MCBU
Molecular and Cell Biology Updates

Series Editors

Prof. Dr. Angelo Azzi
Institut für Biochemie
und Molekularbiologie
Bühlstr. 28
CH–3012 Bern
Switzerland

Prof. Dr. Lester Packer
Dept. of Molecular
and Cell Biology
251 Life Science Addition
Membrane Bioenergetics Group
Berkeley, California 94720-3200
USA

Oxidative Stress in Skeletal Muscle

Edited by
A.Z. Reznick
L. Packer
C.K. Sen
J.O. Holloszy
M.J. Jackson

Birkhäuser Verlag
Basel · Boston · Berlin

Editors

Dr. A.Z. Reznick (Chief Editor)
Senior Lecturer and Researcher
Musculoskeletal Laboratory
Bruce Rappaport Faculty of Medicine
Technion-Israel Institute of Technology
P.O. Box 9649, Bat-Galim
Haifa 31096
Israel

Dr. C.K. Sen [1,2] (Associate Editor)
[1]Dept. of Molecular and Cell Biology
251 Life Science Addition
Membrane Bioenergetics Group
Berkeley, California 94720-3200
USA
[2]Dept. of Physiology, Faculty of Medicine
University of Kuopio
FIN-70211, Kuopio
Finland

Dr. M. Jackson (Associate Editor)
Dept. of Medicine
University of Liverpool
P.O. Box 147
Liverpool, L69 3BX
UK

Prof. Dr. L. Packer (Associate Editor)
Dept. of Molecular and Cell Biology
251 Life Science Addition
Membrane Bioenergetics Group
Berkeley, California 94720-3200
USA

Dr. J.O. Holloszy (Associate Editor)
Dept. of Internal Medicine
Washington University
School of Medicine
St. Louis, MO 63110
USA

Library of Congress Cataloging-in-Publication Data

Oxidative stress in skeletal muscle / edited by A.Z. Reznick ... [et al.].
    p. cm. – (Molecular and cell biology updates)
    Includes bibliographical references and index.
    ISBN 3-7643-5820-3 (hbk. : alk. paper). – ISBN 0-8176-5820-3 (hbk. : alk. paper)
    1. Muscles – Pathophysiology. 2. Stress, Physiology. 3. Oxidation, Physiological.
    I. Reznick, A. Z. (Abraham Z.). 1945– . II. Series.
    RC925.6.O95 1998
    611'.0186 – dc21   88-6332
                                      CIP

Deutsche Bibliothek Cataloging-in-Publication Data

Oxidative stress in skeletal muscle / ed. by A.Z. Reznick ... - Basel
  ; Boston ; Berlin : Birkhäuser, 1998
  (Molecular and cell biology updates)
  ISBN 3-7643-5820-3 (Basel ...)
  ISBN 0-8176-5820-3 (Boston)

© 1998 Birkhäuser Verlag, P.O. Box 133, CH-4010 Basel, Switzerland
Printed on acid-free paper produced from chlorine-free pulp. TFC ∞
Printed in Germany

ISBN 3-7643-5820-3
ISBN 0-8176-5820-3

9 8 7 6 5 4 3 2 1

Table of contents

This book is dedicated to the Krol Foundation of Lakewood NJ USA, for its generous and long-term continuous support of my muscle research work.

Abraham Z. Reznick Ph.D.

List of contributors

J.R. Arthur, Division of Micronutrient and Lipid Metabolism, Rowett Research Institute, Bucksburn, Aberdeen, AB21 9SB, Scotland, U.K

M. Atalay, Department of Physiology, University of Kuopio, FIN-70211 Kuopio, Finland

G.A. Brazeau, Department of Pharmaceutics, College of Pharmacy, University of Florida Box 100494 J.H.M.H.C., Gainesville, Florida 32610, USA

M. Buck, Department of Medicine, Veterans Affairs Medical Center, and Center for Molecular Genetics, University of California, San Diego, CA 92161, USA

M. Chojkier, Department of Medicine, Veterans Affairs Medical Center, and Center for Molecular Genetics, University of California, San Diego, CA 92161, USA

R.H.T. Edwards, Department of Medicine, University of Liverpool, Liverpool L69 3GA, UK

G. Falkous, Neurochemistry Department, Regional Neurosciences Centre, Newcastle General Hospital, Newcastle upon Tyne, NE4 6BE, UK

R. Fielding, Department of Health Sciences, Sargent College of Allied Health Profession, Boston University, Boston, MA 02118, USA

N.V. Gorbunov, Department of Environmental and Occupational Health, University of Pittsburgh, Pittsburgh, PA 15238, USA

S. Goto, Department of Biochemistry, School of Pharmaceutical Sciences, Toho University, Funabashi, Japan

B. Gross, Department of Neurology, Carmel Medical Center, Haifa, Israel

B. Halliwell, International Antioxidant Research Centre, Pharmacology Group, University of London King's College, Manresa Road, London SW3 6LX, UK

O. Hänninen, Department of Physiology, University of Kuopio, FIN-70211 Kuopio, Finland

J.W. Haycock, University Department of Medicine, Clinical Sciences Centre, Northern General Hospital, Sheffield, S5 7AU, UK

A. Hochman, Department of Biochemistry, George S. Wise Faculty of Life Sciences, Tel Aviv University, Tel Aviv, Israel 69978

M.J. Jackson, Muscle Research Centre, Department of Medicine, University of Liverpool, Liverpool L69 3GA, UK

L.L. Ji, Department of Kinesiology and Nutritional Science, University of Wisconsin-Madison, 2000 Observatory Drive, Madison, WI 53706, USA

V.E. Kagan, Department of Environmental and Occupational Health, University of Pittsburgh, Pittsburgh, PA 15238, USA

J. Komulainen, LIKES Research Center for Sport and Health Sciences, University Campus, FIN-40100 Jyväskylä, Finland

H. Kondo, Faculty of Medicine, Kyoto University, Kyoto 606, Japan

J.M. Lawler, Departments of Exercise and Sport Sciences and Physiology, University of Florida, Gainesville, FL 32611, USA

E. Livne, Department of Anatomy and Cell Biology, Bruce Rappaport Faculty of Medicine – Technion-Israel Institute of Technology, Haifa, Israel

D. Mantle, Neurochemistry Department, Regional Neurosciences Centre, Newcastle General Hospital, Newcastle Upon Tyne, NE4 6BE, UK

K.R. Martin, USDA Human Nutrition Research Center on Aging at Tufts University, 711 Washington Street, Boston, MA 02111, USA

E. Menshikova, Department of Cell Biology and Physiology, University of Pittsburgh, Pittsburgh, PA 15238, USA

M. Meydani, USDA Human Nutrition Research Center on Aging at Tufts University, 711 Washington Street, Boston, MA 02111, USA

S. Mokady, Faculty of Food Engineering and Biotechnology – Technion-Israel Institute of Technology, Haifa, Israel

L. Packer, Department of Molecular and Cell Biology, 251 LSA, University of California-Berkeley, Berkeley, CA 94720-3200, USA

T.J. Peters, Department of Clinical Biochemistry, King's College School of Medicine and Dentistry, Bessemer Road, London SE5 9PJ, UK

S.K. Powers, Departments of Exercise and Sport Sciences and Physiology, University of Florida, Gainesville, FL 32611, USA

V.R Preedy, Department of Clinical Biochemistry, King's College School of Medicine and Dentistry, Bessemer Road, London SE5 9PJ, UK

Z. Radák, Laboratory of Exercise Physiology, Univeristy of Physical Education Budapest, Hungary

M.E. Reilly, Department of Clinical Biochemistry, King's College School of Medicine and Dentistry, Bessemer Road, London SE5 9PJ, UK

A.Z. Reznick, Musculoskeletal Laboratory, Department of Morphological Sciences, The Bruce Rappaport Faculty of Medicine, Technion-Israel Institute of Technology, 31096 Haifa, Israel

V.B. Ritov, Department of Environmental and Occupational Health, University of Pittsburgh, Pittsburgh, PA 15238, USA

I. Roisman, Musculoskeletal Laboratory, Department of Morphological Sciences, The Bruce Rappaport Faculty of Medicine, Technion-Israel Institute of Technology, 31096 Haifa, Israel

N. Ronen, Department of Anatomy and Cell Biology, Bruce Rappaport Faculty of Medicine – Technion-Israel Institute of Technology, Haifa, Israel

G. Salama, Department o Cell Biology and Physiology, University of Pittsburgh, Pittsburgh, PA 15238, USA

C.K. Sen, Department of Molecular and Cell Biology, 251 LSA, University of California-Berkeley, Berkeley, CA 94720-3200, USA

V. Vihko, LIKES Research Center for Sport and Health Sciences, University Campus, FIN-40100 Jyväskylä, Finland

H.K. Vincent, Departments of Exercise and Sport Sciences and Physiology, University of Florida, Gainesville, FL 32611, USA

N. Zarzhevsky, Musculoskeletal Laboratory, Department of Morphological Sciences, The Bruce Rappaport Faculty of Medicine, Technion-Israel Institute of Technology, 31096 Haifa, Israel

Preface

This book for the first time allows those interested in muscle physiology and pathophysiology access to the most up-to-date research findings in the field of oxidative stress in skeletal muscle. Understanding the mechanisms associated with oxidative stress in muscle may help to improve the well-being and function of muscles in various physiological conditions, muscle pathologies, and tissue ageing.

Skeletal muscle comprises about 35% of our body weight, making it the largest single type of tissue in the human body. Indeed, the muscle tissue is unique in other ways as well. It has very special requirements for energy and metabolism. Skeletal muscle consumes significant amounts of oxygen, and its oxygen flux increases many times under conditions of exercise and muscle contraction. This makes the muscles vulnerable to oxidative stress as, together with the increase in oxygen flow, there is an increase in oxygen free radicals, which are the byproduct of muscle respiration. A number of studies over the last decade have documented the involvement of oxygen free radicals in exercising muscles. Other studies have tried to counter the oxidative damage by supplementation of antioxidants such as vitamin E, as a means of ameliorating oxidative stress in muscle exercise.

Additional studies over the last few years have shown that oxidative stress and muscle damage may occur in cases of muscular pathologies, such as muscular dystrophy, muscle immobilization, and conditions of ischemia and reperfusion of muscles. Overall, the consequences of muscle oxidative stress have mainly resulted in increased muscle protein oxidation, elevation of lipid peroxidation and depletion of muscle antioxidants. The mechanisms of this oxidative stress are under extensive investigation in laboratories around the world and are discussed in some of the chapters of this book.

Furthermore, some chapters deal with the oxidative stress that is associated with the phenomena of muscle ageing and muscle diseases. Other important contributions include aspects of diet and drugs, and their effects on muscle damage and oxidative status.

A number of chapters have been written in a review style, whereas some report more specific data which provide the reader with recent up-to-date information.

I would like to express my gratitude and appreciation to the associate editors of this book, Drs. Lester Packer, Chandan K. Sen, John O. Holloszy and Malcolm Jackson, for their excellent assistance and cooperation in the planning and reviewing of the chapters.

Abraham Z. Reznick
Haifa, Israel
January 1998

Oxidative Stress in Skeletal Muscle
A.Z. Reznick et al. (eds)
© 1998 Birkhäuser Verlag Basel/Switzerland

Free radicals and oxidative damage in biology and medicine: An introduction

B. Halliwell

International Antioxidant Research Centre, Pharmacology Group, University of London King's College, Manresa Road, London SW3 6LX, UK

Introduction

When living organisms first evolved on the Earth, they did so under an atmosphere containing very little O_2, i.e. they were essentially anaerobes. Anaerobic micro-organisms still survive to this day, but their growth is inhibited and they can often be killed by exposure to 21% O_2, the current atmospheric level. As the O_2 content of the atmosphere rose (as a result of the evolution of organisms with photosynthetic water-splitting capacity), many primitive organisms must have died. Present-day anaerobes are presumably the descendants of those primitive organisms that followed the evolutionary path of "adapting" to rising atmospheric O_2 levels by restricting themselves to environments that the O_2 did not penetrate. However, other organisms began the evolutionary process of evolving antioxidant defence systems to protect against O_2 toxicity. In retrospect, this was a fruitful path to follow. Organisms that tolerated the presence of O_2 could also evolve to use it for metabolic transformations (oxidases, oxygenases and synthases such as nitric oxide synthase [NOS]) and for efficient energy production by using electron transport chains with O_2 as the terminal electron acceptor, such as those present in mitochondria. Human mitochondria make over 80% of the ATP that we need, and the lethal effects of inhibiting this, e.g. by cyanide, show how important the mitochondria are. Mitochondrial defects contribute to the pathology of a wide range of diseases, as shown in subsequent chapters of this book.

Aerobes have developed antioxidant defences to protect against levels of 21% O_2, but not against levels higher than that. Hence all aerobes suffer demonstrable injurious effects if exposed to O_2 at concentrations higher than 21% (reviewed by Balentine, 1982). The nature and speed of onset of these effects depend upon the organism studied, the age, the physiological state, diet and, in cold-blooded animals, temperature. Oxygen toxicity effects have been described in aerobes ranging from fruit flies and nematode worms to humans. For example, the incidence of retrolental fibroplasia (formation of fibrous tissue behind the lens) increased abruptly in the early 1940s among infants born pre-term and often led to blindness. Not until 1954 was it realized that retrolental fibroplasia is associated with the use of high O_2 concentrations in incubators for pre-term babies. More careful control of O_2 concentrations (continuous transcutaneous O_2 monitoring, with supplementary O_2 given only where necessary) and administration of the lipid-soluble antioxidant α-tocopherol have decreased its incidence. However, the problem has not disappeared, because many pre-term infants need increased levels of O_2 in order to survive at all (discussed in Editorial, 1991 and by Ehrenkranz, 1989).

Oxygen toxicity and free radicals

In 1954, Gerschman et al. proposed that the damaging effects of O_2 could be attributed to the formation of oxygen radicals. This hypothesis was popularized and converted into the "superoxide theory of O_2 toxicity" after the discovery of the superoxide dismutase (SOD) enzymes by McCord and Fridovich (reviewed by Fridovich, 1986, 1989). In its simplest form, this theory states that O_2 toxicity results from excess formation of superoxide radical ($O_2^{-\cdot}$) and that the SOD enzymes are important antioxidant defences. Equally important are the enzymes that cooperate with SOD by removing the hydrogen peroxide (H_2O_2) that it produces: a correct balance of SOD activity and H_2O_2-removing systems seems essential. For example, cell transfection experiments suggest that too much SOD in relation to the activity of H_2O_2-removing enzymes is deleterious (Amstad et al., 1991). Such imbalances might contribute to the pathology of Down's syndrome (Avraham et al., 1988) and Parkinson's disease, where the levels of SOD, but not those of H_2O_2-removing enzymes are elevated in the substantia nigra, and levels of glutathione (GSH) are markedly decreased (Jenner, 1994).

Definitions: What radicals are made *in vivo*?

The term "free radical", unlike the term "antioxidant" (see below), is easy to define. In the structure of atoms and molecules, electrons usually associate in pairs, with each pair moving within a defined region of space around the nucleus. This space is referred to as an atomic or molecular orbital. One electron in each pair has a spin quantum number of $+\frac{1}{2}$, the other $-\frac{1}{2}$. A free radical is any species capable of independent existence (hence the term "free") which possesses one or more orbitals containing single electrons (unpaired electrons). The simplest free radical is an atom of the element hydrogen, with one proton and a single electron. A superscript dot is used to denote free radical species. Table 1 gives examples of other free radicals. The spectroscopic technique of electron spin resonance is specific for the detection and measurement of free radicals, recording the energy changes that occur as unpaired electrons respond to a magnetic field (Janzen, 1993), but many other techniques to detect free radicals are available (Packer, 1994).

The chemical reactivity of free radicals varies, and some are metabolically useful in controlled amounts. The prime example is nitric oxide (nitrogen monoxide, NO·), largely synthesized *in vivo* from the amino acid L-arginine by NOS enzymes in vascular endothelial cells, phagocytes, certain neurons and many other cell types. Nitric oxide is a vasodilator and anti-thrombotic agent, a neurotransmitter and a regulator of secretion, among many other physiological roles (Moncada and Higgs, 1993). Nitric oxide may also be involved in the killing of parasites by macrophages in some mammalian species. Some NO· species may also arise *in vivo* by non-enzymic reactions, such as the reaction of nitrite (NO_2^-) in swallowed saliva with gastric acid to produce HNO_2, which can release oxides of nitrogen. Although physiological levels of NO· are useful, excess levels (at or approaching the micromolar range) can be cytotoxic. Indeed, overproduction of NO· (usually involving the inducible nitric oxide synthase, iNOS, enzymes) has been suggested as contributing to the pathology of several chronic inflammatory diseases (Anggard, 1994).

Table 1. Examples of free radicals

Name	Formula	Comments
Hydrogen atom	$H^{\cdot}$	The simplest free radical
Trichloromethyl	$CCl_3^{\cdot}$	A carbon-centred radical (i.e. the unpaired electron resides on carbon). $CCl_3^{\cdot}$ is formed during metabolism of CCl_4 in the liver and contributes to the toxic effects of this solvent. Carbon-centered radicals usually react fast with O_2 to make peroxyl radicals, e.g. $CCl_3^{\cdot} + O_2 \longrightarrow CCl_3O_2^{\cdot}$
Superoxide	$O_2^{\cdot -}$	An oxygen-centred radical. One-electron reduction product of O_2. Selectively reactive
Hydroxyl	$OH^{\cdot}$	A highly reactive oxygen-centred radical. Usually reacts at its site of formation
Thiyl	$RS^{\cdot}$	A group of radicals with an unpaired electron residing on sulphur. Can also combine with O_2 to give thiyl peroxyl ($RSO_2^{\cdot}$) and other oxysulphur radicals
Peroxyl, alkoxyl	$RO_2^{\cdot}$, $RO^{\cdot}$	Oxygen-centred radicals formed (among other routes) during the breakdown of organic peroxides and the reaction of $C^{\cdot}$ radicals with O_2
Oxides of nitrogen	$NO^{\cdot}$, $NO_2^{\cdot}$	Nitric oxide is formed *in vivo* from the amino acid L-arginine. Nitrogen dioxide is made when $NO^{\cdot}$ reacts with O_2. $NO^{\cdot}$ and $NO_2^{\cdot}$ are found in polluted air and smoke from burning organic materials, e.g. cigarette smoke

Superoxide radical ($O_2^{\cdot -}$) is the one-electron reduction product of oxygen. It is produced *in vivo* by phagocytic cells (neutrophils, monocytes, macrophages, eosinophils) and helps them to inactivate viruses, fungi and bacteria (Babior and Woodman, 1990). Activated neutrophils also generate hypochlorous acid, HOCl, a powerful oxidizing and chlorinating agent (Weiss, 1989). Evidence is accumulating that extracellular $O_2^{\cdot -}$ is produced *in vivo* by several cell types other than phagocytes, including B lymphocytes, fibroblasts and possibly vascular endothelial cells (Maly, 1990; Darley-Usmar V and Halliwell, 1996). Superoxide produced by such cells is often thought to be involved in intercellular signalling and growth regulation, and many experiments with cells in culture are consistent with this concept (Burdon et al., 1994). However, it has not yet been proven to occur *in vivo*.

Superoxide *in vivo* is generated not only deliberately but can also arise from what may be called "accidents of chemistry". Superoxide and H_2O_2 are produced by "autoxidation" reactions, in which compounds such as catecholamines, tetrahydrofolates and reduced flavins react directly with O_2 to form $O_2^{\cdot -}$. The $O_2^{\cdot -}$ then oxidizes more of the compound and sets up an autocatalytic free radical-mediated autoxidation. Most "autoxidation" reactions *in vitro* (and presumably also *in vivo*) are not spontaneous reactions with O_2, but are catalyzed by the presence of traces of catalytic transition metal ions (Halliwell and Gutteridge, 1990a).

A major cellular source of $O_2^{\cdot -}$ is the mitochondria. The mitochondrial electron transport chain is a gradient of reduction potential, from the highly reducing $NADH/NAD^+$ couple to the oxidizing $\frac{1}{2}O_2/H_2O$ couple of cytochrome oxidase. Thermodynamically, there is nothing to prevent constituents of the early part of the electron transport chain (e.g. nonhaem iron proteins, quinones, flavoproteins, cytochromes *b*) from reducing O_2 directly to make $O_2^{\cdot -}$. Fortunately, such reactions are restricted, both by the normally low intramitochondrial O_2 tension and by an arrangement of electron carriers in the inner mitochondrial membrane that favours transfer to the next carrier in the chain. Hence, most of the electrons entering the chain arrive at cytochrome oxidase, and only a small percentage (perhaps only 1–3%) may leak away from early chain constituents to form $O_2^{-\cdot}$. Raising O_2 concentration causes more leakage to occur, resulting in more $O_2^{-\cdot}$ formation.

These studies of the leakiness of electron transport chains are based on experiments with bacteria and mitochondria respiring in air-saturated solutions (Fridovich, 1986, 1989) and might be an overestimate. However, if we add in the deliberate $O_2^{-\cdot}$ production from phagocytes, and other sources of $O_2^{-\cdot}$ (such as electron leakage from other electron-transport systems, e.g. cytochromes P450, NOS and xanthine oxidase, and autoxidation reactions), a figure of 1–3% does not seem unreasonable. Although this figure may seem trivial, humans are large animals and breathe in a lot of O_2. Thus, even at rest, we may produce close to 2 kg of $O_2^{-\cdot}$ per year (Box 1). A very important question, to which there is as yet no clear answer, is whether $O_2^{-\cdot}$ production increases proportionally with O_2 uptake during exercise and what the consequences of this might be. The physiological importance of scavenging of mitochondrial $O_2^{-\cdot}$ is graphically illustrated by the pathology of transgenic "knockout" mice lacking manganese SOD (MnSOD), the mitochondrial isoform. They usually die soon after birth from lung problems; survivors develop neurodegeneration (Li et al., 1995; Lebovitz et al., 1996).

The total body generation of NO˙ may also be large, as assessed by measurements of nitrite (NO_2^-) and nitrate (NO_3^-). The major end-product of NO˙ oxidation is NO_2^- which is rapidly oxidized to NO_3^- *in vivo*. For example, in one study which allowed for the confounding effects of dietary NO_3^-, plasma NO_3^- was 29 ± 1 μM corresponding to a basal body production of NO˙ of 840 ± 146 μmol per day in the human body (Wennmalm et al., 1994).

Like the overproduction of NO˙, excess generation of $O_2^{-\cdot}$ can be deleterious, via a variety of mechanisms. Several bacterial enzymes are sensitive to attack by $O_2^{-\cdot}$ and aconitase, a key enzyme of the Krebs cycle, may be a target of attack in mammalian cells (reviewed by Liochev, 1996). Superoxide can release trace amounts of iron, both from ferritin and during its attack upon

Box 1. How much superoxide is made in the human body? (Adapted from Halliwell, 1994).

An adult at rest utilizes about 3.5 ml O_2/kg body mass per minute
or 352.8 l//day (assuming 70 kg body mass)

or 14.7 mol/day

If 1% makes $O_2^{-\cdot}$

this is 0.147 mol/day

or 53.66 mol/year

or about 1.72 kg/year (of $O_2^{-\cdot}$).

During bodily exertion this would increase up to ten-fold, **assuming that the 1% figure still applied.**

some enzymes with iron–sulphur clusters at their active sites (reviewed by Liochev, 1996). Liberation of "catalytic" iron ions in this way can lead to generation of OH˙ by Fenton chemistry (Halliwell and Gutteridge, 1990a). The hydroxyl radical reacts at almost diffusion-controlled rates with all molecules in the human body and its formation probably always causes damage (von Sonntag, 1987). For example, when OH˙ is generated adjacent to DNA it attacks all four DNA bases to produce a multitude of products (Steenken, 1989; Dizdaroglu, 1993). This complex product pattern is believed to be diagnostic for damage to DNA caused by OH˙; the damage pattern produced to the DNA bases by peroxyl radicals, singlet O_2, hypochlorous acid or peroxynitrite is very different and neither O_2^{-} or NO˙ appears to attack DNA bases (Yermilov et al., 1995; Spencer et al., 1996; Halliwell and Dizdaroglu, 1992). Observation of the "OH˙ damage pattern" in cells or tissues subjected to oxidative stress is part of the evidence that this reactive radical is generated *in vivo* (Spencer et al., 1995; reviewed by Halliwell, 1996).

Hydroxyl radical generation *in vivo* might occur by the homolytic fission of water after exposure to ionizing radiation (von Sonntag, 1987),

$$H_2O \longrightarrow OH˙ + H˙$$

by the reaction of O_2^{-} with hypochlorous acid (Candeias et al., 1993),

$$O_2^{-} + HOCl \longrightarrow O_2 + OH˙ + Cl^- \text{ (rate constant } 7.5 \times 10^6 \text{ M}^{-1} \cdot \text{s}^{-1}),$$

by Fenton-type chemistry

$$Fe^{2+} + H_2O_2 \longrightarrow OH˙ + OH^- + Fe^{3+}$$

and by reaction of HOCl with iron (Candeias et al., 1994)

$$Fe^{2+} + HOCl \longrightarrow OH˙ + Cl^- + Fe^{3+}$$

Copper ions also react with H_2O_2 to form OH˙ It has been suggested that reactive species additional to OH˙, such as oxo iron and oxo copper species, are generated but this is a controversial area (discussed by Burkitt, 1993 and Halliwell and Gutteridge, 1990a). Superoxide can assist Fenton-type chemistry by reducing Fe^{3+} to Fe^{2+}, and also by increasing the availability of the necessary iron ions through their displacement from ferritin or iron–sulphur proteins.

Sequestration of metal ions

Iron and copper ions in chemical forms that can decompose H_2O_2 to OH˙, promote lipid peroxidation and catalyse autoxidation reactions are in short supply *in vivo*. The human body ensures that as much iron and copper as possible are kept safely bound to transport or storage proteins. Indeed, this "sequestration" of metal ions is an important antioxidant defence mechanism (Halliwell and Gutteridge, 1990b). Sequestration of metal ions deters the growth of bacteria in human body fluids (Weinberg, 1990) and also ensures that such fluids will, in general, not convert $O_2^{-\cdot}$ and H_2O_2 into $OH^\cdot$. If iron or copper does become available to catalyse free radical reactions in body fluids, as happens in animals that are defective in transferrin synthesis and in certain human metal overload diseases, severe damage to many body tissues occurs. Tissue injury itself causes release of transition metal ions and other catalytic complexes, e.g. haem proteins (Fig. 1).

Hence, a major determinant of the nature of the damage done by excess generation of reactive oxygen species *in vivo* may be the availability and location of metal ion catalysts of $OH^\cdot$ radical formation. If, for example, "catalytic" iron salts are bound to DNA in one cell type and to membrane lipids in another, then excessive formation of H_2O_2 and $O_2^{-\cdot}$ will, in the first case, damage the DNA, and in the second, could initiate lipid peroxidation. Evidence for $OH^\cdot$ formation in the nucleus of cells treated with H_2O_2 has been obtained by showing that all four DNA bases are modified in a way that is characteristic of $OH^\cdot$ attack (e.g. Spencer et al., 1995). If this $OH^\cdot$ is formed by metal ion-dependent reactions, then the "catalytic" metal ions must be bound to the DNA itself. Either they are always present on the DNA, or else the H_2O_2 causes their release within the cell and they then bind to DNA, which avidly binds transition metal ions. For example, an excess of H_2O_2 decomposes haem proteins (Gutteridge, 1986), including myoglobin (Puppo and Halliwell, 1988), to release iron ions, and $O_2^{-\cdot}$ can liberate some iron from ferritin (Bolann and Ulvik, 1990). Rises in chelatable iron in cells subjected to oxidative stress have recently been confirmed by using fluorescent iron ion-specific probes (Breuer et al., 1996).

Free radical reactivity

Radical termination

Free radicals of different types show a wide spectrum of chemical reactivity (Bielski, 1985). If two free radicals meet, they can join their unpaired electrons to form a covalent bond and both radicals are lost. Thus, atomic hydrogen forms diatomic hydrogen:

$$H^\cdot + H^\cdot \longrightarrow H_2.$$

A more biologically-relevant example is the very fast (Huie and Padmaja, 1993) reaction of $NO^\cdot$ and $O_2^{-\cdot}$ to form a non-radical product, peroxynitrite

$$NO^\cdot + O_2^{-\cdot} \longrightarrow ONOO^- \text{ (peroxynitrite)}.$$

At physiological pH, $ONOO^-$ protonates to peroxynitrous acid (ONOOH), which disappears within a few seconds, the end-product being largely nitrate. The chemistry of peroxynitrite/peroxynitrous acid reactions is extremely complex and far from completely understood, but addition of $ONOO^-$ to cells and tissues leads to oxidation and nitration of proteins, DNA and lipids, often resulting in cell injury or death (Beckman et al., 1994; Pryor and Squadrito, 1995; Spencer et al., 1996; Yermilov et al., 1995). Indeed, ONOOH has some reactivities comparable to those of $OH^\cdot$, although it is uncertain whether or not it actually generates $OH^\cdot$ under physiolo-

OXIDATIVE STRESS

↑

INJURY

Ischaemia/reperfusion
Heat
Trauma
Freezing
Exercise to excess
Toxins
Radiation
Infection
Disuse atrophy

↑

- Phagocyte recruitment and activation (makes O_2^-, H_2O_2, NO^-, $HOCl$). **Injury to muscles by strenuous exercise leads to inflammation and oxidative stress.**

- Arachidonic acid release, enzymic peroxide formation (by activation of lipoxygenase, cyclooxygenase enzymes). Decomposition of enzyme-formed and non-enzymically formed peroxides to peroxyl/alkoxyl radicals can spread damage to other lipids/proteins.

- Metal ion release from storage sites (Fe^{2+}, Cu^{2+}), stimulating conversion of H_2O_2 to OH^-, lipid peroxide breakdown to RO_2^-/RO^-, and "autoxidation" reactions. **Metal ion release and oxidative damage have been observed in atrophied muscles, e.g. after prolonged non-usage.**

- Haem protein release (myoglobin, haemoglobin, cytochromes); haem proteins react with peroxides to stimulate free radical damage and (if peroxide is in excess) to release Fe^{2+} and haem, which can decompose peroxides to RO_2^- and RO^-. **Especially important in muscle damage.**

- Interference with antioxidant defence systems, (e.g. GSH and ascorbate loss from cells). Ascorbate loss from extracellular fluids.

- Conversion of xanthine dehydrogenase to oxidase in certain tissues, possible release of oxidase from damaged cells into the circulation to cause systemic damage (e.g. by binding to vascular endothelium), increased hypoxanthine levels due to disrupted energy metabolism.

- Mitochondrial damage, increased leakage of electrons to form O_2^-.

- Raised intracellular Ca^{2+}, stimulating calpains (Ca^{2+}-dependent proteases), Ca^{2+}-dependent nucleases and Ca^{2+}/calmodulin-dependent nitric oxide synthase, giving more NO^- and increased risk of $ONOO^-$ formation.

Items especially relevant to muscle damage are shown in **bold type**.

Figure 1. Some of the reasons why tissue injury causes oxidative stress. Items especially relevant to muscle damage are shown in bold type.

Table 2. Some of the conditions in which the formation of peroxynitrite has been implicated as causing tissue injury

Atherosclerosis	Adult respiratory distress syndrome
Sporadic inclusion-body myositis	Skin inflammation
Rheumatoid arthritis	Gastritis (*Helicobacter pylori* infection)
Inflammatory bowel disease	Cystic fibrosis
Neurodegenerative disease	Endotoxic shock
Acute inflammation	Ageing of skeletal muscle
Carbon monoxide toxicity	Viral infection

For a full list of references see Halliwell (1997).
The evidence for $ONOO^-$ involvement in these diseases usually includes measurement of increased formation of 3-nitrotyrosine, a product generated by reaction of $ONOO^-$-derived nitrating species with free tyrosine or tyrosine residues in proteins (Beckman et al., 1994). However, reliance on this evidence alone may be insufficient to implicate $ONOO^-$, because other reactive nitrogen species can nitrate tyrosine (Eiserich et al., 1995, 1996; van der Vliet et al., 1997).

Box 2. An outline of lipid peroxidation.

Initiation of peroxidation occurs by the generation of a species ($R^{\cdot}$) capable of abstracting hydrogen from a polyunsaturated fatty acid side-chain in a membrane (such fatty acid side-chains are more susceptible to free radical attack than are saturated or monounsaturated side-chains):

$$-CH + R^{\cdot} \longrightarrow -C^{\cdot} + RH$$

Species able to abstract hydrogen include $OH^{\cdot}$ and peroxyl radicals (Tab. 1). The carbon-centred radicals react fast with O_2

$$-C^{\cdot} + O_2 \longrightarrow -CO_2^{\cdot}$$

A fatty acid side chain **peroxyl radical** is formed. This can attack adjacent fatty acid side-chains and **propagate** lipid peroxidation.

$$-CO_2^{\cdot} + -CH \longrightarrow -CO_2H + \cdot C^{\cdot}$$

The chain reaction thus continues and **lipid peroxides** ($-CO_2H$) accumulate in the membrane. Lipid peroxides destabilize membranes and make them "leaky" to ions. Peroxyl radicals can attack not only lipids but also membrane proteins (e.g. damaging enzymes, receptors and signal transduction systems) and oxidize cholesterol. If transition metal ions are present or if lipids are heated to high temperatures, peroxides decompose to peroxyl and alkoxyl ($RO^{\cdot}$) radicals, which can abstract hydrogen and start new peroxidation cycles. A wide range of noxious carbonyl compounds, especially unsaturated aldehydes such as 4-hydroxynonenal, are end-products of peroxide decomposition (Esterbauer et al., 1992; El Ghissassi et al., 1995).

gical conditions (Pryor and Squadrito, 1995; Kaur et al., 1997). Evidence consistent with $ONOO^-$ formation *in vivo* has been presented for a wide range of diseases (Tab. 2). This is perhaps unsurprising, because many chronic diseases result in overproduction of both $O_2^{-\cdot}$ and $NO^\cdot$.

Radical termination reactions do not always produce toxic products. They can be beneficial, e.g. $NO^\cdot$ can inhibit lipid peroxidation by scavenging peroxyl radicals (Rubbo et al., 1994):

$$RO_2^\cdot + NO^\cdot \longrightarrow ROONO$$

although the biological consequences of nitrated lipids (if any) need further investigation.

Radical chain reactions

When a free radical reacts with a non-radical, a new radical results and a chain reaction can also result. As most biological molecules are non-radicals, the generation of reactive radicals such as $OH^\cdot$ *in vivo* usually sets off chain reactions. For example, attack of reactive radicals upon fatty acid side chains in membranes and lipoproteins can abstract hydrogen, leaving a carbon-centred radical and initiating the process of lipid peroxidation (Box 2).

Similarly, if $OH^\cdot$ is generated adjacent to the purine base guanine in DNA, it can undergo an addition reaction to produce an 8-hydroxyguanine radical. This can be oxidized to 8-hydroxyguanine or reduced, eventually yielding a ring-opened product, FaPy guanine (Fig. 2). The ratio of these two products can vary depending on the reaction conditions, so caution must be employed in inferring levels of oxidative DNA damage by measuring only a single product, such as 8-hydroxyguanine. For example, in the substantia nigra of patients with Parkinson's disease there is an elevated level of 8-hydroxyguanine but levels of FaPy guanine are decreased, so it is uncertain if there has been a real increase in free radical damage to DNA (Alam et al., 1997).

Reactive oxygen and nitrogen species

Most of the $O_2^{-\cdot}$ generated *in vivo* probably undergoes a nonenzymatic or SOD-catalysed dismutation reaction, represented by the overall equation:

$$2O_2^{-\cdot} + 2H^+ \longrightarrow H_2O_2 + O_2.$$

This generates H_2O_2, which has no unpaired electrons and so is classified as a non-radical. Other non-radicals include $HOCl$ and $ONOO^-$. Hydrogen peroxide resembles water in its molecular structure and is very diffusible within and between cells. As well as arising from $O_2^{-\cdot}$, H_2O_2 is produced by the action of several oxidase enzymes *in vivo*, including amino acid oxidases and the enzyme xanthine oxidase (Chance et al., 1979; Granger, 1988). Xanthine oxidase catalyses the oxidation of hypoxanthine to xanthine and xanthine to uric acid. Oxygen is simultaneously reduced to both $O_2^{-\cdot}$ and H_2O_2. Low levels of xanthine oxidase are present in many mammalian tissues, especially in the gastrointestinal tract. Levels of xanthine oxidase often increase when tissues are subjected to insult, such as trauma or deprivation of oxygen (Granger, 1988; Grisham, 1994).

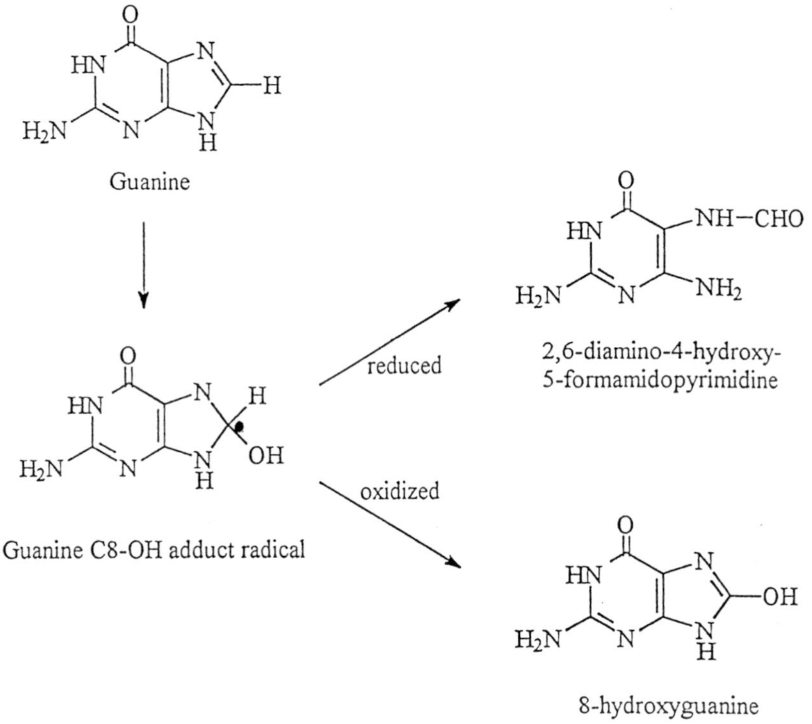

Figure 2. Products of hydroxyl radical attack on the DNA base guanine.

Like NO$^\cdot$ and $O_2^{-\cdot}$, H_2O_2 (at controlled levels) fulfils some metabolic roles. For example, H_2O_2 generated in the thyroid gland is used by a peroxidase enzyme to iodinate the thyroid hormones (Dupuy et al., 1991). In some, but by no means all, cells, H_2O_2 may be a second messenger which mediates (via regulation of the extent of phosphorylation) the displacement of an inhibitory subunit from the cytoplasmic gene transcription factor NF-κB (Schreck et al., 1992). Displacement of the inhibitory subunit causes the active factor to migrate to the nucleus and activate many different genes by binding to specific DNA sequences in enhancer and promoter elements. Thus, H_2O_2 (and other peroxides; Collins, 1993) can upregulate expression of genes controlled by NF-κB, possibly an important event at sites of inflammation and tissue injury generally. As is the case for $O_2^{-\cdot}$ and NO$^\cdot$, whether H_2O_2 is good or bad depends on its level and location.

Hydrogen peroxide at micromolar levels seems poorly reactive, but higher (>50 μM) levels of H_2O_2 can attack certain cellular targets. For example, it can oxidize an essential -SH group on the glycolytic enzyme glyceraldehyde-3-phosphate dehydrogenase, blocking glycolysis. H_2O_2 also interferes with other aspects of cell energy metabolism (Cochrane, 1991). It can lead to OH˙ generation by Fenton-type chemistry. For example, H_2O_2 does not directly modify DNA bases; DNA strand breakage and oxidation of purine and pyrimidine bases, which occur when isolated cells are treated with H_2O_2 (the degree of sensitivity to H_2O_2 depends on the cell type studied, how long it has been in culture, and the composition of the culture medium), appear to be a result of formation of OH˙ from H_2O_2 in the nucleus (Spencer et al., 1995; Halliwell, 1996a).

Reactive oxygen species

Reactive oxygen species (ROS) is a collective term often used by scientists to include not only the oxygen radicals ($O_2^{-˙}$, $RO_2˙$, RO˙ and OH˙) but also H_2O_2, $ONOO^-$, HOCl and even the non-radical ozone (O_3). "Reactive" is, of course, a relative term; neither $O_2^{-˙}$ nor H_2O_2 is particularly reactive in aqueous solution. Hence, some authors use the term "oxygen-derived species" instead. Another popular collective term is "oxidants". However, $O_2^{-˙}$ and H_2O_2 can act as both oxidants and reductants in aqueous solution. The terms "reactive nitrogen species" and "reactive chlorine species" are beginning to enter the literature (Tab. 3).

Antioxidants

Words such as antioxidant and oxidative stress are widely used but surprisingly difficult to define precisely (for a discussion of the latter term see Sies, 1991). For example, the term "antioxidant" as used in the food science literature is often implicitly restricted to chain-breaking antioxidant inhibitors of lipid peroxidation, presumably because food scientists use these compounds mainly to prevent rancidity. However, reactive oxygen, nitrogen and chlorine species (ROS/RNS/RCS) generated *in vivo* damage proteins, DNA and other molecules in addition to lipids. Hence one broader definition is (Halliwell, 1990): "An antioxidant is any substance that, when present at low concentrations compared to those of an oxidizable substrate, significantly delays or prevents oxidation of that substrate."

The term "oxidizable substrate" includes everything found in living cells: proteins, lipids, carbohydrates and DNA. Mechanisms of antioxidant action can include:

- Minimizing O_2 levels (e.g. the packaging of foodstuffs under N_2 or the low intramitochondrial O_2 tension at which these organelles normally operate)
- Scavenging reactive oxygen/nitrogen/chlorine species
- Inhibiting ROS/RNS/RCS formation or scavenging their precursors
- Binding metal ions needed for catalysis of ROS generation
- Upregulation of endogenous antioxidant defences.

Table 3. Reactive species

Radicals	Non-radicals
Reactive oxygen species (ROS)	
Superoxide, $O_2^{-\cdot}$	Hydrogen peroxide, H_2O_2
Hydroxyl, $OH^{\cdot}$	Hypobromous acid, HOBr
Hydroperoxyl, $HO_2^{\cdot}$	Ozone, O_3
Lipid peroxyl, $LO_2^{\cdot}$	Singlet oxygen $O_2{}^1\Delta g$
Lipid alkoxyl, $LO^{\cdot}$	Lipid peroxides, LOOH
Reactive chlorine species (RCS)	
	Hypochlorous acid, HOCl
	Nitryl (nitronium) chloride, NO_2Cl^a
	Chloramines
Reactive nitrogen species (RNS)	
Nitric oxide, $NO^{\cdot}$	Nitrous acid, HNO_2
Nitrogen dioxide, $NO_2^{\cdot}$	Nitrosyl cation, NO^+
	Nitroxyl anion, NO^-
	Dinitrogen tetroxide, N_2O_4
	Dinitrogen trioxide, N_2O_3
	Peroxynitrite, $ONOO^-$
	Peroxynitrous acid, ONOOH
	Nitronium (nitryl) cation, NO_2^+
	Alkyl peroxynitrites, ROONO
	Nitryl (nitronium) chloride, NO_2Cl^a

Reactive oxygen species (ROS) is a collective term that includes both oxygen radicals and certain non-radicals that are oxidizing agents and/or are easily converted into radicals (HOCl, O_3, $ONOO^-$, $^1O_2$, H_2O_2). RNS is also a collective term including nitric oxide and nitrogen dioxide radicals, as well as such non-radicals as HNO_2 and N_2O_4. $ONOO^-$ is often included in both categories. "Reactive" is not always an appropriate term: H_2O_2, $NO^{\cdot}$ and $O_2^{-\cdot}$ react quickly with only a few molecules whereas $OH^{\cdot}$ reacts quickly with almost everything. $RO_2^{\cdot}$, $RO^{\cdot}$, HOCl, $NO_2^{\cdot}$, $ONOO^-$ and O_3 have intermediate reactivities.
HOBr could also be regarded as a "reactive brominating species". [a]NO_2Cl is a chlorinating and nitrating species produced by reaction of HOCl with NO_2^- (Eiserich et al., 1996).

When ROS/RNS/RCS are generated *in vivo*, many antioxidants come into play. Their relative importance as protective agents depends on:

- Which reactive species is generated
- How it is generated
- Where it is generated
- What target of damage is measured

Hence there is no "best antioxidant"; indeed, league tables of antioxidants are meaningless unless the assay system used is defined. For example, if human blood plasma is tested for its ability to inhibit iron ion-dependent lipid peroxidation, the proteins transferrin and ceruloplasmin are found to be the most important protective agents (Gutteridge and Quinlan, 1992). When human blood plasma is exposed to NO_2, uric acid seems to be a major protective antioxidant (Halliwell et al., 1992b) whereas urate appears to play little role as a scavenger of HOCl in plasma (Hu et al., 1992). Similarly, if the oxidative stress is kept the same but a different target of oxidative damage is measured, different answers can result. When plasma is exposed to gas-phase cigarette smoke, lipid peroxidation occurs, which is inhibited by ascorbate (Frei et al., 1991), whereas ascorbate has no effect on the formation of plasma protein carbonyls upon cigarette smoke exposure (Reznick et al., 1992). As an extreme example, the carcinogen diethylstilboestrol can aggravate oxidative DNA damage *in vivo* (Roy and Liehr, 1991) but is a powerful inhibitor of lipid peroxidation *in vitro* (Wiseman and Halliwell, 1993).

Antioxidant defences in the human body

All organisms suffer some exposure to OH˙, because it is generated by homolytic fission of O–H bonds in water driven by background ionizing radiation (von Sonntag, 1987). This radical is so reactive with all biological molecules that it is impossible to evolve (or create in the laboratory) a specific scavenger of it that will work as such *in vivo*. Almost everything in living organisms reacts with OH˙ with second-order rate constants of $10^9 - 10^{10}$ M^{-1} second^{-1}, so that collision of

Table 4. Repair of oxidative damage

Substrate of damage	Repair system
DNA All components of DNA can be attacked by OH˙, whereas singlet O_2 attacks guanine preferentially. H_2O_2, NO˙ and O_2^{-} do not attack DNA. HNO_2 deaminates bases; $ONOO^-$ leads to nitration and deamination; HOCl chlorinates bases	A wide range of enzymes exists that recognize abnormalities in the DNA precursor pool (Mo et al., 1992) or in DNA and remove them, e.g. by excision, resynthesis and rejoining of the DNA strand (Demple and Harrison, 1994).
Proteins Many ROS/RNS/RCS can oxidize -SH and methionine groups. Hydroxyl radicals attack many amino acid residues. Proteins often bind transition metal ions, making them a target of attack by "site-specific" OH˙ generation. RNS can nitrate tryptophan, tyrosine and phenylalanine; HOCl and NO_2Cl can chlorinate aromatics and form chloramines with -NH_2 groups (Dean et al., 1997; Kettle, 1996; Stadtman and Oliver, 1991).	Oxidized methionine residues may be repaired by methionine sulphoxide reductase. Damaged (including oxidized) proteins may be recognized and preferentially destroyed by cellular proteases, especially the proteasome. The fate of nitrated/chlorinated proteins is not yet clear but they are presumably degraded (Ohshima et al., 1990).
Lipids Some ROS/RNS (not including NO˙, O_2^{-} or H_2O_2) can initiate lipid peroxidation to generate cytotoxic products including aldehydes (Esterbauer et al., 1992) and isoprostanes (Morrow and Roberts, 1994).	Chain-breaking antioxidants (especially α-tocopherol) remove chain-propagating peroxyl radicals. Phospholipid hydroperoxide glutathione peroxidase can remove peroxides from membranes, as can some phospholipases. Normal membrane turnover can release damaged lipids.

OH˙ with the molecules almost always results in reaction. Once OH˙ has been formed, damage caused by this radical is probably unavoidable and is dealt with by repair processes (Tab. 4). Thus repair can be regarded as an antioxidant defence mechanism.

Enzymes

A large part of the body's antioxidant defences serves to minimize any additional production of OH˙. Living organisms have evolved proteins to remove excess $O_2^{-˙}$ and H_2O_2. Superoxide dismutase enzymes remove $O_2^{-˙}$ by accelerating its conversion to H_2O_2 by about four orders of magnitude at pH 7.4. The essential nature of MnSOD has already been illustrated from transgenic animal data. A SOD with copper and zinc (CuZnSOD) at the active site is also present in human cells, but largely in the cytosol. Interestingly, transgenic mice lacking CuZnSOD seem to have no acute problems (Reaume et al., 1996).

Superoxide dismutase enzymes work in collaboration with H_2O_2-removing enzymes. Catalases convert H_2O_2 to water and O_2 (Chance et al., 1979):

$$2\,H_2O_2 \longrightarrow 2\,H_2O + O_2$$

but are not usually in the same subcellular location as SOD; they are present in the peroxisomes of most mammalian cells, where they probably serve to destroy H_2O_2 generated by oxidase enzymes located within these organelles. The most important H_2O_2-removing enzymes in mammalian cells are thought to be the selenoprotein glutathione peroxidase (GSHPX) enzymes (Chance et al., 1979; Sies, 1991). A selenocysteine residue, essential for enzyme activity, is present at the active site. These enzymes remove H_2O_2 by using it to oxidize reduced glutathione (GSH) to oxidized glutathione (GSSG) (Fig. 3). However, transgenic mice lacking GSHPX seem normal, although they may be more prone to tissue damage by cytotoxic agents (e.g. anthracyclines) or by ischemia–reperfusion than controls (e.g. Spector et al., 1996). Glutathione reductase, a flavoprotein (FAD-containing) enzyme, regenerates GSH from GSSG, with NADPH as a source of reducing power (Fig. 3).

There may be many other important antioxidant systems of whose significance we are as yet unaware. Examples perhaps include thioredoxin-dependent H_2O_2-removal systems (Yim et al., 1994). In addition, sequestration of metal ions into proteins such as ferritins, transferrins, ceruloplasmin and metallothionein is an important antioxidant defence, as discussed above.

Dietary antioxidants

We obtain certain antioxidants from the diet. The physiological role of some of these is well established (especially vitamin E; reviewed by Gey, 1995 and Diplock, 1997) whereas the role of others is uncertain as yet (Tab. 5). This is an important research area because there is good evidence that endogenous antioxidants do not completely prevent damage by ROS/RNS/RCS in the human body (Tab. 6): the presence of various "biomarkers" of oxidative damage (Fig. 4) indi-

---

GLUTATHIONE PEROXIDASES

$$2GSH + H_2O_2 \rightarrow GSSG + 2H_2O$$

$$2GSH + \text{FATTY ACID-OOH} \rightarrow GSSG + \text{FATTY ACID-OH} + H_2O$$

GLUTATHIONE REDUCTASE

$$GSSG + NADPH + H^+ \rightarrow 2GSH + NADP^+$$

Reduced glutathione (GSH) is a tripeptide, glutamic acid-cysteine-glycine that is present at

millimolar concentrations in most mammalian cells and has multiple metabolic functions

(Anderson, 1997). In oxidized glutathione (GSSG), two tripeptides are linked by a disulphide

bridge. Glutathione peroxidase can also remove fatty acid peroxides by converting them to

alcohols (lipid-OH). Mammalian cells additionally contain a phospholipid hydroperoxide

glutathione peroxidase that performs the same reaction upon lipid peroxides within membranes

(Maiorino *et al* 1991).

Figure 3. The glutathione system.

cates that repair systems are not always 100% effective. This ongoing oxidative damage is widely proposed to contribute to the development of cardiovascular disease (Steinberg et al., 1989), cancer (Totter, 1980; Ames, 1989) and, more recently, neurodegenerative disease (Halliwell, 1992; Ames et al., 1993; Jenner, 1994). Dietary manipulations that decrease steady-state levels of oxidative damage might thus protect against development of these diseases. The use of these oxidative damage markers in nutritional studies is becoming increasingly evident, although there is considerable debate as to which are the best biomarkers to use (Fig. 4; reviewed by Halliwell, 1996a).

Table 5. Dietary antioxidants

Putative antioxidant	Status
Vitamin E	Essential antioxidant in humans, protective against cardiovascular disease (Gey, 1995). Severe deficiency causes neurodegeneration (Muller and Goss-Sampson, 1990) and accelerates atherosclerosis (Diplock, 1997).
Vitamin C	Multiple metabolic roles; antioxidant action only one of its effects but can scavenge many ROS/RNS/RCS (Buettner and Jurkiewicz, 1993; Whiteman and Halliwell, 1996). Can exert pro-oxidant actions *in vitro* by interaction with iron and copper ions; *in vivo* significance of pro-oxidancy unknown (Halliwell, 1996b).
β-Carotene, other carotenoids, related plant pigments	Epidemiological evidence that high body levels are associated with diminished risk of cancer and cardiovascular disease, particularly in smokers (Gey, 1995). Carotenoids are good singlet O_2 quenchers/scavengers and can exert other antioxidant effects *in vitro* (Burton and Ingold, 1984; Rice-Evans et al., 1997). Often simplistically grouped with vitamins E and C as "antioxidant nutrients", it is not yet rigorously proved that any protective effects these pigments exert against human disease are the result of antioxidant action (discussed by Halliwell, 1996a). For example, β-carotene supplementation of the diet did not decrease the elevated urinary excretion of 8-hydroxydeoxyguanosine, a putative index of oxidative DNA damage, in smokers (von Poppel et al., 1995).
Flavonoids, other plant phenols	Many plant phenols inhibit lipid peroxidation and lipoxygenase enzymes *in vitro* and may be important dietary antioxidants (Hertog et al., 1993; Weisburger, 1995). It has been speculated that flavonoids in red wine could explain the "French paradox", although the identity of the phenolics responsible is uncertain. Like ascorbate, some plant phenolics can be prooxidant *in vitro* if mixed with copper or iron ions (Halliwell, 1995a, 1996a). More data are needed on absorption and bioavailability of phenolics, but evidence is growing that some are absorbed. Plant phenols might also scavenge RNS, e.g. preventing tyrosine nitration by $ONOO^-$, but the biological properties of any resulting nitroso- or nitro-phenolics must be considered (Pannala et al., 1997).

A diet rich in fruits, nuts, grains and vegetables is protective against several human diseases. This may be the result of the antioxidants that they contain and/or to the many other compounds present (Block et al., 1992; Johnson et al., 1994; Willett, 1994; Prestera and Talalay, 1995; Verhagen et al., 1995).

Oxidative stress

Oxidative stress is a somewhat vague term, but has been suggested to refer to a serious imbalance between production of ROS/RNS and antioxidant defences (Sies, 1991). It can result from the following:

1. Depletion of endogenous antioxidants (e.g. as a result of inborn errors of metabolism) or of diet-derived antioxidants, caused by malnutrition (Golden, 1994). Dietary constituents such as iron, copper, manganese, sulphur-containing amino acids and riboflavin are also important (e.g. for GSH manufacture or to make the FAD cofactor in glutathione reductase). Folate helps minimize levels of homocysteine in plasma.

Table 6. Evidence that damage by reactive species occurs in the human body

Target of damage	Evidence
DNA	Low baseline levels of DNA base damage products are present in DNA isolated from human cells (Halliwell and Dizdaroglu, 1992; Collins et al., 1993). Urinary excretion of DNA base damage products, presumably resulting from repair of oxidative damage to DNA or DNA precursors (e.g. the nucleotide pool) (Mo et al., 1992).
Protein	Attack of free radicals upon proteins produces protein carbonyls and other modified amino acid residues. Low levels of carbonyls and certain other products (e.g. o-tyrosine) are detected in human tissues and body fluids (Dean et al., 1997). Nitrotyrosines, products of reaction of RNS with tyrosine residues in proteins, have been detected in atherosclerotic lesions; about 40 nM nitrotyrosine is found in human plasma and metabolites are excreted in urine (Ohshima et al., 1990). Bityrosine can be detected in body fluids and atherosclerotic lesions (Dean et al., 1997).
Lipid	Accumulation of "age pigments" in tissues. Lipid peroxidation in atherosclerotic lesions. Presence of end-products of peroxidation in human body fluids, e.g. isoprostanes (Morrow and Roberts, 1994).
Uric acid	Attacked by several ROS/RNS to generate allantoin, cyanuric acid, parabanic acid, oxonic acid and other products, which can be detected in human body fluids. Levels increase during oxidative stress.

For a complete list of references see Halliwell (1996a).

2. Excess production of ROS/RNS, e.g. by exposure to elevated O_2 concentrations, the presence of toxins that are metabolized to produce free radicals, or excessive activation of "natural" radical-producing systems, e.g. inappropriate activation of phagocytic cells in chronic inflammatory diseases, such as rheumatoid arthritis and ulcerative colitis (Grisham, 1994; Halliwell, 1995b).

Cells can tolerate mild oxidative stress, which often results in upregulation of the synthesis of antioxidant defence systems in an attempt to restore the balance. For example, if rats are gradually acclimatized to elevated O_2, they can tolerate pure O_2 for much longer than control rats, apparently as a result of increased synthesis of antioxidant defence enzymes and of GSH in the lung (Iqbal et al., 1989). However, severe oxidative stress can produce major interdependent derangements of cell metabolism, including DNA strand breakage (often an early event), rises in intracellular "free" Ca^{2+}, damage to membrane ion transporters and/or other specific proteins and peroxidation of lipids (Orrenius et al., 1989; Cochrane, 1991). Increases in "free" Ca^{2+} can lead to increased NO˙ production, activation of Ca^{2+}-stimulated proteases (damaging the cytoskeleton and, in some cell types, degrading xanthine dehydrogenase to xanthine oxidase), and stimulation of phospholipase A_2, releasing arachidonic acid. Mechanisms of cell injury by oxidative stress are complex and overlapping. Oxidative stress can also lead to cell death; necrotic cell death leads to the release of transition metal ions able to promote such deleterious events as OH˙ formation (Fig. 5). Reactive oxygen species are not essential for apoptosis to occur but often appear to be involved in this process (Sarafian and Bredesen, 1994).

ANTIOXIDANT DEPLETION
does not prove oxidative damage, only that the
defence system is working

(Total AOX potential, depletion of specific
AOX, measurement of AOX-derived
species e.g. ascorbate radical, urate
oxidation products)

SPECIFIC MARKERS
show that oxidative damage has occurred

LIPID

peroxides   MDA, HNE and other
oxysterols   aldehydes
chlorinated/nitrated lipids   isoprostanes
                                             isoleukotrienes

DNA

strand breaks (e.g. comet assay)
oxidized bases in cells and urine
nitrated/deaminated bases in cells and urine
aldehyde/other base adducts in cells and urine

PROTEIN

-SH oxidation
carbonyl formation
aldehyde adducts
oxidized tyr, trp, his, met, lys, leu, ileu, val
nitrated/chlorinated tyrosine, trp, phe
protein peroxides/hydroxides

INDUCTION OF AOX ENZYMES
Does not prove oxidative damage

ROS/RNS TRAPPING
PBN/other spin traps
Aromatic probes (e.g. salicylate,
phenylalanine) other detectors

ROS/RNS formation does not imply ROS/RNS
importance. If they are important and the traps are
efficient, the traps should be protective

Figure 4. Biomarkers of oxidative damage. AOX, antioxidant; PBN, phenyl-*tert*-butylnitrone; MDA, malondialdehyde; HNE, hydroxynonenal.

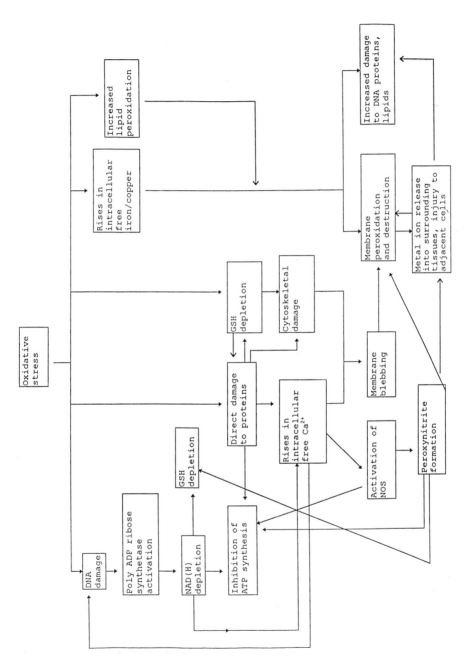

Figure 5. The complexity of cell damage mechanisms by oxidative stress.

Oxidative stress and human disease

Does oxidative damage play a role in human disease? Many of the biological consequences of vitamin E deficiency, selenium deficiency, and excess radiation exposure may be the result of oxidative damage, and oxidative damage may contribute to atherosclerosis, cancer, chronic inflammatory diseases, neurodegenerative diseases and retinopathy of prematurity. There are a multitude of papers in the biomedical literature suggesting a role for oxidative stress in a wide range of other human diseases (Tab. 7).

Table 7. Some of the clinical conditions in which the involvement of ROS/RNS has been suggested

---

Inflammatory–immune injury
    Glomerulonephritis
    Vasculitis[*]
    Autoimmune diseases[*]
    Rheumatoid arthritis[*]
    Hepatitis

Ischemia-reflow states
    Stroke/myocardial infarction/arrhythmias/angina
    Organ transplantation
    Inflamed rheumatoid joint
    Frostbite[*]
    Dupuytren's contracture[*]
    Reattachment of severed limbs[*]

Drug and toxin-induced reactions

Iron overload (tissue and plasma)
    Idiopathic haemochromatosis
    Dietary iron overload (Bantu)
    Thalassaemia and other chronic anaemias treated with multiple blood transfusions
    Nutritional deficiencies (kwashiorkor)
    Alcoholism
    Multiorgan failure
    Cardiopulmonary bypass

Alcoholism
    Includes alcohol-induced iron overload, alcoholic myopathy[*] and alcohol-induced brain damage

Radiation injury
    Nuclear explosions
    Accidental exposure
    Radiotherapy
    Hypoxic cell sensitizers
    Radon gas

Ageing
    Disorders of premature ageing
    Ageing itself
    Age-related diseases, e.g. cancer

Table 7. (continued)

Red blood cells
    Phenylhydrazine
    Primaquine, related drugs
    Lead poisoning
    Protoporphyrin photoxidation
    Malaria
    Sickle-cell anaemia
    Favism
    Fanconi's anaemia
    Haemolytic anaemia of prematurity
    Chemotherapy

Lung
    Cigarette smoke effects
    Other smoke inhalation
    Emphysema (COPD)
    Hyperoxia
    Bronchopulmonary dysplasia
    Air pollutants (O_3, NO_2, SO_2)
    ARDS
    Mineral dust pneumoconiosis
    Asbestos carcinogenicity
    Bleomycin toxicity
    SO_2 toxicity
    Paraquat toxicity
    Skatole toxicity
    Asthma
    Cystic fibrosis

Heart and cardiovascular system
    Alcohol cardiomyopathy
    Keshan disease (selenium deficiency)
    Atherosclerosis
    Claudication[*]
    Anthracycline cardiotoxicity

Kidney
    Autoimmune nephrotic syndromes
    Aminoglycoside nephrotoxicity
    Heavy metal nephrotoxicity (Pb, Cd, Hg)
    Myoglobin/haemoglobin damage[*] (Giulivi and Cadenas, 1994)
    Haemodialysis

Gastrointestinal tract
    Endotoxic liver injury
    Halogenated hydrocarbon liver injury (e.g. bromobenzene, CCl_4, halothane)
    Diabetogenic action of alloxan
    Pancreatitis
    NSAID-induced gastrointestinal tract lesions
    Oral iron poisoning

Table 7. (continued)

Brian/nervous system/neuromuscular disorders
    Hyperbaric oxygen
    Vitamin E deficiency
    Neurotoxins
    Alzheimer's disease
    Parkinson's disease
    Huntington's disease
    Hypertensive cerebrovascular injury
    Neuronal ceroid lipofuscinoses
    Allergic encephalomyelitis and other demyelinating diseases[*]
    Aluminium overload
    Potentiation of traumatic injury
    Muscular dystrophy[*]
    Amyotrophic lateral sclerosis[*]
    Guam dementia

Eye
    Cataractogenesis
    Ocular haemorrhage
    Degenerative retinal damage/macular degeneration
    Retinopathy of prematurity (retrolental fibroplasia)
    Photic retinopathy
    Penetration of metal objects

Skin
    Solar radiation
    Thermal injury
    Porphyria
    Hypericin, other photosensitizers
    Contact dermatitis
    Male pattern baldness

---

[*]Those relevant to muscle function.
ARDS, adult respiratory syndrome; NSAID, non-steroidal anti-inflammatory drug.

Tissue damage by disease, trauma, toxic agents, and other causes usually leads to formation of increased amounts of putative "injury mediators", such as prostaglandins, NO˙, leukotrienes, interleukins, interferons and cytokines, such as tumour necrosis factors (TNFs) (Halliwell et al., 1992a). All of these at various times have been suggested as playing important roles in different human diseases. Reactive oxygen/nitrogen/chlorine species can be placed in the same category, in that tissue damage will usually lead to increased ROS formation and oxidative stress (Fig. 1). Indeed, in most human diseases, oxidative stress may be a secondary phenomenon, a consequence of the disease activity (Fig. 6). Its importance will vary according to the disease state, i.e. the demonstration of oxidative damage does not prove the importance of oxidative damage. However, there is growing evidence that oxidative stress is indeed a major contributor to chronic inflammatory, cardiovascular and neurodegenerative diseases. Criteria to be fulfilled to reach such a conclusion are listed in Table 8. The contribution of oxidative stress to diseases and other patho-physiological conditions affecting skeletal muscle is the subject of the rest of this book.

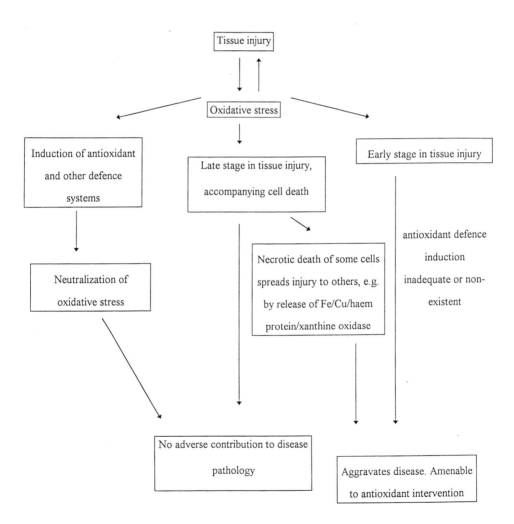

Figure 6. Oxidative stress: significant or not significant in tissue injury?

Table 8. Criteria for implicating ROS/RNS/RCS (or any other agent) as a significant contributor to tissue injury in human disease

---

1. The agent should always be present at the site of injury
2. Its time course of formation should be consistent with the time course of tissue injury
3. Direct application of the agent to the tissue at concentrations within the range found *in vivo* should reproduce most or all of the damage observed
4. Removing the agent or inhibiting its formation should diminish the injury to an extent related to the degree of removal of the agent or inhibition of its formation

---

References

Alam Z, Jenner A, Daniel SE, Lees AJ, Cairns N, Marsden CD, Jenner P and Halliwell B (1997) Oxidative DNA damage in the parkinsonian brain: an apparent selective increase in 8-hydroxyguanine levels in substantia nigra? *J Neurochem* 69: 1196–1203.

Ames BN (1989) Endogenous oxidative DNA damage, aging, and cancer. *Free Radical Res Commun* 7: 121–128.

Ames BN, Shigenaga MK and Hagen TM (1993) Oxidants, antioxidants and the degenerative diseases of aging. *Proc Natl Acad Sci USA* 90: 7915–7922.

Amstad P, Peskin A, Shah G, Mirault ME, Moret R, Zbinden I and Cerutti P (1991) The balance between Cu,Zn-superoxide dismutase and catalase affects the sensitivity of mouse epidermal cells to oxidative stress. *Biochemistry* 30: 9305–9313.

Anderson ME (1997) Glutathione and glutathione delivery compounds. *Adv Pharmacol* 38: 65–78.

Anggard E (1994) Nitric oxide: mediator, murderer, and medicine. *Lancet* 343: 1199–1206.

Avraham KB, Schickler M, Sapoznikov D, Yarom R and Groner Y (1988) Down's syndrome: abnormal neuromuscular function in tongue of transgenic mice with elevated levels of human Cu/Zn-superoxide dismutase. *Cell* 54: 823–829.

Babior BM and Woodman RC (1990) Chronic granulomatous disease. *Semin Hematol* 27: 247–259.

Balentine J (1982) *Pathology of Oxygen Toxicity.* Academic Press, New York.

Beckman JS, Chen J, Ischiropoulos H and Crow JP (1994) Oxidative chemistry of peroxynitrite. *Methods Enzymol* 233: 229–240.

Bielski BHJ (1985) Reactivity of HO_2/O_2^- radicals in aqueous solution. *J Phys Chem Ref Data* 14: 1041–1100.

Block G, Patterson B and Subar A (1992) Fruit, vegetables, and cancer prevention: a review of the epidemiological evidence. *Nutr Cancer* 18: 1–29.

Bolann BJ and Ulvik RJ (1990) On the limited ability of superoxide to release iron from ferritin. *Eur J Biochem* 193: 899–904.

Bowling AC, Schulz JB, Brown RH Jr and Beal MF (1993) Superoxide dismutase activity, oxidative damage and mitochondrial energy metabolism in familial and sporadic amyotrophic lateral sclerosis. *J Neurochem* 61: 2322–2325.

Breuer W, Epsztejn S and Cabantchik ZI (1996) Dynamics of the cytosolic chelatable iron pool of K562 cells. *FEBS Lett* 382: 304–308.

Buettner GR and Jurkiewicz BA (1993) Ascorbate free radical as a marker of oxidative stress: an EPR study. *Free Radical Biol Med* 14: 49–55.

Burdon RH, Alliangana D and Gill V (1994) Endogenously generated active oxygen species and cellular glutathione levels in relation to BHK-21 cell proliferation. *Free Radical Res* 21: 121–134.

Burkitt MJ (1993) ESR spin trapping studies into the nature of the oxidizing species formed in the Fenton reaction: pitfalls associated with the use of 5,5-dimethyl-1-pyrroline-*N*-oxide in the detection of the hydroxyl radical. *Free Radical Res Commun* 18: 43–57.

Burton GW and Ingold KU (1984) β-Carotene, an unusual type of lipid antioxidant. *Science* 224: 569–573.

Candeias LP, Patel KB, Stratford MRL and Wardman P (1993) Free hydroxyl radicals are formed on reaction between the neutrophil-derived species superoxide anion and hypochlorous acid. *FEBS Lett* 333: 151–153.

Candeias LP, Stratford MR and Wardman P (1994) Formation of hydroxyl radicals on reaction of HOCl with ferrocyanide, a model iron(II) complex. *Free Radical Res* 20: 241–249.

Chance B, Sies H and Boveris A (1979) Hydroperoxide metabolism in mammalian organs. *Physiol Rev* 59: 527–605.

Cochrane CG (1991) Mechanisms of oxidant injury of cells. *Mol Aspects Med* 12: 137–147.

Collins AR, Duthie SJ and Dobson VL (1993) Direct enzymic detection of endogenous oxidative base damage in human lymphocyte DNA. *Carcinogenesis* 14: 1733–1735.

Collins T (1993) Endothelial nuclear factor-κB and the initiation of the atherosclerotic lesion. *Lab Invest* 68: 499–508.

Darley-Usmar V and Halliwell B (1996) Blood radicals. *Pharmaceut Res* 13: 649–662.

Dean RT, Fu S, Stocker R and Davies MJ (1997) Biochemistry and pathology of radical-mediated protein oxidation. *Biochem J* 324: 1–18.

Demple B and Harrison L (1994) Repair of oxidative damage to DNA: enzymology and biology. *Annu Rev Biochem* 63: 915–948.

Diplock AT (1997) Will the "good fairies" please prove to us that vitamin E lessens human degenerative disease? *Free Radical Res* 26: 565–583.

Dizdaroglu M (1993) Chemistry of free radical damage to DNA and nucleoproteins. *In*: B Halliwell and OI Aruoma (eds): *DNA and Free Radicals*. Ellis Horwood, Chichester, pp 19–39.

Dupuy C, Virion A, Ohayon R, Kaniewski J, Deme D and Pommier J (1991) Mechanism of hydrogen peroxide formation catalysed by NADPH oxidase in thyroid plasma membrane. *J Biol Chem* 266: 3739–3743.

Editorial (1991) Retinopathy of prematurity. *Lancet* 337: 83–84.

Ehrenkranz RA (1989) Vitamin E and retinopathy of prematurity: still controversial. *J Pediat* 114: 801–803.

Eiserich JP, van der Vliet A, Handelman GJ, Halliwell B and Cross CE (1995) Dietary antioxidants and cigarette smoke-induced biomolecular damage: a complex interaction. *Am J Clin Nutr* 62 (suppl): 14950–15005.

Eiserich JP, Cross CE, Jones AD, Halliwell B and van der Vliet A (1996) Formation of nitrating and chlorinating species by reaction of nitrite with hypochlorous acid. *J Biol Chem* 271: 19199–19208.

El Ghissassi F, Barbin A, Nair J and Bartsch H (1995) Formation of 1,N[6]-ethenoadenine and 3,N[4]-ethenocytosine by lipid peroxidation products and nucleic acid bases. *Chem Res Toxicol* 8: 278–283.

Esterbauer H, Gebicki J, Puhl H and Jurgens G (1992) The role of lipid peroxidation and antioxidants in oxidative modification of LDL. *Free Radical Biol Med* 13: 341–390.

Frei B, Forte TM, Ames BN and Cross CE (1991) Gas phase oxidants of cigarette smoke induce lipid peroxidation and changes in lipoprotein properties in human blood plasma. *Biochem J* 277: 133–138.

Fridovich I (1986) Superoxide dismutases. *Methods Enzymol* 58: 61–97.

Fridovich I (1989) Superoxide dismutases: an adaptation to a paramagnetic gas. *J Biol Chem* 264: 7761–7764.

Gerschman K, Gilbert DL, Nye SW, Dwyer P and Fenn WO (1954) Oxygen poisoning and X-irradiation: a mechanism in common. *Science* 119: 623–626.

Gey KF (1995) Ten-year retrospective on the antioxidant hypothesis of arteriosclerosis. *J Nutr Biochem* 6: 206–236.

Giulivi C and Cadenas E (1994) Ferrylmyoglobin: formation and chemical reactivity toward electron-donating compounds. *Methods Enzymol* 233: 189–202.

Golden M (1994) Free radicals and the aetiology of kwashiorkor. *Biochemist* June/July: 12–15.

Granger DN (1988) Role of xanthine oxidase and granulocytes in ischemia-reperfusion injury. *Am J Physiol* 255: H1269–H1275.

Grisham MB (1994) Oxidants and free radicals in inflammatory bowel disease. *Lancet* 344: 859–861.

Gutteridge JMC (1986) Iron promoters of the Fenton reaction and lipid peroxidation can be released from haemoglobin by peroxides. *FEBS Lett* 201: 291–295.

Gutteridge JMC and Quinlan GJ (1992) Antioxidant protection against organic and inorganic oxygen radicals by normal human plasma: the important primary role for iron-binding and iron-oxidizing proteins. *Biochim Biophys Acta* 1159: 248–254.

Halliwell B (1990) How to characterize a biological antioxidant. *Free Radical Res Commun* 9: 1–32.

Halliwell B (1992) Reactive oxygen species and the central nervous system. *J Neurochem* 59: 1609–1623.

Halliwell B (1994) Free radicals and antioxidants: a personal view. *Nutr Rev* 52: 253–265.

Halliwell B (1995a) Antioxidant characterization, methodology and mechanism. *Biochem Pharmacol* 49: 1341–1348.

Halliwell B (1995b) Oxygen radicals, nitric oxide and human inflammatory joint disease. *Ann Rheum Dis* 54: 505–510.

Halliwell B (1996a) Oxidative stress, nutrition and health. Experimental strategies for optimization of nutritional antioxidant intake in humans. *Free Radical Res* 25: 57–74.

Halliwell B (1996b) Vitamin C: antioxidant or pro-oxidant *in vivo*? *Free Radical Res* 25: 439–454.

Halliwell B (1997) What nitrates tyrosine? Is nitrotyrosine specific as a biomarker of peroxynitrite formation *in vivo*. *FEBS Lett* 411: 157–160.

Halliwell B and Dizdaroglu M (1992) The measurement of oxidative damage to DNA by HPLC and GC/MS techniques. *Free Radical Res Commun* 16: 75–87.

Halliwell B and Gutteridge JMC (1990a) Role of free radicals and catalytic metal ions in human disease: an overview. *Methods Enzymol* 186: 1–85.

Halliwell B and Gutteridge JMC (1990b) The antioxidants of human extracellular fluids. *Arch Biochem Biophys* 280: 1–8.

Halliwell B, Gutteridge JMC and Cross CE (1992a) Free radicals, antioxidants, and human disease: where are we now? *J Lab Clin Med* 119: 598–620.

Halliwell B, Hu ML, Louie S, Duvall TR, Tarkington BR, Motchnik P and Cross CE (1992b) Interaction of nitrogen dioxide with human plasma. *FEBS Lett* 313: 62–66.

Hertog MGL, Feskens EJM, Hollman PCH et al. (1993) Dietary antioxidant flavonoids and risk of coronary heart disease: the Zutphen elderly study. *Lancet* 342: 1007–1011.

Hu ML, Louie S, Cross CE, Motchnik P and Halliwell B (1992) Antioxidant protection against hypochlorous acid in human plasma. *J Lab Clin Med* 121: 257–262.

Huie RE and Padmaja S (1993) The reaction of NO with superoxide. *Free Radical Res Commun* 18 195–199.

Iqbal J, Clerch LB, Hass MA, Frank L and Massaro D (1989) Endotoxin increases lung Cu, Zn superoxide dismutase mRNA: O_2 raises enzyme synthesis. *Am J Physiol* 257: L61–L64.

Janzen E (ed.) (1993) Third international symposium on spin-trapping and aminoxyl radical chemistry. *Free Radical Res Commun* 19 (Suppl. 1): S1–230.

Jenner P (1994) Oxidative damage in neurodegenerative disease. *Lancet* 344: 796–798.

Johnson IT, Williamson G and Musk SRR (1994) Anticarcinogenic factors in plant foods. A new class of nutrients? *Nutr Res* 7: 175–204.

Kaur H, Whiteman M and Halliwell B (1997) Peroxynitrite-dependent aromatic hydroxylation and nitration of salicylate and phenylalanine. Is hydroxyl radical involved? *Free Radical Res* 26: 71–82.

Kettle AJ (1996) Neutrophils convert tyrosyl residues in albumin to chlorotyrosine. *FEBS Lett* 379: 103–106.

Laires A, Gaspar J, Borba H, Proença M et al. (1993) Genotoxicity of nitrosated red wine and of the nitrosatable phenolic compounds present in wine: tyramine, quercetin and malvidine-3-glucose. *Food Chem Toxicol* 31: 989–994.

Lebovitz RM, Zhang H, Vogel H et al. (1996) Neurodegeneration, myocardial injury, and perinatal death in mitochondrial superoxide dismutase-deficient mice. *Proc Natl Acad Sci USA* 93: 9782–9787.

Li Y, Huang TT, Carlson EJ et al. (1995) Dilated cardiomyopathy and neonatal lethality in mutant mice lacking manganese superoxide dismutase. *Nat Genet* 11: 376–381.

Liochev S (1996) The role of iron-sulfur clusters in *in vivo* hydroxyl radical production. *Free Radical Res* 25: 369–384.

Maiorino M, Chu FF, Ursini F, Davies KJ, Doroshow JH and Esworthy RS (1991) Phospholipid hydroperoxide glutathione peroxidase is the 18-kDa selenoprotein expressed in human tumor cell lines. *J Biol Chem* 266: 7728–7732.

Maly FE (1990) The B-lymphocyte: a newly-recognized source of reactive oxygen species with immunoregulatory potential. *Free Radical Res Commun* 8: 143–148.

Mo JY, Maki H and Sekiguchi M (1992) Hydrolytic elimination of a mutagenic nucleotide, 8-oxodGTP, by human 18-kilodalton protein: sanitization of nucleotide pool. *Proc Natl Acad Sci USA* 89: 11021–11025.

Moncada S and Higgs A (1993) The L-arginine-nitric oxide pathway. *N Engl J Med* 329: 2002–2011.

Morrow JD and Roberts LJII (1994) Mass spectrometry of prostanoids: F_2-isoprostanes produced by non-cyclooxygenase free radical-catalysed mechanism. *Methods Enzymol* 233: 163–174.

Muller DPR and Goss-Sampson MA (1990) Neurochemical, neurophysiological, and neuropathological studies in vitamin E deficiency. *Crit Rev Neurobiol* 5: 239–263.

Ohshima H, Friesen M, Brouet I and Bartsch H (1990) Nitro-tyrosine as a new marker for endogenous nitrosation and nitration of proteins. *Food Chem Toxicol* 28: 647–652.

Orrenius S, McConkey DJ, Bellomo G and Nicotera P (1989) Role of Ca^{2+} in toxic cell killing. *Trends Pharmacol Sci* 10: 281–285.

Packer L (ed.) (1994) Oxygen radicals in biological systems. Part C. *Methods Enzymol* 233: 1–711.

Pannala AS, Rice-Evans CA, Halliwell B and Singh S (1997) Inhibition of peroxynitrite-mediated tyrosine nitration by catechin polyphenols. *Biochem Biophys Res Commun* 232: 164–168.

Prestera T and Talalay P (1995) Electrophile and antioxidant regulation of enzymes that detoxify carcinogens. *Proc Natl Acad Sci USA* 92: 8965–8969.

Pryor WA and Squadrito GL (1995) The chemistry of peroxynitrite: a product from the reaction of nitric oxide with superoxide. *Am J Physiol* 268: L699–L722.

Puppo A and Halliwell B (1988) Formation of hydroxyl radicals in biological systems. Does myoglobin stimulate hydroxyl radical production from hydrogen peroxide? *Free Radical Res Commun* 4: 415–422.

Reaume AG, Elliott J, Hoffman EK et al. (1996) Motor neurons in Cu/Zn superoxide dismutase-deficient mice develop normally but exhibit enhanced cell death after axonal injury. *Nat Genet* 13: 43–47.

Reznick AZ, Cross CE, Hu M, Suzuki YJ, Khwaja S, Safadi A, Motchnik PA, Packer L and Halliwell B (1992) Modification of plasma proteins by cigarette smoke as measured by protein carbonyl formation. *Biochem J* 286: 607–611.

Rice-Evans C, Sampson J, Bramley PM and Holloway DE (1997) Why do we expect carotenoids to be antioxidants *in vivo*? *Free Radical Res* 26: 381–398.

Roy D and Liehr JG (1991) Elevated 8-hydroxydeoxy-guanosine levels in DNA of diethylstilboestrol-treated syrian hamsters: covalent DNA damage by free radicals generated by redox cycling of diethylstilboestrol. *Cancer Res* 51: 3882–3885.

Rubbo H, Radi R, Trujillo M et al. (1994) Nitric oxide regulation of superoxide and peroxynitrite-dependent lipid peroxidation. *J Biol Chem* 269: 26066–26075.

Sarafian TA and Bredesen DE (1994) Is apoptosis mediated by reactive oxygen species? *Free Radical Res* 21: 1–8.

Schreck R, Albermann KAJ and Baeuerle PA (1992) Nuclear factor κB: an oxidative stress-responsive transcription factor of eukaryotic cells (a review). *Free Radical Res Commun* 17: 221–237.

Sies H (ed.) (1991) *Oxidative Stress, Oxidants and Antioxidants.* Academic Press, London.

Spector A, Yang Y, Ho YS et al. (1996) Variation in cellular glutathione peroxidase activity in lens epithelial cells, transgenics and knockouts does not significantly change the response to H_2O_2 stress. *Exp Eye Res* 62: 521–540.

Spencer JPE, Jenner A, Chimel K, Aruoma OI, Cross CE, Wu R and Halliwell B (1995) DNA strand breakage and base modification induced by hydrogen peroxide treatment of human respiratory tract epithelial cells. *FEBS Lett* 374: 233–236.

Spencer JPE, Wong J, Jenner A, Aruoma OI, Cross CE and Halliwell B (1996) Base modification and strand breakage in isolated calf thymus DNA and in DNA from human skin epidermal keratinocytes exposed to peroxynitrite or 3-morpholinosydnonimine. *Chem Res Toxicol* 9: 1152–1158.

Stadtman ER and Oliver CN (1991) Metal-catalysed oxidation of proteins. Physiological consequences. *J Biol Chem* 266: 2005–2008.

Steenken S (1989) Purine bases, nucleosides and nucleotides: aqueous solution redox chemistry and transformation reactions of their radical cations and e^- and OH· adducts. *Chem Rev* 89: 503–520.

Steinberg D, Parthasarathy S, Carew TE et al. (1989) Beyond cholesterol. Modifications of low-density lipoprotein that increase its atherogenicity. *N Engl J Med* 320; 915–924.

Totter JR (1980) Spontaneous cancer and its possible relationship to oxygen metabolism. *Proc Natl Acad Sci USA* 77: 1763–1767.

van der Vliet A, Eiserich JP, Halliwell B and Cross CE (1997) Formation of reactive nitrogen species during peroxidase-catalysed oxidation of nitrite. *J Biol Chem* 272: 7617–7625.

Verhagen V, Poulsen HE Loft S et al. (1995) Reduction of oxidative DNA-damage in humans by Brussels sprouts. *Carcinogenesis* 16: 969–970.

von Poppel G, Poulsen H, Loft S and Verhagen H (1995) No influence of beta carotene on oxidative DNA damage in male smokers. *J Nat Cancer Inst* 87: 310–311.

von Sonntag C (1987) *The Chemical Basis of Radiation Biology.* Taylor and Francis, London.

Weinberg ED (1990) Cellular iron metabolism in health and disease. *Drug Metab Rev* 22: 531–579.

Weisburger JH (1995) Tea antioxidants and health. *In*: E Cadenas and L Packer (eds): *Handbook of Antioxidants.* Marcel Dekker Inc., New York.

Weiss SJ (1989) Tissue destruction by neutrophils. *N Engl J Med* 320: 365–376.

Wennmalm A, Benthin G, Jungersten L, Edlund A and Petersson AS (1994) Nitric oxide formation in man as reflected by plasma levels of nitrate, with special focus on kinetics, confounding factors and response to immunological challenge. *In*: S Moncada, M Feelish, R Busse and EA Higgs (eds): *The Biology of Nitric Oxide,* Vol. 4. Portland Press, UK, pp 474–476.

Whiteman M and Halliwell B (1996) Protection against peroxynitrite-dependent tyrosine nitration and α_1-antiproteinase inactivation by ascorbic acid. A comparison with other biological antioxidants. *Free Radical Res* 25: 275–283.

Willett WC (1994) Diet and health: what should we eat? *Science* 264: 532–537.

Wiseman H and Halliwell B (1993) Carcinogenic antioxidants: diethylstilboestrol, hexoestrol and 17α-ethynyl-oestradiol. *FEBS Lett* 332: 159–163.

Wiseman H and Halliwell B (1996) Damage to DNA by reactive oxygen and nitrogen species: Role in inflammatory disease and progression to cancer. *Biochem J* 313: 17–29.

Yermilov V, Rubio J and Ohshima H (1995) Formation of 8-nitroguanine in DNA treated with peroxynitrite *in vitro* and its rapid removal from DNA by depurination. *FEBS Lett* 376: 207–210.

Yim MB, Chae HZ, Rhee SG, Chock PB and Stadtman ER (1994) On the protective mechanism of the thiol-specific anti-oxidant enzyme against the oxidative damage of biomolecules. *J Biol Chem* 269: 1621–1626.

Oxidative Stress in Skeletal Muscle
A.Z. Reznick et al. (eds)
© 1998 Birkhäuser Verlag Basel/Switzerland

Oxidative metabolism in skeletal muscle

O. Hänninen and M. Atalay

Department of Physiology, University of Kuopio, FIN-70211 Kuopio, Finland

Introduction

Skeletal muscle is unique because in addition to its aerobic capacity, it is adapted for short-term anaerobic activity, allowing for both extended physical activity of a lower intensity and short-term high-energy output. The dynamic range for the change in rate of ATP utilization is large, in excess of 100-fold for skeletal muscle. Changes in ATP use require compensatory adjustments of circulatory, cardiac and respiratory functions. In humans at rest, skeletal muscle receives about 5 ml blood/100 g tissue. During heavy exercise the share of cardiac output of the muscle tissue can increase in trained subjects up to 80% of the total cardiac output or even more (Fig. 1). The extraction of the oxygen also increases, as evidenced by the increasing arteriovenous difference

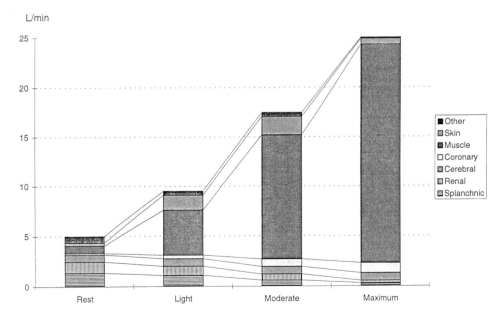

Figure 1. Distribution of cardiac output (l/min)expressed as blood flow to various tissues at rest and during light, moderate and maximum exercise (Modified from Anderson, 1968).

from 25% at rest to 80% or even more in maximal exercise. Thus, oxygen consumption in exercising human muscle can increase about a hundredfold, which is in fact quite modest in comparison to some animals, in which the increase can be a thousandfold.

In skeletal muscle, contractions are normally associated with depolarization of the plasma membrane which initiates the release of calcium ions from intracellular stores within the sarcoplasmic reticulum. Calcium ions bind to troponin C, a regulatory protein associated with the thin filaments, producing a change in protein conformation. This change of shape is transmitted to the other thin filament components (troponin T, troponin I, tropomyosin and actin), with the result that the actin subunits of the thin filaments are permitted to interact with the neighbouring myosin molecules. The contraction ceases when calcium ions are taken up by the sarcoplasmic reticulum, through the operation of an ATP-driven pump, commonly known as Ca^{2+} ATPase. In skeletal muscle, the bulk of the ATP produced through aerobic and anaerobic metabolism is consumed by this pump.

Chemical energy in skeletal muscle is derived mostly from glucose and fatty acids, which are also stored in significant amounts as glycogen and triglycerides, respectively, in the muscle fibres. A stationary equilibrium between the production and breakdown of ATP must be reached for

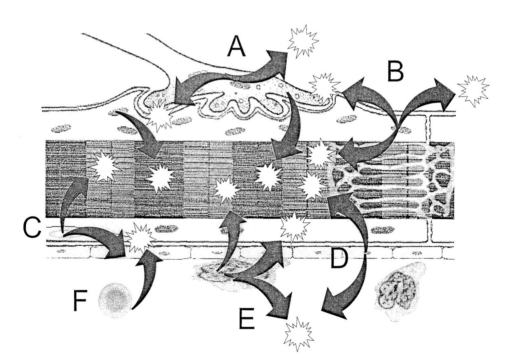

Figure 2. Free radical production in mitochondria of (A) neuromuscular junction, (B) skeletal muscle fibre and (C) fibroblast, as well as in (D) endothelium, (E) neutrophils and (F) erythrocytes.

sustained aerobic metabolism (Spurway, 1992). ATP and creatine phosphate concentrations are therefore fairly constant (about 5 mM and 30 mM, respectively). The stores of ATP itself are sufficient to provide energy for a few seconds, after which they are secured through three main mechanisms: short-term high-energy phosphates (creatine phosphate), and a medium-term (via anaerobic glycolysis) and a long-term energy supply (via oxidative phosphorylation of glucose and fatty acids to water and CO_2) (Brooks and Fahey, 1984; Astrand and Rodahl, 1986).

Skeletal muscle is also readily subjected to anaerobiosis, which allows for a short-term performance that far exceeds levels that can be handled aerobically. This can hamper the function of neuromuscular junctions, muscle fibres themselves, cells of the connective tissue and also the vessels, but it is a stimulus for adaptive changes in metabolisms which is an important element in training, e.g. for sports.

The intensive use of O_2 also results in profound generation of different forms of O_2, including reactive oxygen species (ROS) (Fig. 2). Reactive oxygen species promote muscle fatigue and tissue damage. Muscle tissue must therefore be able not only to generate energy, but also to handle secondary reactions and the oxidants generated. Skeletal muscle is capable of synthesizing glutathione (GSH) (Sen et al., 1992, 1993) which has a central role in the maintenance of antioxidant defence. It is an oxidizable substrate itself and helps to maintain vitamins C (in the aqueous phase) and E (in the lipid phase) in a reduced state (Sen, 1995; Sen et al., 1997). Enzymes of the glutathione system, such as glutathione peroxidase and glutathione S-transferases, complement the enzyme catalase in peroxide metabolism (Sen and Hänninen, 1994; Atalay et al., 1996b).

Mitochondria and aerobic metabolism

Citric acid cycle

In skeletal muscle mitochondria consume most of the oxygen and serve as the primary source of metabolic energy for sustained work. The citric acid cycle (also known as the tricarboxylic acid cycle) consists of a series of reactions in the mitochondria which bring about the catabolism of acetyl-coenzyme A (acetyl-CoA), liberating hydrogen equivalents that are used in oxidative phosphorylation for generation of ATP from ADP. The citric acid cycle plays a pivotal role in aerobic metabolism. It is the common final pathway for glucose (glycolysis), lipid (fatty acid β-oxidation) and protein catabolism. Both at rest and during exercise, fatty acid β-oxidation and glycolysis provide over 95% of acetyl-CoA entering the citric acid cycle.

In the citric acid cycle, acetyl-CoA condenses with oxaloacetate to form citrate. In subsequent reactions, oxaloacetate is again formed. In the process, three molecules of NADH, one of $FADH_2$, one of GTP and two of CO_2 are produced. The hydrogen equivalents subsequently undergo oxidative phosphorylation. The three molecules of NADH and one $FADH_2$ yield 11 ATP in the process. Thus, one turn of the citric acid cycle generates 12 molecules of ATP.

Mitochondrial oxidative phosphorylation

Mitochondrial oxidative phosphorylation couples respiration with generation of high-energy ATP. In oxidative phosphorylation, the reducing equivalents formed through the citric acid cycle flow through the respiratory chain in order of increasing redox potential, from the more electronegative components to the more electropositive oxygen. Coupling of the respiratory chain is not perfect, however. About 5% of O_2 consumed in the respiratory chain at rest has been estimated to escape as superoxide (Halliwell and Gutteridge, 1989). Such generation of ROS may have implications both in the resting state (Halliwell and Gutteridge, 1989) and with exercise (Sen and Hänninen, 1994; Sen et al., 1994a,b; Laaksonen et al., 1996).

In cells at high oxygen pressures, energy demand determines the rate of mitochondrial respiration, but substrate supply determines the cellular energy level at which that rate is attained. The pool of ATP and creatine phosphate is small compared with the energy required by active cells. As a result of the small size of the high-energy phosphate pool, the rate of ATP use can be more or less than the rate of ATP synthesis for only very short periods of time. Synthesis of ATP must therefore occur at a rate which, on average, is equal to that at which it is hydrolysed by cellular processes. Mitochondrial oxidative phosphorylation is thus tightly coupled to several different metabolic pathways and quickly responds to changes in tissue demands for ATP. The mitochondrial membrane is impermeable to NADH, and the reducing equivalents are therefore transported by energy-consuming shuttle systems. The glutamate/aspartate exchange across the mitochondrial membrane, which is essential for the function of the malate/aspartate shuttle, has been shown to be electrogenic, and thus the difference in NADH concentrations between the mitochondria and the cytosol is influenced by the mitochondrial membrane potential (Davis et al., 1980). An increase in mitochondrial NADH, or alternatively a decrease in mitochondrial membrane potential which could result from either a decrease in the ATP/ADP ratio (Davis et al., 1980) or a decrease in the O_2 availability (Chance, 1976) thus leads to increased cytosolic NADH. Increased cytosolic NADH relative to mitochondrial NADH is also necessary to create a driving force for the increased flux of reducing equivalents into the mitochondria, which occurs during high mitochondrial respiratory loads such as in aerobic exercise.

In recent years, ubiquinone or coenzyme Q has gained a great deal of attention. Ubiquinone is an electron carrier in the mitochondrial respiratory chain, and it links the flavoproteins to cytochrome *b*, the member of the cytochrome chain with the lowest redox potential. Ubiquinone exists in mitochondria as oxidized quinone under aerobic conditions and in the reduced quinol form under anaerobic conditions; it is a constituent of the mitochondrial lipids. The structure of ubiquinone is analogous to vitamins K and E. There is a large stoichiometric excess of ubiquinone in the mitochondria compared with other members of the respiratory chain. This suggests that ubiquinone is a mobile component of the respiratory chain and collects electrons from the more fixed flavoprotein complexes, passing them on to the cytochromes (Beyer, 1990).

Uncoupling protein-3 has recently been found in skeletal muscle. This protein accelerates O_2 consumption without ATP synthesis. An increase of ROS production can also be expected as a side reaction of the uncoupling-3 protein function. At present not much is known about the function or regulation of this protein (Boss et al., 1997).

Metabolism of glucose and glycogen in muscle fibres

Skeletal muscle derives glucose from glycogenolysis or by transport from the blood. Glucose can be stored as glycogen up to the level of 4–5% of the wet weight of the muscle tissue. Glycogen is the major supply of glucose during moderate exercise, and is a limiting factor in endurance events such as marathon runs. Muscle extracts glucose from the blood via insulin-dependent mechanisms (DeFronzo et al., 1981). Exercise increases skeletal muscle insulin sensitivity (Henriksson, 1995). During exercise, muscle also increases glucose uptake (Berger et al., 1976; DeFronzo et al., 1981; Katz et al., 1986; Richter et al., 1988) via the contraction-induced increase in membrane permeability to glucose (Richter et al., 1988), as well as via enhanced metabolic rate.

Other regulatory mechanisms, such as increased glycogenolysis or higher resting glycogen concentration, have been shown to inhibit glucose uptake (Jansson et al., 1986; Richter et al., 1988). Glucose uptake during exercise may also be decreased by increased concentrations of free fatty acids, although there is controversy about this question (Berger et al., 1976; Rennie and Holloszy, 1977; Tuominen et al., 1996). Muscle GLUT4 (glucose transporter) levels, an important limiting factor of glucose utilization, and glycogen synthase activity increase in response to exercise training (Rodnick et al., 1992; Henriksson, 1995). Increased GLUT4 levels may not necessarily mean increased glucose uptake, however (Idstrom et al., 1986).

Glycolysis – catabolism of glucose to acetyl-CoA or lactate

Glycolysis is the pathway for the catabolism of glucose, occurring in the cytosol. Glycolysis is unique in that it can use O_2 if available (pyruvate $\longrightarrow$ acetyl-CoA), or function without it (pyruvate $\longrightarrow$ lactate). The relative role of glycolysis as a source of energy varies between tissues (e.g. slight in heart, and major in the brain and red blood cells). In skeletal muscle, glycolysis permits high performance when aerobic metabolism alone is not sufficient. In skeletal muscle at rest, glycolysis provides almost half of the acetyl-CoA used in the citric acid cycle. In the process, six-carbon glucose is catabolized to three-carbon pyruvate and then to acetyl-CoA, resulting in a net production of two NADH and two ATP molecules. NADH formed through glycolysis is transported via the malate shuttle into the mitochondria and oxidized in the respiratory chain, with a net yield of two ATP molecules per NADH. Thus, in the complete oxidation of 1 mole of glucose under aerobic conditions, glycolysis yields eight ATP and the citric acid cycle 30 ATP molecules.

During sufficiently intense physical exercise, energy requirements exceed the capacity of skeletal muscle to generate energy aerobically. Muscle NADH decreases during low-intensity exercise, but it increases above resting values during high-intensity exercise. The increase in muscle NADH can result from limited availability of O_2 in the contracting muscle (Ren et al., 1988). During intense exercise, the increase in cytosolic NADH inhibits pyruvate dehydrogenase, resulting in increased reduction of pyruvate into lactate via extraction of a hydrogen atom from the NADH. Oxidized NAD (NAD^+) can act as a hydrogen acceptor, allowing glycolysis to proceed and provide delivery of energy for the reconstitution of energy-rich phosphates. Anaerobic formation of ATP is, however, at a high cost. Oxidation of 1 mol of glucose results in a net yield of only two ATP molecules.

Aerobic and anaerobic thresholds

Aerobic and anaerobic thresholds are commonly used terms in exercise physiology. In exercise testing, the aerobic threshold is the point at which lactate production in skeletal muscle exceeds peripheral or local catabolism, such that lactate levels start to accumulate in the systemic circulation, frequently arbitrarily defined as the point when blood lactate levels exceed 2 mM. The anaerobic threshold occurs when lactate levels start to accumulate at a steeper rate, frequently defined as when blood lactate levels exceed 4 mM. The aerobic threshold occurs around 55% and the anaerobic threshold around 65% of maximal O_2 consumption, although there is wide individual variation. The magnitude of increase in the aerobic and anaerobic thresholds brought about by training is greater than the increase in maximal O_2 consumption. Despite the concept of thresholds, the formation of lactate or acetyl-CoA from pyruvate in skeletal muscle is not an all-or-nothing reaction, but rather an equilibrium that shifts in favour of one or other in a tightly regulated fashion.

Some NADH molecules may be transhydrogenated by an energy-dependent mechanism to form NADPH. The physiological significance of transhydrogenation of NADH to NADPH is still unclear. Glutathione reductase and nitric oxide synthase, among other important enzymes, are NADPH dependent. It has been suggested that NADH can contribute to the reducing capacity by sparing NADPH (Lund et al., 1994). However, in addition to stimulating glycolysis, a high $NADH:NAD^+$ ratio impairs lipid metabolism and induces superoxide and peroxynitrite production, resulting in cell damage (Dawson et al., 1993; Williamson et al., 1993; Roy et al., 1997).

Fatty acids and triglycerides as an energy source

In the resting state, oxidation of free fatty acids provides more than 50% energy of skeletal muscle. During sustained exercise, their share depends on the intensity of exercise and the fitness of the subject, during mild-to-moderate exercise, the proportion may increase slightly, whereas during intense exercise, it decreases, although endurance athletes are capable of deriving a higher share of energy from fatty acids (van der Vusse and Reneman, 1996).

Fatty acids oxidized in skeletal muscle arise from blood triglycerides and fatty acids, adipose tissue and triglyceride stored within the skeletal muscle. Moderate exercise results in an increase in free fatty acids, arising mainly from adipose stores. The adipose tissue constitutes a long-term fat store of the order of 300 000 kJ (Guezennec, 1992; Ramsay, 1996). The release of free fatty acids is regulated by a number of hormones such as epinephrine, norepinephrine and adrenocorticotropic hormone which increase with physical and mental activity (van der Vusse and Reneman, 1996). Lipoprotein lipase synthesized in muscle fibres and partially transferred to the endothelium helps skeletal muscle to obtain free fatty acids. Muscle tissue contains triglycerides that have an energy value exceeding that of glycogen (van der Vusse and Reneman, 1996).

The sarcolemma of skeletal muscle contains fatty acid transporter proteins. In the fibres, fatty acid binding-protein carries the fatty acids to their intracellular sites of use. It is of interest that the dissociation of fatty acids is promoted by lowering of the pH, i.e. when the metabolic activity of the muscle fibres increases (Shiau and Levine, 1980; van der Horst et al., 1993; Calles-Escandon et al., 1996).

Fatty acid oxidation

Mitochondrial β-oxidation is the dominant form of fatty acid oxidation. A modified form of β-oxidation of very-long-chain fatty acids also takes place in the peroxisomes. Information about the contribution of peroxisomal β-oxidation to overall fatty acid metabolism is scanty. At rest in rat quadriceps femoris muscle its share of O_2 consumption has been reported to be about 15% (Chance et al., 1979). Owing to their high oxidase content peroxisomes are a significant site of H_2O_2 production.

Free fatty acid activation to acyl-CoA takes place mainly in the outer mitochondrial membrane. Long-chain acyl-CoA (and inactive fatty acids) cannot penetrate the inner mitochondrial membrane. Transport of long-chain acyl-CoA into the inner mitochondrial space occurs by combination with carnitine. Once across the inner mitochondrial membrane, acyl-carnitine is converted back to acyl-CoA.

Mitochondrial β-oxidation occurs in the inner mitochondrial space. In β-oxidation, the acyl chain is successively shortened by cleavage between the α and β carbons to form acetyl-CoA, a free fatty acid that is two carbons shorter than at the start of the cycle, a reduced flavoprotein and one NADH. Free fatty acids with an even number of carbons are oxidized completely to acetyl-CoA, whereas those with an odd number are oxidized to propionyl-CoA. Propionyl-CoA is converted to succinyl-CoA, which is the only product of fatty acid catabolism that can be used for glucogenesis. Complete oxidation of the eight carbon fatty acid palmitate requires two ATP molecules for fatty acid activation and yields 35 ATP molecules in the successive production of acetyl-CoA and 96 from the eight acetyl-CoA molecules that enter the citric acid cycle, giving a net yield of 129 ATP molecules.

Numerous reports indicate that endurance training increases the capacity of mitochondrial fatty acid oxidation in muscle (Fig. 3). Recent studies indicate that chronic muscle stimuli can up-regulate the gene that encodes peroxisomal fatty acid oxidation, suggesting that training may also up-regulate peroxisomal β-oxidation (Cresci et al., 1996). During acute exercise, however, there is indirect evidence to suggest that, compared with the resting state, the relative role of peroxisomal β-oxidation decreases (Meydani et al., 1993; Sen et al., 1997).

Skeletal muscle fibre type, and aerobic and glycolytic capacity

Motor units in skeletal muscle contain fibres that have differing metabolic or functional capabilities. Type 1 fibres have a low contraction speed and low glycolytic capacity, and are rich in mitochondria and myoglobin. As a result of high myoglobin content, they have a red appearance.

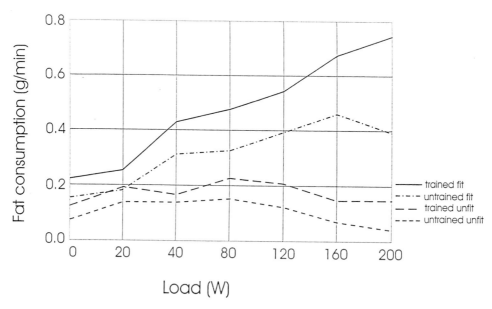

Figure 3. Fat consumption in fit and unfit obese young adults, prior to and after 8 weeks of light, endurance training, 4 days/week (R.K. Koskelo and O.H. Hänninen, unpublished data).

Type 1 fibres are adapted for steady, continuous work output with slow development of fatigue, and they operate largely on fat-based metabolism (Brooks and Fahey, 1984; van der Vusse and Reneman, 1996). Type 1 fibres are well supplied with capillaries, and their extracts after physical exercise have been shown to induce capillary growth *in vitro* (Oh-ishi et al., 1996).

Type 2A fibres combine high contraction speeds with an abundance of mitochondria and a high glycolytic capacity. They also have a red colour, and marked resistance to fatigue (Fitts, 1994). Type 1 and 2a fibres have higher lipoprotein lipase activity and more fatty acid transporter proteins than white fibres.

Type 2B fibres have high glycolytic capacity, but they are low in mitochondria and myoglobin, which leads to a pale appearance and low resistance to fatigue. They rely mainly on stored glycogen for energy production, and their maximal work output can be very high for a short period of time (Astrand and Rodahl, 1986). Lactate production is fast in these large-diameter fibres during exhaustive exercise. They have a relatively poor capillary blood supply. The ability of type 2B fibres to extract and consume fatty acids is also poor (Brooks and Fahey, 1984; van der Vusse and Reneman, 1996).

Among these muscle fibre types, type 2B fibres are dependent on glycogen stores as metabolic fuel (Griffiths and Rahim, 1978). A wide variety of substrates (free fatty acids, ketone bodies, triglycerides, branched-chain amino acids, pyruvate, lactate and glucose) can be used as energy source in types 1 and 2A fibres. The availability of these substrates depends on the dietary and hormonal status of the organism (Rennie and Holloszy, 1977). The oxidation ratio between fats and carbohydrates can be tilted towards carbohydrates by circulating insulin, partly through improved glucose transport into the tissue (DeFronzo et al., 1981).

Muscular fatigue and mitochondrial respiration

A rapid decrease in endurance time of skeletal muscle takes place at contraction levels above 15–20% of the maximum voluntary contraction for both sustained isometric and dynamic exercise (Hagberg, 1981). A muscle or muscle group may fatigue because of failure of any one or all of the different neuromuscular mechanisms involved in muscular contraction. Exposure of muscle to free radicals has been shown to accelerate fatigue (Barclay and Hansel, 1991). Davies et al. (1982) reported the first electron proton resonance (EPR) data showing that exhaustive exercise-induced increase of ROS production was associated with decreased mitochondrial respiration. Biochemical changes observed in fatigued muscle *in vivo*, such as increased production of thio-barbituric acid-reactive substances (Anzueto et al., 1992) and GSH oxidation (Cooper et al., 1986; Anzueto et al., 1992) are indeed major markers of oxidative stress (Sen et al., 1994a, 1994b, 1997; Sen, 1995).

Non-specific antioxidants used to test the involvement of ROS in fatigue have yielded mixed results (Barclay and Hansel, 1991; Shrier et al., 1991). However, Reid et al. (1994) demonstrated that *N*-acetylcysteine, a thiol group antioxidant, delays fatigue in human muscle without effecting contractility in unfatigued muscle. Thus antioxidants can be a therapeutic tool in delaying muscular fatigue and enhancing exercise performance.

Mitochondria, vasculature and NO metabolism as sources of free radicals in skeletal muscle

The main generation of radicals in skeletal muscle takes place in mitochondria. A moderately active person consumes about 22 mol O_2 in a day. It is estimated that about 3–5% of O_2 consumed escapes the normal metabolic routes to water and other products, and ends up as ROS (Halliwell and Gutteridge, 1989). Reactive oxygen species are formed during activity of the mitochondrial electron transfer chain through univalent reduction of O_2. The rate of superoxide formation is directly proportional to the rate of mitochondrial O_2 use (Boveris and Chance, 1973). During metabolic stress conditions, such as ischemia–reperfusion and heavy physical exercise which are associated with increased oxygen use, electron pressure in the electron transport chain increases, and cytochrome oxidase activity decreases, resulting in the requirements of alternative electron acceptors such as coenzyme Q (Sjodin et al., 1990; Soussi et al., 1990). Boveris et al. (1976) and Forman and Boveris (1982) have suggested that semiquinone is the major mitochondrial autooxidizable component, which is univalently reduced from coenzyme Q (CoQ) by electrons from NADH or succinate in the presence of NADH–CoQ reductase and succinate dehydrogenase. Using electron spin resonance (ESR) spectrographic data, Ksenzenko et al. (1983) have demonstrated that the superoxide is formed from the coenzyme Q semiquinone under experimental conditions in which the submitochondrial electron transfer chain is completely blocked by inhibitors that interfere with the oxidant ferricytochrome *b*-566. The other sites of superoxide production in the respiratory chain are located within NADH dehydrogenase between the mercurial-sensitive and rotenone-sensitive sites, most probably a non-haem iron–sulphur function (Boveris and Chance, 1973; Takeshige and Minakami, 1979).

Muscle blood flow increases exponentially with metabolism. Reactive oxygen species arising from the vasculature may therefore play an especially important role during exercise. Xanthine dehydrogenase, located in endothelial cells in most tissues, including muscle, is an important enzyme for degradation of purine metabolites. Xanthine dehydrogenase uses NAD^+ as an electron acceptor. During metabolic stress conditions, resulting from proteolytic activities (Della Corte and Stirpe, 1968) and oxidation of sulphhydryl groups (Della Corte and Stirpe, 1972), xanthine dehydrogenase is converted to xanthine oxidase, which uses molecular O_2 as an electron acceptor and generates superoxide. In skeletal muscle the vascular endothelium is rich in xanthine dehydrogenase. This enzyme can be converted into xanthine oxidase under anaerobic conditions. Xanthine oxidase activity is an important source of ROS in muscle. Activated neutrophils also contribute to this conversion (Wakabayashi et al., 1995).

Increased xanthine oxidase activity in muscle microvascular walls and in leukocytes that have migrated to muscle contributes to ROS production in muscle during exercise, especially eccentric-type exercise (Hellsten et al., 1997). It is well established that in ischemia–reperfusion injury excess ROS are produced through xanthine oxidase activity. Skeletal muscle is subject to partial ischemia and reperfusion during heavy physical exercise (Soussi et al., 1990).

Acute heavy physical exercise induces as acute phase immune response characterized by activation and mobilization of neutrophils, as in ischemia and reperfusion (McCord and Roy, 1982; Komatsu et al., 1992). In heavy physical exercise, especially eccentric loading, the neutrophils migrate to the site of injury in the muscles. Activated neutrophils accumulate at the site of injury and phagocytose cellular fragments and micro-organisms via proteolytic enzymes and with the help of ROS (Cannon and Blumberg, 1994; Jones, 1994). During the respiratory burst, activated neutrophils and other phagocytes produce an excess of superoxide by an NADPH oxidase, which is subsequently converted to strong physiological oxidants, e.g. hydrogen peroxide and hypochlorite (Chance et al., 1979; Weiss, 1989; Segal and Abo, 1993; Jones, 1994). Reactive oxygen species formed during the oxidative burst contribute to pathogen killing and wound healing (Besner et al., 1992). Under certain conditions, such as physical exercise, ROS formed during the oxidative burst may also contribute to oxidative host tissue damage (Weiss, 1989; Ward and Michigan, 1991; Atalay et al., 1996a).

Skeletal muscle expresses constitutive nitric oxide synthase (NOS) and produces nitric oxide (NO) (Brenman et al., 1995), which modulates vascular control (King et al., 1994; Ward and Hussain, 1994), glucose uptake (Balon and Nadler, 1994) mitochondrial O_2 consumption (King et al., 1994; Kobzik et al., 1995), cyclic nucleotide metabolism (Kobzik et al., 1994), and contractile function (Kobzik et al., 1994; Murrant et al., 1994) of skeletal muscle. NOS activity is detectable in a variety of limb and respiratory muscles, and it differs markedly among muscles. In the absence of exogenous interventions, rat skeletal muscle expresses neuronal NOS endothelial NOS, or both (Kobzik et al., 1994, 1995). It has recently been demonstrated that NO is synthesized in skeletal muscle by neuronal-type NOS (nNOS), especially in fast-twitch fibres.

Nitric oxide may influence the contractile function of muscles that rely heavily on oxidative metabolism. Recent observations have shown that NOS inhibitors augmented, whereas NO donors diminished, skeletal muscle contraction (Kobzik et al., 1994). The amount of NOS activity within different muscles correlated with their contractile properties. However, endogenous NO is also suggested to be essential for optimal myofilament function during active shortening (Morrison et al., 1996). NO selectively facilitates detachment of a slower cross-bridge population. This action would minimize the internal load against which muscle shortens, thereby increasing both velocity and external power (Claflin and Faulkner, 1985; Josephson and Edman, 1988; Morrison et al., 1996).

In skeletal muscle, motor end-plates are very rich in mitochondria. They undergo active aerobic metabolism. It has recently been shown that, with chronic overloading of the soleus muscle, oxidative enzyme activities increase specifically in the motor end-plate areas (Campbell et al., 1996). Motor end-plates of fast-twitch fibres are very rich in neural-type NOS (Brenman et al., 1995). Nitric oxide reacts with superoxide to form peroxynitrite, which is strongly neurotoxic (Lipton et al., 1993) and attacks side chains of cysteine and tyrosine in proteins (Beckman et al., 1993). Thus, neuromuscular junctions may be considered to be a significant location of free radical production in the skeletal muscle. It has been discovered very recently by Nagase et al. (1997) that NO can be generated non-enzymatically, by the reaction between arginine and H_2O_2.

Alternatively, NO might also act as a brake on respiration through direct inhibition of cytochromes, or to control other physiological functions that are regulated by mitochondrial Ca^{2+} release which follows cytochrome inhibition (Richter et al., 1994). NO and NO donors have been observed to modulate oxygen consumption directly in intact skeletal muscle and to inhibit mitochondrial function (Schweizer and Richter, 1994). This inhibition may be blocked by pre-treatment with NOS inhibitors. As oxygen free radicals can promote the development of fatigue (Reid et al., 1992), the rapid inactivation of mitochondria-derived superoxide radical by NO may serve a protective role.

Conclusion

The mass of the locomotor system makes up about two-thirds of the human body. At rest its share of the cardiac output is one-sixth of the total, and equal to that of the brain. At maximal activation in aerobic work, the O_2 consumption of muscle dominates, and its blood circulation corresponds to 80% of the cardiac output. The main sources of energy for muscle are glucose and fatty acids, the consumption of which depends on the load and fitness of the subject as well as the availability of O_2. ATP production from cytosolic glycolysis, mitochondrial β-fatty acid oxidation and the citric acid cycle is tightly regulated and responds quickly to muscle demands for more ATP. When energy demands exceed the capacity of muscle to provide ATP through the citric acid cycle, glycolysis is stimulated and lactic acid is produced, yielding ATP anaerobically. The generation of reactive intermediates of oxygen metabolism and of other radicals has been observed in the skeletal muscle. Free radicals thus generated can cause harm and contribute to fatigue development and to slowing down of muscle activity.

References

Anderson KL (1968) The Cardiovascular System In Exercise. *In*: HB Falls (ed.): *Exercise Physiology*. Academic Press, New York.

Anzueto A, Andrade FH, Maxwell LC, Levine SM, Lawrence RA, Gibbons WJ and Jenkinson SG (1992) Resistive breathing activates the glutathione redox cycle and impairs performance of rat diaphragm. *J Appl Physiol* 72: 529–534.

Astrand PO and Rodahl K (1986) *Textbook of Work Physiology*, 3rd ed. McGrawHill, NewYork.

Atalay M, Marnila P, Lilius EM, Hänninen O and Sen CK (1996a) Glutathione dependent modulation of exhaustive exercise induced changes in neutrophil function. *Eur J Appl Physiol Occup Physiol* 74: 342–347.

Atalay M, Seene T, Hänninen O and Sen CK (1996b) Skeletal muscle and heart antioxidant defences in response to sprint training. *Acta Physiol Scand* 158: 129–134.

Balon TW and Nadler JL (1994) NO release is present from incubated skeletal muscle preparations *J Appl Physiol* 77: 2519–2521.

Barclay JK and Hansel M (1991) Free radicals may contribute to oxidative skeletal muscle fatigue. *Can J Physiol Pharmacol* 69: 279–284.

Beckman JS, Carson M, Smith CD and Koppenol WH (1993) ALS, SOD and peroxynitrite [letter]. *Nature* 364: 584.

Berger M, Hagg SA, Goodman MN and Ruderman NB (1976) Glucose metabolism in perfused skeletal muscle. Effects of starvation, diabetes, fatty acids, acetoacetate, insulin and exercise on glucose uptake and disposition. *Biochem J* 158: 191–202.

Besner GE, Glick PL, Karp MP, Wang WC, Lobe TE, White CR and Cooney D (1992) Recombinant human granulocyte colony-stimulating factor promotes wound healing in a patient with congenital neutropenia. *J Pediat Surg* 27: 288–291.

Beyer RE (1990) The participation of coenzyme Q in free radical production and antioxidation. *Free Radical Biol Med* 8: 545–565.

Boss O, Samec S, Paoloni Giacobino A, Rossier C, Dulloo A, Seydoux J, Muzzin P and Giacobino JP (1997) Uncoupling protein-3: a new member of the mitochondrial carrier family with tissue-specific expression. *FEBS Lett* 408: 39–42.

Boveris A, Cadenas E and Stoppani AO (1976) Role of ubiquinone in the mitochondrial generation of hydrogen peroxide. *Biochem J* 156: 435–444.

Boveris A and Chance B (1973) The mitochondrial generation of hydrogen peroxide. General properties and effect of hyperbaric oxygen. *Biochem J* 134: 707–716.

Brenman JE, Chao DS, Xia H, Aldape K and Bredt DS (1995) Nitric oxide synthase complexed with dystrophin and absent from skeletal muscle sarcolemma in Duchenne muscular dystrophy. *Cell* 82: 743–752.

Brooks GA and Fahey TD (1984) *Exercise Physiology: Human Bioenergetics and its Applications*. Macmillan Publishing Company New York.

Calles-Escandon J, Sweet L, Ljungqvist O and Hirshman MF (1996) The membrane-associated 40 KD fatty acid binding protein (Berk's protein), a putative fatty acid transporter is present in human skeletal muscle. *Life Sci* 58: 19–28.

Campbell RJ, Jasmin BJ and Michel RN (1996) Succinate dehydrogenase activity within synaptic and extrasynaptic compartments of functionally overloaded rat skeletal muscle fibres. *Pflügers Arch* 431: 797–799.

Cannon JG and Blumberg JB (1994) Acute phase immune response in exercise. *In*: CK Sen, L Packer and O Hänninen (eds): *Exercise and Oxygen Toxicity*. Elsevier, Amsterdam, pp 89–126.

Chance B (1976) Pyridine nucleotide as an indicator of the oxygen requirements for energy-linked functions of mitochondria. *Circ Res* 38(suppl 1): I31–I38.

Chance B, Sies H and Boveris A (1979) Hydroperoxide metabolism in mammalian organs. *Physiol Rev* 59: 527–605.

Claflin DR and Faulkner JA (1985) Shortening velocity extrapolated to zero load and unloaded shortening velocity of whole rat skeletal muscle. *J Physiol London* 359: 357–363.

Cooper MB, Jones DA, Edwards RH, Corbucci GC, Montanari G and Trevisani C (1986) The effect of marathon running on carnitine metabolism and on some aspects of muscle mitochondrial activities and antioxidant mechanisms. *J Sport Sci* 4: 79–87.

Cresci S, Wright LD, Spratt JA, Briggs FN and Kelly DP (1996) Activation of a novel metabolic gene regulatory pathway by chronic stimulation of skeletal muscle. *Am J Physiol* 270: C1413–C1420.

Davis EJ, Bremer J and Akerman KE (1980) Thermodynamic aspects of translocation of reducing equivalents by mitochondria. *J Biol Chem* 255: 2277–2283.

Davies KJ, Quintanilha AT, Brooks GA and Packer L (1982) Free radicals and tissue damage produced by exercise. *Biochem Biophys Res Commun* 107: 1198–1205.

Dawson TL, Gores GJ, Nieminen AL, Herman B and Lemasters JJ (1993) Mitochondria as a source of reactive oxygen species during reductive stress in rat hepatocytes. *Am J Physiol* 264: C961–C967.

DeFronzo RA, Ferrannini E, Sato Y, Felig P and Wahren J (1981) Synergistic interaction between exercise and insulin on peripheral glucose uptake. *J Clin Invest* 68: 1468–1474.

Della Corte E and Stirpe F (1968) The regulation of rat-liver xanthine oxidase, activation by proteolytic enzymes. *FEBS Lett* 2: 83–84.

Della Corte E and Stirpe F (1972) The regulation of rat-liver xanthine oxidase. Involvement of thiol groups in the conversion of the enzyme from dehydrogenase (typeD) into oxidase (type O) and the purification of the enzyme. *Biochem J* 126: 739–745.

Fitts RH (1994) Cellular mechanisms of muscle fatigue. *Physiol Rev* 74: 49–94.

Forman HJ and Boveris A (1982) Superoxide radical and hydrogen peroxide in mitochondrial. *In*: WA Pryor (ed.): *Free Radicals in Biology*. Academic Press, New York, pp 65–90.

Griffiths JR and Rahim ZH (1978) Glycogen as a fuel for skeletal muscle. *Biochem Soc Trans* 6: 530–534.

Guezennec CY (1992) Role of lipids on endurance capacity in man. *Int J Sport Med* 13 Suppl 1: S114–S118.

Hagberg M (1981) Muscular endurance and surface electromyogram in isometric and dynamic exercise. *J Appl Physiol* 5: 1–7.

Hagberg M (1984) Occupational musculoskeletal stress and disorders of the neck and shoulder: a review of possible pathophysiology. *Int Arch Occup Environ Health* 53: 269–278.

Halliwell B and Gutteridge JMC (1989) *Free Radicals in Biology and Medicine*. Clarendon Press, Oxford.

Hellsten Y, Frandsen U, Orthenblad N, Sjodin B and Richter EA (1997) Xanthine oxidase in human skeletal muscle following eccentric exercise: a role in inflammation. *J Physiol London* 498 (Pt 1): 239–248.

Henriksson J (1995) Influence of exercise on insulin sensitivity. *J Cardiovasc Risk* 2: 303–309.

Idstrom JP, Elander A, Soussi B, Schersten T and Bylund-Fellenius AC (1986) Influence of endurance training on glucose transport and uptake in rat skeletal muscle. *Am J Physiol* 251: H903–H907.

Jansson E, Hjemdahl P and Kaijser L (1986) Epinephrine-induced changes in muscle carbohydrate metabolism during exercise in male subjects. *J Appl Physiol* 60: 1466–1470.

Jones OT (1994) The regulation of superoxide production by the NADPH oxidase of neutrophils and other mammalian cells. *Bioessays* 16: 919–923.

Josephson RK and Edman KA (1988) The consequences of fibre heterogeneity on the force-velocity relation of skeletal muscle. *Acta Physiol Scand* 132: 341–352.

Katz A, Broberg S, Sahlin K and Wahren J (1986) Leg glucose uptake during maximal dynamic exercise in humans. *Am J Physiol* 251: E65–E70.

King CE, Melinyshyn MJ, Mewburn JD, Curtis SE, Winn MJ, Cain SM and Chapler CK (1994) Canine hindlimb blood flow and O_2 uptake after inhibition of EDRF/NO synthesis. *J Appl Physiol* 76: 1166–1171.

Kobzik L, Reid MB, Bredt DS and Stamler JS (1994) Nitric oxide in skeletal muscle. *Nature* 372: 546–548.

Kobzik L, Stringer B, Balligand JL, Reid MB and Stamler JS (1995) Endothelial type nitric oxide synthase in skeletal muscle fibres: mitochondrial relationships. *Biochem Biophys Res Commun* 211: 375–381.

Komatsu H, Koo A, Ghadishah E, Zeng H, Kuhlenkamp JF, Inoue M, Guth PH and Kaplowitz N (1992) Neutrophil accumulation in ischemic reperfused rat liver: evidence for a role for superoxide free radicals. *Am J Physiol* 262: G669–676.

Ksenzenko M, Konstantinov AA, Khomutov GB, Tikhonov AN and Ruuge EK (1983) Effect of electron transfer inhibitors on superoxide generation in the cytochrome bc1 site of the mitochondrial respiratory chain. *FEBS Lett* 155: 19–24.

Laaksonen DE, Atalay M, Niskanen L, Uusitupa M, Hänninen O and Sen CK (1996) Increased resting and exercise-induced oxidative stress in young IDDM men. *Diabetes Care* 19: 569–574.

Lipton SA, Choi YB, Pan ZH, Lei SZ, Chen HS, Sucher NJ, Loscalzo J, Singel DJ and Stamler JS (1993) A redox-based mechanism for the neuroprotective and neurodestructive effects of nitric oxide and related nitroso compounds *Nature* 364: 626–632.

Lund LG, Paraidathathu T, Kehrer JP (1994) Reduction of glutathione disulfide and the maintenance of reducing equivalents in hypoxic hearts after the infusion of diamide. *Toxicology* 93: 249–262.

McCord JM and Roy RS (1982) The pathophysiology of superoxide: roles in inflammation and ischemia. *Can J Physiol Pharmacol* 60: 1346–1352.

Meydani M, Evans WJ, Handelman G, Biddle L, Fielding RA, Meydani SN, Burrill J, Fiatarone MA, Blumberg JB and Cannon JG (1993) Protective effect of vitamin E on exercise-induced oxidative damage in young and older adults. *Am J Physiol* 264: R992–R998.

Morrison RJ, Miller CC and Reid MB (1996) Nitric oxide effects on shortening velocity and power production in the rat diaphragm. *J Appl Physiol* 80: 1065–1069.

Murrant CL, Woodley NE and Barclay JK (1994) Effect of nitroprusside and endothelium-derived products on slow twitch skeletal muscle function *in vitro*. *Can J Physiol Pharmacol* 72: 1089–1093.

Nagase S, Takemura K, Ueda A, Hirayama A, Aoyagi K, Kondoh M and Koyama A (1997) A novel nonenzymatic pathway for the generation of nitric oxide by the reaction of hydrogen peroxide and D- or L-arginine. *Biochem Biophys Res Commun* 233: 150–153.

Ohishi S, Yamashita H, Kizaki T, Nagata N, Sen CK, Hänninen O, Izawa T, Sakurai T and Ohno H (1996) A single bout of exercise induces capillary growth in skeletal muscle through basic fibroblast growth factor (bFGF). *Pathophysiology* 3: 197–201.

Ramsay TG (1996) Fat cells. *Endocrinol Metab Clin N Amer* 25: 847–870.

Reid MB, Haack KE, Franchek KM, Valberg PA, Kobzik L and West MS (1992) Reactive oxygen in skeletal muscle. I. Intracellular oxidant kinetics and fatigue *in vitro*. *J Appl Physiol* 73: 1797–1704.

Reid MB, Stokic DS, Koch SM, Khawli FA and Leis AA (1994) N-acetylcysteine inhibits muscle fatigue in humans. *J Clin Invest* 94: 2468–2474.

Ren JM, Henriksson J, Katz A and Sahlin K (1988) NADH content in type I and type II human muscle fibres after dynamic exercise. *Biochem J* 251: 183–187.

Rennie MJ and Holloszy JO (1977) Inhibition of glucose uptake and glycogenolysis by availability of oleate in well-oxygenated perfused skeletal muscle. *Biochem J* 168: 161–170.

Richter EA, Kiens B, Saltin B, Christensen NJ and Savard G (1988) Skeletal muscle glucose uptake during dynamic exercise in humans: role of muscle mass. *Am J Physiol* 254: E555–E561.

Richter C, Gogvadze V, Schlapbach R, Schweizer M and Schlegel J (1994) Nitric oxide kills hepatocytes by mobilizing mitochondrial calcium. *Biochem Biophys Res Commun* 205: 1143–1150.

Rodnick KJ, Piper RC, Slot JW and James DE (1992) Interaction of insulin and exercise on glucose transport in muscle. *Diabetes Care* 15: 1679–1689.

Roy S, Sen CK, Tritschler HJ and Packer L (1997) Modulation of cellular reducing equivalent homeostasis by alpha-lipoic acid. Mechanisms and implications for diabetes and ischemic injury. *Biochem Pharmacol* 53: 393–399.

Schweizer M and Richter C (1994) Nitric oxide potently and reversibly deenergizes mitochondria at low oxygen tension. *Biochem Biophys Res Commun* 204: 169–175.

Segal AW and Abo A (1993) The biochemical basis of the NADPH oxidase of phagocytes. *Trends Biochem Sci* 18: 43–47.

Sen CK, Marin E, Kretzschmar M and Hänninen (1992) O Skeletal muscle and liver glutathione homeostasis in response to training, exercise, and immobilization. *J Appl Physiol* 73: .

Sen CK, Rahkila P and Hänninen O (1993) Glutathione metaboilsm in skeletal muscle derived cells of the L6 line. *Acta Phyisiol Scand* (1) 21–26.

Sen CK and Hänninen O (1994) *In:* CK Sen, L Packer and O Hänninen (eds): *Physiological Antioxidants in Exercise and Oxygen Toxicity.* Elsevier, Amsterdam, pp 89–126.

Sen CK, Atalay M and Hänninen O (1994) Exercise-induced oxidative stress: glutathione supplementation and deficiency. *J Appl Physiol* 77: 2177–2187.

Sen CK, Rankinen T, Vaisanen S and Rauramaa R (1994) Oxidative stress after human exercise: effect of N-acetylcysteine supplementation. *J Appl Physiol* 76: 2570–2577.

Sen CK (1995) Oxidants and antioxidants in exercise. *J Appl Physiol* 79: 675–686.

Sen CK and Packer L (1996) Antioxidant and redox regulation of gene transcription. *FASEB J* 10: 709–720.

Sen CK, Atalay M, Agren J, Laaksonen D, Roy S and Hänninen O (1997) Regulation of oxidative stress by fish oil and vitamin E supplementation at rest and after physical exercise. *J Appl Physiol* 83: 189–195.

Shiau YF and Levine GM (1980) pH dependence of micellar diffusion and dissociation. *Am J Physiol* 239: G177–G182.

Shrier I, Hussain S and Magder S (1991) Failure of oxygen radical scavengers to modify fatigue in electrically stimulated muscle. *Can J Physiol Pharmacol* 69: 1470–1475.

Sjodin B, Hellsten Westing Y and Apple FS (1990) Biochemical mechanisms for oxygen free radical formation during exercise. *Sport Med* 10: 236–254.

Soussi B, Idstrom JP, Schersten T and Bylund-Fellenius AC (1990) Cytochrome c oxidase and cardiolipin alterations in response to skeletal muscle ischemia and reperfusion. *Acta Physiol Scand* 138: 107–114.

Spurway NC (1992) Aerobic exercise, anaerobic exercise and the lactate threshold. *Brit Med Bull* 48: 569–591.

Takeshige K and Minakami S (1979) Mitochondrial H_2O_2 production occurs during state 4 respiration, and it ceases upon transition to state 3. *Biochem J* 180: 129–135.

Tuominen JA, Ebeling P, Bourey R, Koranyi L, Lamminen A, Rapola J, Sane T, Vuorinen-Markkola H and Koivisto VA (1996) Postmarathon paradox: insulin resistance in the face of glycogen depletion. *Am J Physiol* 270: E336–E343.

van der Horst DJ, van Doorn JM, Passier PC, Vork MM and Glatz JF (1993) Role of fatty acid-binding protein in lipid metabolism of insect flight muscle. *Mol Cell Biochem* 123: 145–152.

van der Vusse GJ and Reneman RS (1996) Lipid metabolism in muscle *In:* LB Rowell and JT Shepherd (eds): *Handbook of Physiology,* Section 12: *exercise: regulation and integration of multiple systems.* Oxford University Press, New York.

Wakabayashi Y, Fujita H, Morita I, Kawaguchi H and Murota S (1995) Conversion of xanthine dehydrogeanse to xanthine oxidase in bovine carotid artery endothelial cells induced by activated neutrophils: involvement of adhesion molecules. *Biochim Biophys Acta* 1265: 103–109.

Ward PA and Michigan AA (1991) Mechanisms of endothelial cell killing by H_2O_2. *Am J Med* 91: 89S–94S.

Ward ME and Hussain SN (1994) Diaphragmatic pressure flow relationship during hemorrhagic shock: role of nitric oxide. *J Appl Physiol* 77: 2244–2249.

Weiss SJ (1989) Tissue destruction by neutrophils. *N Engl J Med* 320: 365–376.

Williamson JR, Chang K, Frangos M, Hasan KS, Ido Y, Kawamura T, Nyengaard JR, van den Enden M, Kilo C and Tilton RG (1993) Hyperglycemic pseudohypoxia and diabetic complications. *Diabetes* 42: 801–813.

Oxidative Stress in Skeletal Muscle
A.Z. Reznick et al. (eds)
© 1998 Birkhäuser Verlag Basel/Switzerland

Strategies to assess oxidative stress

A.Z. Reznick[1], L. Packer[2] and C.K. Sen[2]

[1]*Musculoskeletal Laboratory, Department of Morphological Sciences, Technion Faculty of Medicine, Haifa, Israel 31096*
[2]*Department of Molecular and Cell Biology, 251 LSA, University of California-Berkeley, Berkeley, CA 94720-3200, USA*

Summary. While oxygen is vital to the existence of all aerobic organisms, under certain circumstances it forms reactive species and derivative oxidants that are noxious to biological molecules. However, nature has developed a coordinated antioxidant defence system against such oxidants. An imbalance between oxidant insult and antioxidant defences in favor of the former constitutes the basis for oxidative stress and for conditions leading towards oxidative injury.
In order to ascertain the validity and significance of oxidative stress in biological reactions, it is essential to characterize and quantitate the various substances and enzymes in this balance between oxidants and antioxidants. On the one hand, it is possible to identify, quantify, and localize the various oxidants and antioxidants. But on the other hand, and equally important, it is imperative to assess the oxidative damage inflicted upon biomolecules, e.g. lipids, proteins, and DNA by the various oxidants. The strategies and methodologies currently available to tackle such issues are reviewed in the present work.

Introduction

The interaction of oxidants, biomolecules and antioxidants can be depicted in the form of a golden triangle as shown in Figure 1.

This figure shows that antioxidants and oxidants interact with each other. However, when the level of oxidants exceeds the capacity of antioxidant defence, the result is oxidative stress and damage to biological systems.

The main sources of oxidants can be divided into the following four categories:

1. Reactive oxygen species (ROS)
2. Reactive nitrogen oxide species (RNOS)
3. Reactive aldehydes (RA)
4. Oxidant-generating enzymes (OE).

Examples of ROS are species such as the superoxide ion ($O_2^{-\cdot}$), hydroxyl radical ($OH^{-\cdot}$) and hydrogen peroxide (H_2O_2), whereas RNOS is represented by the ubiquitous nitric oxide ($NO^{-\cdot}$) and peroxynitrite ($ONOO^-$).

Reactive aldehydes, including 4-hydroxynonenal (4-HNE) and malondialdehyde (MDA), are generated primarily as products of lipid peroxidation. Reactive unsaturated aldehydes such as acrolein in the gas phase of cigarette smoke may also prove to be noxious. Some examples of oxidant generating enzymes are monoamine oxidase (MAO), xanthine oxidase, NADPH oxidase, myeloperoxidase, cytochrome 450 system, and nitric oxide synthase (NOS).

The conditions that may trigger oxidative stress in biological systems are summarized in Figure 2. Under normal physiological levels, some oxidants have essential functions in biological

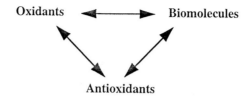

Figure 1. The interaction of oxidants, antioxidants, and biomolecules.

systems, for example, NO⁻ has been implicated as a blood vessel relaxant (vasodilator) affecting the endothelial and smooth muscles of arteries. It has also been found to function as a neurotrans-

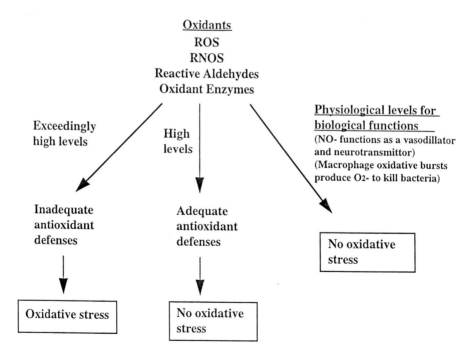

Figure 2. Conditions for occurrence of oxidative stress.

mitter in the central nervous system in parts of the brain that are associated with cognition and memory (Halliwell and Gutteridge, 1993). Another example of the important physiological role of free radicals is the phenomenon known as the respiratory or "oxidative" burst in which macrophages transiently increase the production of $O_2^{-\cdot}$ radicals as a means of eradicating bacteria and other pathogens (Moncada and Higgs, 1993). In the above cases, the effects of these oxidants are local and temporary, and do not constitute oxidative stress resulting in damage to cells and tissues. A situation may arise where oxidants are produced above physiological levels, although, if the antioxidant defences are adequate, oxidative stress may be circumvented. When oxidant levels are high enough to overwhelm the biological antioxidant defence systems, it will lead to conditions of oxidative stress, and the possible consequences are outlined in Figure 3.

The concept of oxidative stress therefore implies that oxidative damage can affect small as well large biomolecules in the human body. Thus, oxidants can alter the structure and function of lipids, proteins and nucleic acids as shown in Figure 3, which will eventually cause cell and tissue damage. To assess oxidative stress, it is necessary to identify and establish initial baseline physio-

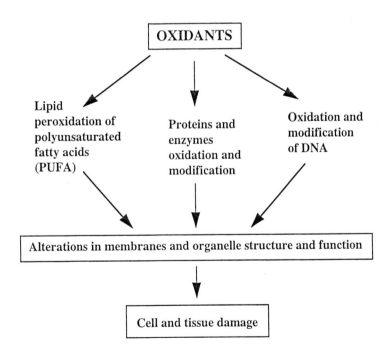

Figure 3. Influence of oxidants on the major biomolecules of biological systems.

logical values for each of the three groups depicted in Figure 1. Any deviation from the baseline levels of these normative values may indicate the occurrence of oxidative stress. For example, in extreme conditions of oxidative assault one would expect a marked depletion of various antioxidant levels. This has been shown when human plasma was exposed, *in vitro*, to cigarette smoke, for example, plasma ascorbate, uric acid and vitamin E were depleted considerably upon exposure to cigarette smoke (Cross et al., 1993). Another example is the depletion of tissue antioxidants under conditions of exhaustive exercise which may be overcome by exogenous antioxidant supplementation(Reznick et al., 1992b; Sen, 1995).

Hence, one, two or even all of the following can be measured and characterized:

1. The nature and quantity of various oxidants
2. The level of antioxidants
3. The level of oxidatively modified biomolecules.

The various approaches to assess these factors are outlined in the following.

Assess the nature and quantity of various oxidants

1. ROS: using spin trap for measurement of $O_2^{-\cdot}$ and $OH^{-\cdot}$. Measure H_2O_2 biochemically.
2. RNOS: measure $NO^{-\cdot}$ and peroxynitrite $ONOO^-$. Assess nitrates and nitrites in biological systems.
3. Aldehydes: measure MDA, 4-HNE and acrolein.
4. Oxidant enzymes: measure the activities of enzymes such as monoamine oxidase, xanthine oxidase, NOS, NADPH oxidase and myeloperoxidase.

Measuring reactive oxygen species (ROS)

The major reactive oxygen species that play a significant role in biological systems are the hydroxyl radical ($OH^{-\cdot}$), superoxide radical ($O^{-\cdot}$), singlet oxygen ($^1O_2$), and hydrogen peroxide (H_2O_2). Other than H_2O_2, which at 37 °C has a half-life in the order of minutes, the half-lives of $OH^{-\cdot}$, $O_2^{-\cdot}$, and $^1O_2$ are 10^{-9}, 10^{-5} and 10^{-6} seconds, respectively. As a result of their relatively short life and low concentration under normal physiological conditions, the detection and the quantitation of ROS is a cumbersome task. However, using direct electron spin resonance (ESR) spectroscopy, which detects the presence of unpaired electrons, it is possible to ascertain the characteristic spectra of oxygen free radicals. As ROS, as stated above, are short lived, it is quite difficult to observe the appearance of these species with conventional ESR.

To overcome this hurdle, free radicals can be reacted with nitroso compounds (R˙NO) to produce longer-lived radicals. This technique, known as spin-trapping, generates highly characteristic ESR spectra for various radicals (Halliwell and Gutteridge, 1993). Thus spin-traps, such as α-phenyl-*tert*-butylnitrone (PBN), *tert*-nitroso-butane (NtB), and 5,5-dimethyl pyroline-*N*-oxide (DMPO), are commonly used to detect superoxide and hydroxyl radicals. Unfortunately, the above methodology suffers from rather poor sensitivity. It has also been shown that spin-traps of

$O_2^{-\cdot}$ are still quite unstable and often decompose rapidly. In addition, certain hydroxyl spin-traps such as DMPO-OH may undergo reduction *in vivo* producing diamagnetic molecules that are incapable of being detected by the ESR technique (Floyd, 1983).

Another approach to measure hydroxyl radicals *in vivo*, has been the hydroxylation of aromatic compounds such as 2-hydroxybenzoic acid to trap hydroxyl radicals. As salicylate is less toxic than other spin-traps and reacts quite fast (1.6×10^9 mole/sec) to yield stable products, this has proved to be a successful technique for measuring the appearance of hydroxyl radical. One of the main products of the reaction of salicylate with $OH^{-\cdot}$ is 2,3-dihydrobenzoic acid (DHBA) which does not occur endogenously in biological systems. It can be detected by gas–liquid chromatography (GLC), or by high-performance liquid chromatography (HPLC) using electrochemical detection (ECD) (Giovanni et al., 1995). A similar way to measure the appearance of $OH^{-\cdot}$ is the reaction of $OH^{-\cdot}$ with 2-hydroxybenzoate. This yields a product, 2,3-dihydroxy-benzoate which is excited at 305 nm and is highly fluorescent at 407 nm. This last technique proved to be highly sensitive.

Singlet oxygen ($^1O_2$) can react with several biological molecules, especially with amino acids such as histidine, tryptophan, and methionine, which cause oxidative damage to proteins. Measuring $^1O_2$ in biological materials has proved to be a complicated matter. Most $^1O_2$ scavengers also react with hydroxyl radicals. However, the products of $^1O_2$ reaction with cholesterol and tryptophan are different from those obtained in reaction with $OH^{-\cdot}$. Thus, characterization of these reaction products can distinguish between $^1O_2$ and $OH^{-\cdot}$. Another approach that detects $^1O_2$ in biological systems is the measurement of chemiluminescence (light emission), using detection by photomultipliers. Usually, individual molecules of $^1O_2$ emit light in the infrared range. As other biochemical reactions can produce light, however, it is essential to measure the $^1O_2$ spectrum to ascertain the appearance of this ROS (Halliwell and Gutteridge, 1993).

One of the most common ways to measure superoxide radical is its reduction of nitroblue tetrazolium ion to form a tetrazolium radical. Two of these tetrazolium radicals combine to give the blue dye monoformazan, which can be detected spectrophotometrically (Halliwell and Gutteridge, 1993). This has been a popular approach to measuring $O_2^{-\cdot}$ production in experimental aqueous solutions. However, in biological preparations, $O_2^{-\cdot}$ levels may depress formazan production and make this technique difficult to apply.

A number of methodologies have been devised to detect H_2O_2 production in biological systems. Halliwell and Gutteridge (1993) described eight different approaches to quantify H_2O_2 production. These methods were described in a variety of systems, from bacteria and plants to specific organs in animals, for example, if catalase is present in small amounts, then adding a large excess of catalase results in oxygen release, which can be measured by an oxygen electrode. If a biochemical reaction requires H_2O_2 for its progression, its inhibition by a large excess of catalase or other peroxidases can indicate how much H_2O_2 is present. One of the most traditional ways of measuring H_2O_2 production has been the inhibition of catalase by aminotriazole. As the inhibition is dependent on the presence of H_2O_2, the amount of H_2O_2 present in the unknown sample can be calculated from the rate of inhibition.

Detection and quantification of reactive nitrogen oxide species (RNOS)

The discovery of nitrogen monoxide or nitric oxide (NO^-) and nitric oxide synthase (NOS), and the realization that NO^- reacts readily with oxygen to form a variety of nitrogen oxide species, has received a large amount of attention over the past decade (Packer, 1996). A number of methodologies have been described in the literature which have attempted to identify and quantify NO^- in biological systems. The main methods that have been used to detect and measure NO^- are summarized in Table 1.

The methods described in Table 1 are different in their sensitivity for NO^- detection. In addition, they also differ in their specificity and interference factors. However, it is not within the scope of this review to elaborate on the intricacies of these methodologies. The reader is referred to a recent issue of *Methods in Enzymology* (Packer, 1996) for further reading.

Detection of reactive aldehydes in biological systems

Malondialdehyde (MDA) and 4-hydroxynonenal (4-HNE) are the end-products of peroxidation of PUFAs. They are considered to be reactive species because they are capable of reacting with proteins, for example, 4-HNE, which has a single double bond, may react with protein-SH groups through a Michael-type addition reaction. Similar Michael-type addition reactions occur where

Table 1. Mehods for detecting NO^- production in biological systems

Method	Principle of method	Reference and system
Direct methods		
Electrochemical detection	Oxidation of NO^- on solid electrodes	Shibuki (1990) Clark type probe Malinski and Taha (1992) Phorphyrin sensor
Reaction with oxy haemoglobin	Detection of nitrate (NO_3^-) spectrophotometrically	Archer (1993)
Reaction with deoxyhaemoglobin	EPR spectroscopy	Wennmalm et al. (1990)
Chemiluminescence detection	Reaction with ozone to form excited NO_2 which decay for ground state	Brien et al. (1996)
Indirect methods		
Measurements of nitrite (NO_2^-) and nitrate (NO_3^-)	Reaction with Griess reagent Quantification by spectrophotometer	Archer (1993)
Measurement of nitrite ion by fluorescence	Using 2,3-diaminonaphthonic reagent	Misko et al. (1993)

EPR, electron proton resonance.

Table 2. List of some compounds reacting with thiobarbituric acid to form chromogens

Compound	Examples
Aldehydes	Acetaldehyde
Carbohydrates in their aldehyde form	Deoxyribose, deoxyglucose, ascorbate
Amino acids	Glutamic acid, proline, arginine, methionine

unsaturated aldehydes such as acrolein and crotonaldehyde present in cigarette smoke react with plasma proteins to form carbonyl derivatives (O'Neil et al., 1994). Malondialdehyde can react with protein amines to form intra- and interprotein cross-links (Halliwell and Gutteridge, 1993).

The quantification of MDA and 4-HNE constitutes an effective strategy for assessing the extent of lipid peroxidation. The most common assay for measurement of MDA has been the test with thiobarbituric acid (TBA), in which TBA reacts with MDA under acidic conditions with boiling to form a pink solution that is read spectrophotometrically at 532 nm. However, the specificity of this reaction of TBA in biological samples has been questioned by many workers, because TBA may react with a variety of substances (Tab. 2). Thus, a lipid extraction step is a necessary procedure, and it is also recommended that, as a control, an additional measure of lipid peroxidation, such as measuring conjugated dienes, be performed to establish the validity of the TBA test for assessing lipid peroxidation (LPX).

An elaborate method for detecting 4-HNE has been described by Goldering et al. (1993). In this procedure 4-HNE is derivatized with 2, 4-dinitrophenylhydrazine (DNPH). After a step of purification on solid thin-layer chromatography (TCL), the DNPH derivative was analyzed using the HPLC system with electrochemical detection (Goldering et al., 1993).

Measuring oxidant enzyme activities

The products of oxidase reactions are usually $O_2^{-\cdot}$ or H_2O_2. As in the case of monoamine oxidase (MAO), elevated activity of the enzymes involved contributes to an increased level of oxidative stress. Also, the level of the various forms of nitric oxide synthases (NOS enzymes) can be monitored which may indicate an increased production of $NO^{-\cdot}$ in certain physiological and pathological conditions. Some of these enzymes are abundant in specific tissues, for example, MAO is present in the dopaminergic neurons of the midbrain, NADPH oxidase activity is a feature of neutrophils and macrophages, and neural NOS (nNOS) is typically present in the neurons of the central nervous system. Assays of these enzyme activities in their characteristic location is an important indication of an increase in production of various ROS.

NOS enzymes have at least three isoforms: nNOS (neuronal form); eNOS (endothelial NOS) which consist of constitutive enzymes; and iNOS (the inducible form) which is present in many tissues and is sensitive to induction by cytokines and other active biofactors. A general biochemical assay for assessing the level of NOS activity in tissue such as muscle has been described (Kobzik et al., 1994). In this assay, muscle homogenates were incubated in the presence of

[^{3}H]arginine (the substrate for NOS) and using separation on Dowex AG SOWX-8, the amount of [^{3}H]citrulline produced was quantified by liquid scintillation counting. However, as the constitutive and inducible forms, and sometimes nNOS and eNOS, may be present in brain and other tissues, it is important to use specific inhibitors of the various isoforms of NOS. Such a variety of inhibitors has been described and can distinguish quite clearly between nNOS, eNOS, and iNOS (Garvey et al., 1996). Another approach for localizing NOS in tissues and ascertaining the various forms is to use histochemical and immunohistochemical techniques that employ specific antibodies against the three isoforms of NOS. This approach has gained considerable popularity in recent years (Norris et al., 1995).

Characterize and quantify the levels of antioxidants

1. Measure water-soluble antioxidants: biothiols including glutathione (GSH), dihydrolipoic acid (DHLA) and *N*-acetylcysteine (NAC), uric acid and ascorbate, natural metal chelators, antioxidant enzymes such as superoxide dismutase (SOD), catalase, glutathione peroxidase and glutathione reductase.
2. Measure levels of lipid soluble antioxidants: tocopherols, ubiquinone, ubiquinol and carotenoids.

Water-soluble antioxidants

Water-soluble antioxidants are a group of very diversified small biomolecules, proteins, and enzymes. Among the small biomolecules that are potent antioxidants are uric acid, ascorbate, and biothiols such as GSH, *N*-acetylcysteine (NAC), and lipoic acid. The importance of ascorbate (vitamin C) as an antioxidant in the physiology of health and disease has been illustrated in a recent book (Packer and Fuchs, 1997). Strategies to measure ascorbate in biological preparations are numerous and range from spectroscopic methods where ascorbate has strong absorption in the ultraviolet region, to electrochemical, fluorometric, and mass spectrometric methods. Table 3 summarizes some of the principal approaches to determining the amount of ascorbate in biological samples.

It is not feasible to describe, in this chapter, the multitude of procedures that have been developed to measure biothiols. The reader is referred to a comprehensive volume (Packer, 1995) for a detailed discussion on the determination of biothiols. The volume outlines several methodologies to quantify monothiols, bithiols, and protein thiols. In addition, several aspects of thiol-associated enzymology are also discussed in Packer (1995). Recently, a powerful modified flow cytometric method for the differential study of specific biological thiols in human cells has also been described (Sen et al., 1997a).

Measurement of the activities of the major antioxidant enzymes, including catalase, superoxide dismutase, glutathione peroxidate, and glutathione reductase, has been described in length by Halliwell and Gutteridge (1993). The catalase assay is a rather straightforward determination. If the level of H_2O_2 is fixed and in excess, the initial decomposition of H_2O_2 is proportional to the

Table 3. General approaches for the analysis of ascorbate and ascorbate derivatives

Method	System	Reference
Colorimetric methods	UV absorption or visible absorption with derivatization of DNP	Saciberlich et al. (1982)
Electrochemical detection	HPLC analysis of ascorbate and ascorbate derivatives	Tsao and Young (1985)
Fluorometric method	Derivatized with fluorescent probe	Deutsch and Weeks (1965)
Mass spectroscopy	Using isotope ratio mass spectrometry	Gensler et al. (1995)
Chemiluminescence	Ascorbate sensitized by oxidation	Peretz-Ruiz et al. (1995)

DNP, 2,4-dinotrophenyl hydrazine.

presence of catalase in a biological sample. Removal of H_2O_2 can be monitored by the loss of absorption at 240 nm or by measuring the release of oxygen using an oxygen electrode. The biological activity and relevance of catalase can also be assessed by introducing a specific inhibitor for catalase, aminotriazole, both *in vivo* and *in vitro*.

Three isoenzymes of SOD have been described: CuZnSOD, the cytosolic isoform, MnSOD, the mitochondrial isoform, and FeSOD which is found mainly in certain bacteria. The measurements of SOD activities in cells and tissues have employed both direct and indirect approaches. One can directly measure the loss of ultraviolet absorbance of $O_2^{-\cdot}$ when KO_2 is added to aqueous solution at basic pH (Grankvist et al., 1981). However, the direct approaches have proved difficult to execute because of their alkaline conditions which were incompatible with the physiological pH of 7.4. In contrast, several indirect methods were developed which in principle have a superoxide-generating system mixed with a detector system, for example, a xanthine–xanthine oxidase system is reacted with nitroblue tetrazolium (NBT). The latter is reduced by the generated $O_2^{-\cdot}$ to form formazan, which has a deep-blue color. The quantification of inhibition of formazan formation by SOD has been used as a means of defining SOD activity (Halliwell and Gutteridge, 1993). In addition, to distinguish the various isoforms of SOD, there have been reports on a traditional approach using specific antibodies and specific inhibitors to the various forms (Ohno et al., 1994).

Finally, the importance of metal ions, such as copper and iron, in generating OH^- radicals and the sequestration of these metal ions in biological fluids by proteins, such as ceruloplasmin (copper), and transferrin (iron), are discussed in this volume (Halliwell).

Quantification of lipid soluble antioxidants

Probably the most efficient and time-saving procedure for identifying and quantifying lipid-soluble antioxidants is to use a single electrochemical detector and thereto separate these antioxidants on HPLC (Motchnik et al., 1994). In this procedure, it is possible to carry out simultaneous determination of α- and γ-tocopherols, β-carotene, lycopene, ubiquinol, and ubiquinone from lipoproteins of human plasma. Eventually, the above methodology was also applied to homogenates of other tissues, e.g. plasma, liver, and muscle (Lang et al., 1986). In this HPLC procedure, the initial step is to extract the lipid antioxidants using several steps to dissolve the lipids in organic solvents such as ethanol, hexane, and methanol, after which the sample is dried under nitrogen, dissolved again in methanol, and injected into an HPLC column. Specific peaks are identified on the chromatograph by comparing the retention time of the corresponding HPLC standards (Motchnik et al., 1994). The sensitivity of the above procedure ranges from about 1 to 50 μM. However, because of the wide range of concentrations of the various lipid antioxidants, (e.g. α-tocopherol, 25–30 μM and ubiquinol 0.7 μM, special adjustment of baseline and experimental conditions is needed to display all of these lipids on the same run.

Assess the levels of oxidized or oxidatively modified biomolecules

1. Peroxidation of polyunsaturated fatty acids (PUFAs), and change in tissue free fatty acid (FFA) profile.
2. Oxidation and other oxidative modification of proteins and enzymes.
3. Determination of DNA oxidation products.

The three major classes of biomolecules that are generally affected by oxidation are PUFAs, proteins, and DNA. There are numerous methodologies for detecting oxidation of PUFA proteins and DNA. A brief description of the most commonly used methods is outlined below.

Peroxidation of PUFAs

In such a general review, it would be impossible to cover the multitude of approaches and techniques that have been developed to identify and quantify parameters indicative of lipid peroxidation (LPX). Nevertheless, in Table 4 a few representative methodologies of such assays are described.

Obviously, when the question arises of which assay (in Tab. 4) is the most suitable, a simple answer cannot be found. Each technique measures something different in the LPX chain reaction. Each assay has different factors that may interfere with measurement, and this needs to be taken into consideration. To validate LPX *in vivo*, a single approach may not suffice and must be supported by one or two other assays of LPX.

Oxidation and modification of proteins and enzymes

As the specific and proper structure of proteins and enzymes is the basis for the phenotypic expression of most biological systems, their oxidation and modification may lead to crucial loss of cellular and physiological functions. There are several known categories of protein oxidations and modifications, and methods for their assessment have been described in the literature; a brief discussion is presented below. Major categories include:

Table 4. Representative methodologies for measurement of lipid peroxidation

Method	Principle and system	Reference
Diene conjugation	Early stage in LPX UV absorption at 235 nm Some interference may occur in biological systems.	Vossen et al. (1993)
Measurement of lipid peroxides and hydroperoxides	Iodine liberation Lipid peroxides oxidize I^- to I_2 HPLC can detect with ECD lipid peroxides and lipid hydroperoxide	Wieland et al. (1992)
Hydrocarbon gas measurement	Measuring pentane and ethane, by gas–liquid chromatography	Jeejeehoy (1991)
Chemiluminescence and thermochemiluminescence	Peroxyradicals convert to cyclic peroxides (dioxetanes) which decompose to excited carbonyls. Light photons are detected by a photomultiplier.	Murphy and Sies (1990)
Thiobarbituric acid (TBA) test	Malondialdehyde reacts with TBA to form pink chromagen absorbed at 532nm	Halliwell and Gutteridge (1993)

1. Accumulation of carbonyl derivatives:
   (a) by metal-catalyzed oxidation
   (b) by Michael-type addition of unsaturated aldehydes
   (c) by glycation of proteins
2. Loss of protein sulfhydryl groups
3. Oxidation of methionine to methionine sulfoxide and methionine sulfone
4. Measure of dityrosine as a marker of oxidatively modified proteins
5. Oxidation by RNOS to form nitrotyrosine
6. Reaction of aldehydes with protein amines to form Schiff-base adducts
7. S-thiolated proteins – reaction of small molecular thiols with protein-SH
8. Glycation and glycooxidation-covalent binding of sugars and other ketoaldehydes to form Amadori products and advanced glycosylated end-products (AGE).

A popular and effective technique for measuring protein oxidation has been the employment of the reagent dinitrophenol-hydrazine (DNPH) which reacts with protein carbonyls to form a yellow adduct of protein hydrazone, which, in turn, can be detected spectrophotometrically (Reznick and Packer, 1994). In addition, they can be separated on HPLC and detected by electrochemical detection. Another approach to identifying DNPH–protein adducts has been to use anti-DNPH-specific antibodies and to quantify the amount of DNPH using Western blotting (Levine et al., 1994; Nakamura and Goto, 1996). Elevated protein carbonyls were detected quite conclusively in: plasma proteins exposed to cigarette smoke (Reznick et al., 1992a); exercise (Reznick et al., 1992b; Sen et al., 1997b); Alzheimer's disease (Smith et al., 1996); brain motor neuron disease (Lyras ét al., 1996); and as a function of ageing (Stadtman, 1992).

Another very common approach used to assess protein oxidation is the monitoring of the disappearance of protein-SH or the oxidation of protein thiols. A spectrophotometric assay based on its reactions with 2,2-dithiodinitrobenzoic acid (DTNB), also known as Ellman's reagent, is commonly used to quantitate protein-SH groups (Hu, 1994). Loss of plasma protein-SH, as a result of exposure to gas-phase cigarette smoke, correlated with an increase of protein carbonyls (O'Neil et al., 1994).

A recent method for analysis of methionine sulfoxide in proteins has been described using a cyanogen bromide procedure. In this approach amino acids are separated from acid hydrolyzates of proteins treated with cyanogen bromide using reverse phase HPLC. This enables the simultaneous determination of homoserine, corresponding to nonoxidized methionine, and residual methione, corresponding to oxidized methione (Maier et al., 1995). The appearance of a dityrosine adduct in proteins as a result of tyrosyl radical reaction and oxidative modification has been described in the literature in the last three decades and a useful method for detection of dityrosine using HPLC with fluorescence has been reported (Giulivi and Davies, 1994).

Probably the oxidation product studied most using RNOS is nitrotyrosine. An HPLC method for analysis of nitrotyrosine using UV detection has been reported by van der Vliet et al. (1996). However, this procedure lacked sensitivity and specificity for repeated determination of nitrotyrosine *in vivo*. Nevertheless, evidence for tyrosine nitration *in vivo* has been provided by immunochemistry using monoclonal antibodies against nitrotyrosine (Beckman et al., 1994). Finally, using HPLC with increased sensitivity – employing multiple electrodes of ECD – an *in vivo* assay of nitrotyrosine, has been shown to be effective (Schulz et al., 1995).

A procedure for the detection of protein-S-thiolation using radioactive S-labeled amino acids has been described by Thomas et al. (1994). In this procedure, cells are labeled with ^{35}S-labeled precursors for short periods of time in the presence of either *N*-ethylmaleimide (NEM) or dithiothreitol (DTT). The difference between NEM and DTT samples represents the number of radioactive thiols bound to proteins by reducible bonds (S-thiolated proteins). The nonenzymatic glycosylation of proteins and the identification of advance glycolated end-products (AGE) has been the focus of extensive studies (Makita et al., 1992; Makita et al., 1995).

It should also be noted that oxidation of certain enzymes may lead to their inactivation and loss of biological activity, for example, glutamine synthetase was shown to be inactivated partially when brain proteins were damaged by the oxidative stress of ischemia and reperfusion (Oliver et al., 1990) Another example is the inactivation of glucose-6-phosphate dehydrogenase by covalent linking of 4-hydroxynonenal (Szweda et al., 1993). Thus chemical changes and formation of oxidative adducts, together with alterations in biological activity, are the best indications for oxidative modifications of proteins.

Determination of DNA oxidation products

In cancer, during the process of ageing and in several other degenerative chronic diseases, free radicals have been implicated in DNA damage. It would be therefore of great interest to assess the extent of DNA damage in biological systems exposed to oxidative stress. In general, there are two main types of DNA damage: DNA strand breakage and oxidative base modifications. A fluorometric analysis of DNA binding as a measure of strand breaks in mammalian cells has been described (Birnboim, 1990). In this procedure, the rate of unwinding of the two strands of DNA in alkaline solution has been used for assessing the number of breaks occurring in the DNA backbone. In cells and tissues where radioactive labeling is possible, a more sensitive procedure involving a DNA labeling technique has been used to measure strand breaks (Birnboim and Jevcak, 1981). Exercise-induced DNA damage in human leukocytes has been detected using this technique (Sen et al., 1994).

Oxidative damage to DNA bases has been shown to result in mutagenic lesions to DNA (Loeb et al., 1988). Thus, measuring oxidative base modifications is an important tool in evaluating the level of oxidative damage to biological systems. Over 20 different modified DNA bases have been identified using gas chromatography/mass spectroscopy (GC/MS) by Dizdaroglu (1994). However, several modified bases such as thymine glycol, hydroxymethyluracil and 8-hydroxy-2-deoxygluanosine (8-OHdG) have been used quite often to assess oxidative changes in DNA; 8-OHdG can be measured with great sensitivity using HPLC separation coupled with electrochemical detection (Floyd et al., 1990). Another application of the HPLC approach has been used to detect DNA damaged biomarkers, e.g. 8-OHdG and 8-oxognamine in biological fluids such as urine and plasma (Shigenaga et al., 1994). Altogether, measuring oxidative modifications to DNA bases has been a very useful strategy in the assessment of oxidative damage to biological systems.

Conclusion

Assessing oxidative parameters of stress is not a simple matter. The myriad reactions of oxygen and its derivatives, and the multitude of enzymatic pathways in which oxygen is involved, add considerable complexity to the challenge. However, if the various components that play pivotal roles in biological oxidations are isolated, it is possible to identify the main players in the interactions of oxidants, antioxidants, and the biomolecules involved. In the past two decades, many methodologies and various diverse approaches have been developed, which enable us to measure and quantify with a great degree of accuracy the levels of different oxidants, antioxidants, and oxidized biomolecules.

Acknowledgement
This work has been suported by the Krol Foundation, Nutley, NJ, USA and Technion VPR Grant No. 181–603.

References

Archer S (1993) Measurements of NO in biological models. *FASEB J* 7: 349.
Beckman JS, Ye YZ, Anderson PG, Chen J, Accavitti MA, Tarpey MM, White CR (1994) Extensive Nitration of protein tyrosines in human atherosclerosis detected by immunohistochemistry. *Biol Chem Hoppe-Seyler* 375: 81–89.
Birnboim HC and Jevcak JJ (1981) Fluorometric method for rapid detection of DNA strand breaks in human white blood cells produced by low doses of radiation. *Cancer Res* 41: 1889–1892.

Birnboim HC (1990) Fluorometric Analysis of DNA unwinding to study strand breaks and repair in mammalian cells. *In*: L Packer and AN Glazer (eds): *Oxygen Radicals in Biological Systems (Part B): Methods in Enzymology*. Academic Press, 186: pp 550–555.

Brien JF, McLaughlin BE, Nakatsu K and Marks GS (1996) Chemiluminescence headspace-gas analysis for determination of nitric oxide formation in biological systems. *In*: L Packer (ed.): *Nitric Oxide (Part A): Methods in Enzymology*. Academic Press, 268: pp 83–92.

Cross CE, O'Neill CA, Reznick AZ, Hu ML, Marcocci L, Frei B (1993) Cigarette smoke oxidation of human plasma constituents. *Ann N Y Acad Sci* 686: 72–90.

Deutsch MJ and Weeks CE (1965) Microfluorometric assay for vitamin C. *J Assn Offic Analyt Chem* 48: 1248–1256.

Dizdaroglu M (1994) Chemical determination of oxidative DNA damage by gas chromatography-mass spectrometry. *In*: L Packer (ed.): *Oxygen Radicals in Biological Systems (Part D): Methods in Enzymology*. Academic Press, 234: pp 1–16.

Floyd RA, West MS, Eneff KL, Schneider JE, Wong PK, Tingey DT and Hogsett WE (1990) Conditions influencing yield and analysis of 8-hydroxy-2'-deoxyguanosine in oxidatively damaged DNA. *Anal Biochem* 188: 155–158.

Floyd RA (1983) The spin trap 5,5-dimethyl-1-pyrroline-1-oxide (DMPO) reacts with hydroxyl radicals but becomes reduced *in vivo*. *Biochim Biophys Acta* 756: 201–216.

Garvey EP, Furfine ES and Sherman PA (1996) Purification and inhibitor screening of human nitric oxide synthase isozymes. *In*: L Packer (ed.): *Nitric Oxide (Part A): Methods in Enzymology)*. Academic Press, 268: pp 339–349.

Gensler M, Rossman A and Schmidt H-L (1995) Detection of added L-ascorbic acid in fruit juices by isotope ratio mass spectrometry. *J Agr Food Chem* 43: 2662–2666.

Giovanni A, Liang LP, Hastings TG and Zigmond MJ (1995) Measuring cortical tissue levels of hydroxyl free radicals following systemic and intraventricular salicylate administration. *J Neurochem* 64: 1819–1825.

Giulivi C and Davies KJA (1994) Dityrosine: A marker for oxidatively modified proteins and selective proteolysis. *In*: L Packer (ed.): *Oxygen Radicals in Biological Systems, (Part C), Methods in Enzymology*. Academic Press, 233: pp 357–363.

Goldering C, Casini A, Maellero E, Del Bello B and Comporti M (1993) Determination of 4-hydroxynonemal by HPLC with ECD. *Lipids* 28: 141–145.

Grankvist K, Marklund SL, Taljedal IB (1981) CuZn-superoxide dismutase, Mn-superoxide dismutase, catalase and glutathione peroxidase in pancreatic islets and other tissues in the mouse. *Biochem J* 199: 393–398.

Halliwell B and Gutteridge JMC (1993) *Free Radicals in Biology and Medicine,* 2nd edn. Clarendon Press, Oxford.

Hu M-L (1994) Measurement of protein thiol groups and glutathione in plasma. *In*: L Packer (ed.): *Oxygen Radicals in Biological Systems (Part C), Methods in Enzymology*. Academic Press, 233: pp 380–385.

Janata J (ed.) (1989) *Principles of Chemical Sensors*. Plenum, New York.

Jeejeebhoy KN (1991) *In vivo* breath alkane as an index of lipid peroxidation. *Free Radical Biol Med* 10: 191–3.

Kobzik L, Reid MB, Bredt DS and Stamler JS (1994) Nitric oxide in skeletal muscle. *Nature* 372: 546–548.

Lang JK, Gohil K, Packer L (1986) Simultaneous determination of tocopherols, ubiquinols, and ubiquinones in blood, plasma, tissue homogenates, and subcellular fractions. *Anal Biochem* 157: 106–116.

Levine RL, Williams JA, Stadtman ER and Shacter E (1994) Carbonyl assays for determination of oxidatively modified proteins. *In*: L Packer (ed.): *Oxygen Radicals in Biological Chemistry (Part C) Methods in Enzymology*. Academic Press, 233: pp 346–357.

Loeb LA, James EA, Waltersdorph AM and Klebanoff SJ (1988) Mutagenesis by the autoxidation of iron with isolated DNA. *Proc Natl Acad Sci USA* 85: 3918–3922.

Lyras L, Evans PJ, Shaw PJ, Ince PG and Halliwell B (1996) Oxidative damage and motor neuron disease: difficulties in measurement of protein carbonyls in human brain tissue. *Free Radical Res* 24: 397–406.

Maier DL, Lenz AG, Beck-Speier and Costabel U (1995) Analysis of methionine sulfoxide in proteins. *In*: L Packer (ed.): *Biothiols, Part A: Methods in Enzymology*. Academic Press, 251: pp 455–461.

Makita Z, Vlassara H, Cerami A and Bucala R (1992) Immunochemical detection of advanced glycosylation end products *in vivo*. *J Biol Chem* 267: 5133–5138.

Makita A, Yanagisawa K, Kuwajima S, Yoshioka N, Atsumi T, Hasnuma Y and Koike T (1995) Advanced glycation end products and diabetic nephropathy. *J Diabetes Complication* 9: 265–268.

Malinski T and Taha Z (1992) Nitric Oxide from a single cell measured insitu by a porphyrinic-based microsensor. *Nature* 358: 676–678.

Misko TP, Schilling RJ, Salvemini D, Moore WM and Currie MG (1993) A fluorometric assay for the measurement of nitrite in biological samples. *Anal Biochem* 214: 11–16.

Moncada S and Higgs A (1993) Mechanisms of disease – the L-arginine nitric oxide pathway. *N Engl J Med* 329: 2002–2012.

Motchnik PA, Frei B and Ames BN (1994) Measurement of antioxidants in human blood plasma. *In*: L Packer (ed.): *Oxygen Radicals in Biological Systems (Part D): Methods in Enzymology*. Academic Press, 234: pp 269–293.

Murphy ME and Sies H (1990) Visible-range low level chemiluminescence in biological systems. *In*: L Packer (ed.): *Oxygen Radicals in Biological Systems (Part B): Methods in Enzymology.* Academic Press, 186: pp 595–610.

Murphy ME and Noack E (1994) Nitric oxide assay using hemoglobin method. *In*: L Packer (ed.): *Oxygen Radicals in Biological Systems* (Part C): *Methods in Enzymology* Academic Press, 233: p. 240.

Nakamura A and Goto S (1996) Analysis of protein carbonyls with 2,4–dinitrophenylhydrazine and its antibodies by immunoblot in two-dimensional gel electrophoresis. *J Biochem* 119: 768–774.

Norris PJ, Charles IG, Scorer CA and Emson PC (1995) Studies on the localiztion and expression of nitric oxide synthase using histochemical techniques. *Histochem J* 10: 745–56.

Ohno H, Suzuki K, Fujii J, Yamashita H, Kizaki T, Oh-Ishi S and Taniguchi N (1994) Superoxide dismutases in exercise and disease. *In*: CK Sen, L Packer and O Hänninen (eds): *Exercise and Oxygen Toxicity*. Elsevier Science, Amsterdam, pp 127–161.

Oliver CN, Starke-Reed PE, Stadtman ER, Liu GJ, Carney JM and Floyd RA (1990) Oxidative damage to brain proteins, loss of glutamine synthetase activity, and production of free radicals during ischemia/reperfusion-induced injury to gerbil brain. *Proc Natl Acad Sci USA* 87: 5144–5147.

O'Neill CA, Halliwell B, van der Vliet A, Davis PA, Packer L, Tritschler H, Strohman WJ, Rieland T, Cross CE and Reznick AZ (1994) Aldehyde induced protein modifications in human plasma: protection by glutathione and dihydrolipoic acid. *J Lab Clin Med* 124: 359–370.

Packer L (ed.) (1995) *Biothiols (Part A) Methods Enzymol 251.* Academic Press.

Packer L (ed.) (1996) *Nitric Oxide (Part A): Methods Enzymol 268.* Academic Press.

Packer L and Fuchs J (eds) (1997) *Vitamin C in Health and Disease.* Marcel Dekker, Inc., New York.

Perez-Ruiz T, Martinez-Lozano C and Sanz A (1995) Flow-injection chemiluminometric determination of ascorbic acid based on its sensitized photooxidation. *Anal Chim Acta* 308: 299–307.

Reznick AZ, Cross CE, Hu M-L, Suzuki YJ, Khwaja S, Safadi A, Motchnick PA, Packer L and Halliwell B (1992a) Modification of plasma proteins by cigarette smoke as measured by protein carbonyl formation. *Biochem J* 286: 607–611.

Reznick AZ, Witt E, Matsumoto M and Packer L (1992b) Vitamin E inhibits protein oxidation in skeletal muscle of resting and exercising rats. *Biochemical and Biophysical Research Communications.* 189: 801–806.

Reznick AZ and Packer L (1994) Oxidative damage to proteins: Spectrophotometic method for carbonyl assay. *In*: L. Packer (ed.): *Oxygen Radicals in Biological Systems (Part C): Methods in Enzymology.* Academic Press, 233: pp 357–363.

Sauberlich HE, Green MD and Omaye ST (1982) Determinatin of ascorbic acid and dehydroascorbic acid. *In*: PA Seib and BM Tolbert (eds): *Ascorbic Acid: Chemistry, Metabolism and Uses.* Advances in Chemistry Series, American Chemical Society, Washington DC, pp 199–221.

Schulz JB, Matthews RT, Muquit NK, Browne SE and Beal MF (1995) Inhibition of neuronal nitric oxide synthatse by 7–nitroindazole protects against MPTP-induced neurotoxicity in mice. *J Neurochem* 64: 936–939.

Sen CK, Rankinen T, Vaisanen S, Rauramaa R (1994) Oxidative stress after human exercise: effect of N-acetylcysteine supplementation. *J Appl Physiol* 76: 2570–2577.

Sen CK (1995) Oxidants and antioxidants in exercise. *J Appl Physiol* 79: 675–676.

Sen CK Roy S, Han D and Packer L (1997a) Regulation of cellular thiols in human lymphocytes by alpha-lipoic acid: a flow cytometric analysis. *Free Radical Biol Med* 22: 1241–1257.

Sen CK, Atalay M, Agren J, Laaksonen DE, Roy S, Hänninen O (1997b) Fish oil and vitamin E supplementation in oxidative stress at rest and after physical exercise. *J Appl Physiol* 83: 189–195.

Shibuki K (1990) An electrochemical microprobe for detecting NO release in brain tissue. *Neurosci Res* 9: 69–76.

Shigenaga MK, Aboujaoude EN, Chen Q and Ames BN (1994) Assays of oxidative DNA damage biomarkers 8-oxo-2'-deoxyguanosine and 8-oxoguanine in nuclear DNA and biological fluids by high-performance liquid chromatography with electrochemical detection. *Methods Enzymology* 234: 16–33.

Smith MA, Perry G, Richey PL, Sayne LM, Anderson VE, Beal MF and Kowall N (1996) Oxidative damage in Alzheimer's. *Nature* 382: 120–121.

Stadtman ER (1992) Protein oxidation and aging. *Science* 257: 1220–1224.

Szweda LI, Uchida K, Tsai L and Stadtman ER (1993) Inactivation of glucose-6-phosphate dehydrogenase by 4-hydroxy-2-nonenal. *J Biol Chem* 268: 3342–3347.

Thomas JA, Chai YC and Jung CH (1994) Thiolation and detection. *In*: L Packer (ed.): *Oxygen Radicals in Biological Systems (Part C): Methods in Enzymology.* Academic Press, 233: pp 385–393.

Tsao CS and Young M (1985) Analysis of ascorbic acid derivatives by high performance liquid chromatography with electrochemical detection. *J Chromatogr* 13: 855–856.

Uchida K and Stadtman ER (1994) Quantitation of 4–hyroxynonenal protein adducts. *In*: L Packer (ed.): *Oxygen Radicals in Biological Systems (Part D): Methods in Enzymology.* Academic Press, 233: pp 371–380.

Van der Vliet A, Eiserich JP, Kaur H, Cross CE, Halliwell B (1996) Nitrotyrosine as a biomarker for reactive nitrogen species. *In*: L Packer (ed.): *Nitric Oxide Part B: Methods in Enzymology.* Academic Press, 269: pp 185–194.

Vossen RC, van Dam-Mieras MC, Hornstra G and Zwaal RF (1993) Continuous monitoring of lipid peroxidation by measuring conjugated diene formation in an aqueous liposome suspension. *Lipids* 28: 857–861.

Wennmalm A, Lanne B and Petersson A-S (1990) Detection of endothelial-derived relaxing factor in human plasma in the basal state and following ischemia using electron paramagnetic resonance spectrometry. *Anal Biochem* 187: 359–363.
Wieland E, Schettler V, Diedrich F, Schuff-Werner P and Oellerich M (1992) Determination of lipid hydroperoxides in serum iodometry and high performance liquid chromatography compared. *Eur J Clin Chem Clin Biochem* 30: 363–369.

Oxidative Stress in Skeletal Muscle
A.Z. Reznick et al. (eds)
© 1998 Birkhäuser Verlag Basel/Switzerland

The course of exercise-induced skeletal muscle fibre injury

J. Komulainen and V. Vihko

LIKES Research Center for Sport and Health Sciences, University Campus, FIN-40100 Jyväskylä, Finland

Summary. Excessive physical exercise, such as prolonged running, causes injuries in the skeletal muscle fibres of humans and animals unaccustomed to such exercise. This chapter describes the ultrastructural and histopathological changes seen in fibre injury, and the susceptibility of fibre and muscle types to lethal injury. Also, some aspects of injury mechanisms, particularly the role of the eccentric component of muscle functioning, as well as training-induced protection against muscle fibre injury are highlighted.

Introduction

To distinguish the injury processes known as exercise-induced delayed muscle damage from the more severe forms of soft tissue injury, such as crush injury, its pathology has been referred to simply as microinjury (Vihko et al., 1978a; Armstrong, 1990). This is justifiable because the initial lesions are subcellular and segmental (Armstrong et al., 1983; Fridén et al., 1983b; Kuipers et al., 1983; Newham et al., 1983), and mainly occur in a small proportion of fibres in an affected skeletal muscle (Vihko et al., 1978a; Armstrong et al., 1983; Kuipers et al., 1983). The initial and early events in the muscle fibre injury process, which are observable during the first hours after exercise, are autogenic, i.e. they are indigenous to the muscle fibres and occur before phagocytic cells invade the affected site (see, for examples, Armstrong, 1990).

Early structural changes

The course of the histopathological changes in exercise-induced muscle fibre injury has been studied after several types of exercise both in humans (Fridén et al., 1983b; Newham et al., 1983; Fridén, 1984) and in animal studies on different species such as the mouse (Vihko et al., 1978a; Komulainen et al., 1993), rat (Altland and Highman, 1961; Armstrong et al., 1983; Komulainen et al., 1994) and rabbit (Lieber and Fridén, 1988). The early ultrastructural changes are observable within a few hours of termination of exertion. Surprisingly, Gibala et al. (1995) found fibre disruption to the same extent both immediately and 48 h after acute concentric and eccentric exercises in the biceps brachii muscles. The early histopathological changes consist of streaming, disarray and disruption in the myofibrillar banding pattern of the injured fibre (Fridén et al., 1981, 1983b; Armstrong et al., 1983), cytoskeletal changes (Fridén, 1984; Duncan, 1987), broken fibres (Stauber et al., 1988), swollen mitochondria (Gollnick and King, 1969; Vihko and Arstila; 1974; Hikida et al., 1983), and subsarcolemmal vacuolization (Kuipers et al., 1983). Disruption of the sarcolemma has also been suggested (McNeil and Khakee, 1992). Ogilvie et al. (1988) described two main types of ultrastructural lesions: focal disruption of the A-band and localized dissolution of Z-lines. The distribution of lesions showed that A-band lesions were the main pathological change, whereas the I-band lesions were far less frequent. The Z-lines appeared to be mostly normal, sometimes broadened and, in severely disrupted regions, they could even be absent (Ogilvie et al., 1988).

Fridén et al. (1981, 1983b) suggested that the weakest link in the integrity of the sarcomere and in the chain of contractile units is the Z-band in eccentric exercise. The cytoskeletal intermediate filament desmin has been found between adjacent myofibrils and acts as a link. Desmin also links the contractile apparatus via the Z-bands to the sarcolemma, mitochondria and nucleus (Tokuyasu et al., 1982), and thus plays an important role in the integrity of muscle fibre structure and synchronization of contraction. Exercise may initiate the disruption of this linking system. In a recent study, Lieber et al. (1996) found desmin dissolution in rabbit tibialis anterior muscle as early as 15 min after the start of eccentric exercise. Komulainen et al. (1996) observed a loss of desmin and discontinuous dystrophin (submembrane structural protein) staining in some swollen muscle fibres immediately after forced lengthening of rat tibialis anterior muscle. The loss of submembrane and cytoskeletal structural protein stainings occurred before the staining of disorganized actin, i.e. before the disruption of the contractile apparatus. On the other hand, Mcpherson et al. (1996), using the model of a single contracting muscle fibre, proposed that the force developed by the muscle during activation directly affects the degree of heterogeneity in the sarcomere, and thus becomes the determining factor in the magnitude of the injury. This is in accordance with the finding of Lynn and Morgan (1994), who noticed that the damage was caused by sarcomere length instabilities in such a way that longer sarcomeres, when stretched beyond filament overlap, are damaged.

Fridén et al. (1991) demonstrated cores heavily stained for fibronectin in large fibres of eccentrically exercised rabbit muscles, which showed no other abnormalities, except for giant size, in haematoxylin and eosin (H&E) stained sections. Fibronectin was thus located intracellularly, indicating perhaps that the alteration may reflect autophagic vacuolization of membrane components (Fridén et al., 1991). McNeil and Khakee (1992) showed by immunohistochemical albumin staining that albumin was also present intracellularly in certain fibres immediately after downhill running of rats that was eccentrically biased. This intracellular staining identifies muscle fibres injured at their sarcolemma which are therefore rendered transiently or permanently permeable to extracellular albumin. Twenty-one per cent of the fibres were intracellularly stained. After downhill running, up to 5% of fibres are necrotized (Armstrong et al., 1983; Komulainen et al., 1994). Hence, the changes in sarcolemmal permeability do not necessarily correlate with the number of lethally injured fibres. McNeil and Khakee (1992) proposed that the change in permeability of the sarcolemma does not occur in all fibres, but at random in the muscle cell population; they considered plasma membrane wounding to be a common but previously undetected event in the lifespan of skeletal muscle fibre. Kasper (1996), using the same technique as McNeil and Khakee (1992), also observed that the incidence of sarcolemmal disruption exceeded the occurrence of muscle necrosis and degeneration in rats. She also concluded that sarcolemmal disruption does not always lead to fibre death.

Secondary changes: Degeneration and necrosis

From 2 to 6 hours after the induction of injury, secondary changes, such as fibre autophagy and heterophagy by macrophages (Vihko et al., 1978a; Vihko and Salminen, 1986), are already observable in the muscle fibres. This stage continues for the next 2–4 days (Salminen, 1985; Vihko et al., 1978a). Secondary changes are associated with the autogenic and phagocytic stages. In the former, the proteolytic and lipolytic systems start the process of degrading cellular structures such as myofibrils and the muscle membrane, respectively (Armstrong, 1990). In the phagocytic stage, injured muscle fibres show basophilic subsarcolemmal staining (Kuipers et al., 1983), muscle fibre swelling (Peeze Binkhorst et al., 1989; Komulainen et al., 1993) and an increase in lysosomal hydrolytic activity (Vihko et al., 1978b; Salminen and Kihlström, 1985). Neutrophils phagocytose tissue debris and can release cytotoxic factors such as elastase and oxygen radicals (Babior et al., 1973). These factors increase vascular permeability by affecting basement membranes near the site of injury (Movat et al., 1987), and thus promote the invasion of other inflammatory cells. The role of free oxygen radicals in exercise-induced damage is discussed in detail in the next chapter (M.J. Jackson, this volume). The neutrophil response in injured muscle usually occurs within 24 h of exercise (Hikida et al., 1983). However, Fielding et al. (1993) observed a significant accumulation of neutrophils in muscle biopsies taken both 45 min and 5 days after downhill running in men. This accumulation of neutrophilic cells was positively correlated with intracellular Z-band damage and interleukin-1β increase in damaged muscle.

In addition, monocytes invade the sites of injury and undergo morphological and functional differentation within the tissue, becoming macrophages. Their invasion into the muscle interstitium has been observed within 24 h of strenuous exercise in both humans (Hikida et al., 1983) and experimental rodents (Vihko et al., 1978a; Armstrong et al., 1983; Kuipers et al., 1983).

Functionally distinct subclasses of macrophages may play distinct roles in a muscle's response to injury. ED1[+] macrophages are present in circulating populations of macrophages and monocytes, which are supposed to invade the injured area (Heuff et al., 1993; Forrester et al., 1994). ED2[+] macrophages, on the other hand, are viewed as resident tissue macrophages that may respond early to muscle injury, which is later accompanied by immigration of neutrophils and provides the additional signals for the late chemotaxis of inflammatory cells from the circulation to the injured site of the muscle. The above considerations of the role of macrophages, and the overall response of inflammatory cells to acute muscle injury, have been discussed in detail in a review by Tidball (1995). In contrast, Stauber et al. (1988) found no support for the predominance of macrophages or mast cells in the injured area after forced muscle lengthening. They mainly observed the invasion of cells of myogenic origin, whereas the remainder of the invading cells appeared to be lymphoidal.

Regeneration

Although there is no clear demarcation between the degenerative and regenerative phases in the muscle injury–repair process, in rodents within 4–6 days of exercise the regenerative processes have started (Vihko et al., 1978a; Armstrong et al., 1983; Salminen, 1985). It is interesting that processes that degrade damaged material may also be involved in regeneration. Phagocytosis of necrotic fibres and autophagocytosis in the surrounding surviving fibres may produce material for regeneration, as suggested by Salminen (1985). The attenuation of inflammation and the simultaneous appearance of regenerating myotubes are prominent 4–5 days after exercise in mice (Vihko and Salminen, 1986). Collagen synthesis is also increased during the repair phase (Myllylä et al., 1986).

The influence of inflammatory cells on muscle regeneration results from their phagocytic role in the lysis of myofibril debris, as well as from their ability to release growth factors (Nathan, 1987). Growth factors can control the replication and differentiation of the satellite cells to myoblasts, as well as their fusion to myotubes, which ultimately mature to myofibres (Florini et al., 1991). Repair of damaged skeletal muscle starts with the activation of normally quiescent satellite cells, which are located between the basal lamina and the sarcolemma. In the soleus muscle of adult rats, the proliferative response of the satellite cells reaches a maximum within 24 h of an acute bout of eccentrically biased running, and the response is related to the amount of degeneration and fibre necrosis (Darr and Schultz, 1987). Simultaneously, however, no damage is observed in the extensor digitorum longus muscles, but a clear response of the satellite cells occurs. Thus, the pathological state of the muscle would be accurately estimated by serial sectioning through the length of the muscle, because even small lesions in a fibre are capable of activating all the satellite cells associated with the fibre (Schultz, 1989).

Estimation of muscle damage

In addition to direct histological and biochemical studies on muscle samples, a number of indirect approaches to estimate exercise-induced muscle damage have been widely used. These estimates have been considered also to reflect muscle fibre injuries. They include estimation of pain (delayed-onset muscle soreness – DOMS), the decrease of maximal muscle power (Clarkson et al., 1992), impaired range of motion (Stauber et al., 1990), shortened muscle length (Clarkson and Tremblay, 1988) and increase of myocellular proteins in the serum (Noakes, 1987; Hortobágyi and Denahan, 1989; Amelink, 1990). Several studies have measured force-producing capability (Warren et al., 1994a) as an early indicator of damage, and it has been proposed as the best quantitative, indirect measure of the amount of damage (McCully and Faulkner, 1986; Balnave and Allen, 1995). The signal intensity of magnetic resonance imaging (MRI) was used to estimate water content in muscle in the study by Nurenberg et al. (1992), who found prominent increases in signal intensity after eccentric exercise which correlated well with the extent of histologically verified muscle fibre injury.

It has been suggested that a loss of muscle cell membrane integrity plays a triggering role in the pathogenesis of exercise-induced damage and, hence, a general assumption has been made that there is a release of myocellular proteins initially into the extracellular space, and later into the circulation. The increased activity or concentrations of serum myocellular proteins have, consequently, been used in several studies as specific and quantitative markers of muscle damage. However, in recent studies this assumption has been questioned (Evans and Cannon, 1991; van der Meulen et al., 1991; Komulainen and Vihko, 1994). There are experiments that have failed to show the relationship between increases of serum myocellular proteins (e.g. creatine kinase, carbonic anhydrase III) and histological verification of necrotic fibre damage (Komulainen et al., 1994, 1995). In unaccustomed running exercise in rats, essentially minor increases (one- to tenfold) in serum creatine kinase activity and exercise-induced necrotic fibre necrosis are not chronologically in concert but may coexist (Komulainen et al., 1995). After swimming exercise, a clear increase in serum creatine kinase occurs, although no fibre necrosis is observed (Komulainen et al., 1995). The authors offered a lymph hypothesis (Lindena et al., 1984) to explain the serum changes, i.e. that an exercise-induced creatine kinase increase may reflect its enhanced wash-out via the lymphatics. After eccentric contractions of forearm flexors using high force, Nosaka and Clarkson (1996) found that the exercise-induced changes in indirect muscle damage indicators, such as force generation, range of motion and muscle swelling measured by MRI, all correlated with changes in serum creatine kinase activity (100-fold maximum increase).

An exact method of studying cellular and subcellular muscle damage is possible using light and electron microscopic methods. In such studies, good quantification of the amount of damage is problematic. The frequency of necrotic fibres has been below 5% (Vihko et al., 1978a; Kuipers et al., 1983), whereas ultrastructural changes have been observed (Gibala et al., 1995) in 37% and 80% of forearm flexor fibres after acute concentric or eccentric resistance exercise, respectively. Evidently, a part of the ultrastructural changes is sublethal (Vihko et al., 1978a) and does not cause fibre necrosis. Specific immunohistochemical stainings for certain structural elements of myofibre microarchitecture (Lieber et al., 1994) have proved to be an excellent tool when also estimating the damage in the early phase before frank histological changes.

The amount of muscle injury after exercise has also been measured biochemically by analysing the activities of glucose-6-phosphate dehydrogenase (Armstrong et al., 1983) or β-glucuronidase (Vihko et al., 1978b; Salminen and Kihlström, 1985). Glucose-6-phosphate dehydrogenase is used as an indicator of inflammation, because inflammatory cells possess high levels of this enzyme (Armstrong et al., 1983). Histochemical studies have shown that the increased activity of β-glucuronidase in the injured muscles of exercised mice (Vihko et al., 1978b; Salminen, 1985) results from both invading phagocytes rich in acid hydrolases and the increase in lysosomal activity in the surviving muscle fibres adjacent to necrotic foci (Vihko et al., 1978a). Salminen (1985) also observed a dose–response relationship between the duration of running exercise and the increase in β-glucuronidase activity measured 5 days after the termination of exercise. The increases in the activity of both these enzymes are thought partly to reflect muscle damage in the same way: the larger the number of inflammatory cells, the larger the number of injured fibres. In addition, total β-glucuronidase activity reflects overall muscle damage because it also accounts for the increased activity originating from the affected but surviving muscle fibres (Salminen, 1985). Salminen and Kihlström (1985) showed that the β-glucuronidase activity correlated significantly with the amount of histopathological changes.

Susceptibility to injuries

Muscle and fibre type

Three main types of skeletal muscle fibres have generally been identified histochemically: slow-twitch oxidative (SO), fast-twitch oxidative glycolytic (FOG) and fast-twitch glycolytic (FG) fibres (Peter et al., 1972). These fibres are generally referred to as types I, IIA and IIB, respectively. Myosin is the major component of the contractile apparatus in skeletal muscle fibre, and several combinations of slow and fast heavy chain and light chain isoforms of myosin exist. Thus these varying proportions reflect a continuum of fibre types rather than a threefold categorization, as well as the dynamic nature of skeletal muscle (Pette and Staron, 1993; Staron and Johnson, 1993).

In a study by Vihko et al. (1978b), the largest increases in certain lysosomal acid hydrolases, after exhaustive running of mice, were found to occur mainly in the deep red portions of the quadriceps femoris muscle, which is composed predominantly of oxidative "red" fibres. Several other rodent studies have also demonstrated this selective damage between different fibre types, suggesting that oxidative fibres are more susceptible to exercise-induced damage than glycolytic fibres (Armstrong et al., 1983; Schwane and Armstrong, 1983; Ogilvie et al., 1988; Peeze Binkhorst et al., 1990) as they also are to ischemic injury (Jennische et al., 1979) or vitamin E deficiency (Ruth and van Vleet, 1974).

Studies using models for precise muscle stimulation during forced lengthening in the rabbit have shown the predominance of fast glycolytic fibre damage (Lieber and Fridén, 1988; Lieber et al., 1991). However, a similar stimulation of rat tibialis anterior muscle caused greater damage to type IIA than to type IIB fibres (van der Meulen, 1991). In humans, exercise-induced pathological changes have been shown to be restricted mainly to type II fibres (Fridén et al., 1983a, 1988; Jones et al., 1986). In the study by Mcpherson et al. (1996), susceptibility to contraction-induced injury was investigated in single permeable muscle fibre segments from the extensor digitorum longus (fast) and soleus (slow) muscles of rats that had been made. Strain of a similar order produced a greater force deficit in fast fibres than in slow fibres. If the force deficit was the same, ultrastructural damage was notably similar in both muscles. Different proportions of isoforms of structural proteins have been hypothesized as contributing to the difference in susceptibility to damage between muscle fibre types (Mcpherson et al., 1996). For instance, titin has been shown to be smaller and less compliant in fast fibres than in slow fibres (Horowits, 1992). Kuipers (1994) proposed instead that recruitment rather than selective vulnerability is responsible for the susceptibility of skeletal muscle fibre to exercise injuries.

In the extensor muscle groups, SO fibres are predominant in the deep portion, whereas FG fibres predominate in the superficial layers. However, this preferential predominance of certain fibres is not apparent in the flexor groups (Armstrong and Phelps, 1984). The degree of exercise-induced damage varies in different muscles, and is probably associated with the different recruitment of muscles during running and possibly with their anatomical location in separate compartments (Salminen and Vihko, 1983). In addition, injuries in the hindlimbs have been observed more frequently in the extensor than the flexor muscles after prolonged exercise (Schumann, 1972). In hindlimb unloading followed by reloading, Kasper (1996) found a significantly different time course of disruption in the sarcolemma between rat plantaris and soleus muscles. Warren et al. (1994a) observed that the extensor digitorum longus muscle was more susceptible than the soleus muscle to eccentric contraction-induced injury in normal weight-bearing mice. However, they postulated that part of the explanation for the difference appeared to be a greater prior loading of the soleus muscle than the fibre type composition itself.

Animal studies have shown that exercise-induced muscle damage is restricted to relatively short segments of a few fibres (Vihko et al., 1978a; Armstrong et al., 1983; Kuipers et al., 1983). The location of the damage can vary even in different anatomical portions of the same muscle. Ogilvie et al. (1988) found a different percentage of necrotic changes in the proximal and distal parts of rat soleus muscle according to whether the running exercise was level or downhill.

Eccentric component of exercise

The nature of total body exercise is complex, and most types of exercise are composed of some combination of concentric, isometric and eccentric muscle contractions. Eccentric contraction occurs when a given muscle elongates while active tension is being produced. The phenomenon of greater injury to muscles in rodents caused by the amount of eccentric work was first demonstrated by Armstrong et al. (1983) in exercising rats, and later by McCully and Faulkner (1986) in mice with *in situ* stimulation of the lengthened muscles. Eccentric contractions generate relatively high forces per active cross-sectional area compared with concentric contractions (Asmussen, 1956).

Responses vary after different types of exercise in both humans and experimental animals. The responses have been compared in many studies using, for example, eccentrically versus concentrically biased running models (Armstrong et al., 1983), electrical stimulations of muscle in lengthened, isometric or shortened positions (McCully and Faulkner, 1986; Lieber and Fridén, 1988), step tests (Newham et al., 1983) or high force eccentric actions of, for example, forearm flexors (Clarkson et al., 1992; Gibala et al., 1995). The muscle exercised eccentrically by stepping (Newham et al., 1983) exhibited immediate damage, but biopsies taken 24 or 48 h after exercise exhibited marked damage, which was not detected in biopsies from concentrically exercised muscle. Gibala et al. (1995) found extensive myofibrillar disruption of the biceps brachii muscle fibres in untrained subjects after a single bout of arm-curl resistance exercise. Significantly greater fibre injury occurred during eccentric contraction. However, fibre disruption was also observed after concentric contraction.

Running-induced muscle injury, at least in rats, is related to the eccentric component of muscle functioning (Armstrong et al., 1983). When compared with running, the muscles do not work against gravity during swimming. Swimming exercise by untrained mice and rats (Komulainen and Vihko, 1992; Komulainen et al., 1995) does not cause muscle fibre necrosis. Swimming exercise lacks the eccentric component of muscle functioning as well as impact work and high muscle tensions, which may explain the lack of damage to the skeletal muscle fibres.

Lieber et al. (1991), who studied the chronology of the decline in tension after eccentric contraction-induced injury in rabbit muscle, observed that the decline in force occurred during the first few minutes of exercise. Lieber and Fridén (1993) have proposed the active strain hypothesis, which describes the interaction of the myofibrillar cytoskeleton, sarcomere and sarcolemma. They have suggested that it is not the high force itself that causes muscle damage in eccentric contraction, but the magnitude of the active strain.

Proposed injury mechanisms

Initiating mechanisms

The exact mechanism that triggers exercise-induced muscle damage is not known. Hypothetical formulations of initiating mechanisms have been offered (Armstrong, 1986; Armstrong et al., 1991; Ebbeling and Clarkson, 1989; Appell et al., 1992), and these can be divided into physical (mechanically or temperature induced) and metabolic hypotheses. However, this distinction may not be adequate in many cases.

Several studies and reviews have suggested that the initiator of damage is dependent on the eccentric component of muscle contraction. Armstrong et al. (1991) listed three main factors involved in eccentric exercise which support a mechanically induced initiating mechanism. First, force production during eccentric contraction may exceed maximal isometric force by 50–100% (Woledge et al., 1985); second, the force per active fibre ratio is the greatest during eccentric contractions; and third, the number of attached cross-bridges decreases as lengthening velocity increases (McMahon, 1984). The metabolic hypothesis assumes that the demand for ATP exceeds its production. This would lead to calcium overloading in the cell (Armstrong et al., 1991). Support for a metabolic cause is provided by studies in which ischemia in non-exercise situations has led to muscle fibre injury similar to that caused by damaging exercise (Mäkitie and Teräväinen, 1977; Appell et al., 1993). The hypothesized, temperature-induced mechanisms suggest that local muscle temperature is higher during eccentric contractions, thus predisposing the muscle fibre to structural and/or metabolic changes (Armstrong et al., 1991).

During prolonged running, the leg extensor muscles perform eccentric actions for several hours and many studies have suggested that this eccentric phase is associated with muscle damage (Armstrong et al., 1983; Schwane and Armstrong, 1983; Evans, 1987). A higher amount of damage in the proximal part of the soleus muscle of rats after level running exercise (Ogilvie et al., 1988; Jacobs et al., 1995) may point to a mechanical rather than a metabolic cause of injury, and may be explained by a higher tension per fibre in the proximal part. In a study by Komulainen et al. (1995), the energy production for contractile work was greater in the rats in the combined exercise (swim and run) group, than in the rats that only ran. However, the degree of muscle damage was similar in both groups, suggesting that, in this case, muscle damage was obviously less dependent on metabolic demands. In addition, the morphometric findings on capillary morphology (Peeze Binkhorst et al., 1989), after uphill running in the soleus muscle of rats, do not support the hypothesis that local ischemia exists in exercise-induced damage. In the second study, it was observed that, 24 h after exertion, the capillaries had an increased cross-sectional luminal area and an increased luminal circumference adjacent to the degenerative muscle fibres.

Lieber et al. (1994), who found cellular infiltration uniquely associated with damaging cyclic eccentric contractions, hypothesized that eccentric exercise initiates events that result in extravasation of leukocytes and monocytes from the circulation, and infiltration of these cells into the tissue. This cellular infiltration results in further tissue degradation by the release of proteolytic enzymes, which enable tissue remodelling. So, if this assumption is correct, post-exercise infiltration and the associated release of proteolytic enzymes were responsible for as much injury as the initial mechanical (and/or metabolic) insults.

Calcium

Regardless of the primary cause of exercise-induced muscle damage, the next step in the injury process is thought to be an elevation of intracellular Ca^{2+} concentration at the injury site (Armstrong et al., 1991). Duan et al. (1990a, b) found increased Ca^{2+} levels in the damaged muscles of rats after downhill walking. Armstrong et al. (1993), studying the effects of passive stretch of the rat soleus muscle, found evidence to support the idea that stretch plays a role in elevations of muscle Ca^{2+} which occur during eccentric exercise. The importance of extracellular Ca^{2+} to the injury process is indicated by studies showing that the removal of Ca^{2+} from the incubation medium markedly reduces muscle damage (Jackson et al., 1984; Jackson and Edwards, 1986) and attenuates tissue degradation (Baracos et al., 1986).

Carpenter and Karpati (1989) used micropuncture of muscle fibres to cause segmental necrosis of the fibres. This model seems to initiate lesions which are quite comparable to changes that have been observed in muscle fibres after eccentric exercise (Ogilvie et al., 1988). The necrotic area was demarcated, and the adjacent sarcomeres were protected from further degenerative processes. The separated area was heavily stained for precipitated Ca^{2+} (Carpenter and Karpati, 1989).

The mechanisms postulated as leading to increased free cytosolic Ca^{2+} are the disruption of the sarcolemma and dysfunction of the sarcoplasmic reticulum (Armstrong et al., 1991). In muscle fibres, mitochondria can accumulate large quantities of free cytosolic Ca^{2+} under pathological conditions (Gillis, 1985), which leads to suppression of mitochondrial function (Wrogemann and Pena, 1976). Calcium also activates neutral proteases (Reddy et al., 1975) and phospholipase A_2, which can degrade cytoskeletal structures and membranes, respectively.

Swelling

One of the most common and visible responses to various injuries is an acute type of inflammation. A typical feature in the sequence of events in the inflammatory response is the accumulation of fluid in the tissues. Before advanced exercise-induced necrosis sets in, the muscles are oedematous. The fibres are swollen, with a more rounded appearance, when compared with the normal polyhedral shape. Also, the interstitial area is enlarged when compared with the tightly packed arrangement of fibres in control samples (Peeze Binkhorst et al., 1990; Komulainen et al., 1993). Biopsies from human tibialis anterior muscle show greater fibre size and higher muscle water content after eccentric compared with concentric exercise (Fridén et al., 1988).

The regulation of the volume of interstitial fluid is a complex phenomenon, which depends on the transcapillary movements of proteins and water, the physicochemical properties of interstitial fluid and the function of the lymphatics. Increased muscle water content (muscle swelling) results from the increased permeability of small blood vessels, which allow fluid, including serum proteins, to leak into the tissue of the affected area (Hurley, 1983). Stauber et al. (1988) demonstrated an increase in vascular permeability in injured rat soleus muscle after forced muscle lengthening.

The contribution of muscle swelling to muscle damage was clearly shown by Komulainen and Vihko (1994). They found that muscle swelling in exercise-induced damage preceded increase in β-glucuronidase activity and was a prerequisite fibre necrosis, even in a quantitative sense, because the stronger the damage, the earlier and the larger the increase in water content. The maintainance of increased water content in the damaged muscles 96 h after termination of downhill running exercise (Komulainen et al., 1994) may reflect the increased synthesis of connective tissue associated with the process of regeneration as proposed by Smith (1991).

It has been suggested that swelling is involved in the generation of delayed onset muscle soreness (DOMS) caused by an increase in local tissue pressure (Bobbert et al., 1986; Fridén et al., 1986). Although significant increases in intramuscular pressure have been observed in a tight compartment (Fridén et al., 1986), no significant increases were seen either in humans (Newham and Jones, 1985) or in rats (Peeze Binkhorst et al., 1990) in a less restricted compartment. In rats, swelling has been observed microscopically in muscle fibres as well in the interstitium (Peeze Binkhorst et al., 1990). Mair et al. (1992) examined quadriceps muscles involved in eccentric contraction by MRI. They attributed MRI signal changes partly to oedema of the injured slow fibres because of the increased plasma concentration of the heavy chain fragment of the myosin slow-twitch-type. Bär et al. (1994) observed that delayed muscle soreness preceded the development of muscle oedema after eccentric exercise, and thereafter postulated that oedema is obviously not the cause of DOMS. Warren and co-workers (1994b), using confocal laser scanning microscopy, observed the segmental swelling of muscle fibres with an elevated cytosolic Ca^{2+} concentration in mice after eccentric exercise.

In conclusion, muscle swelling could result from the muscle fibre injury and the release of osmotically active material to the interstitium. The calcium influx into a damaged fibre could be connected to accumulation of water within the muscle fibre. Disruption of the extracellular matrix and an increase in its hydration state may also play a role in muscle swelling. Despite the fact that muscle swelling is a common phenomenon connected with exercise injuries, only a few systematic studies about the role of muscle water content in such injuries have been done.

Training-induced protection

Adequate training improves physical performance. One of the training-induced adaptations in skeletal muscle is the property to resist exercise-induced injury. Prior training induces a protection against muscle damage after exercise in experimental animals, and similar protective effect has also been observed in several human studies. Physical conditioning results in a reduced amount (Fridén et al., 1983a) or absence of morphological changes (Vihko et al., 1979), unchanged muscle water content (Komulainen and Vihko, 1995), and unchanged β-glucuronidase (Salminen et al., 1984) or glucose-6-phosphate dehydrogenase (Schwane and Armstrong, 1983) activity in the trained muscle after sudden strenuous exercise, which is damaging in unconditioned state. Other training effects on the indices of muscle damage are a reduction in, if not total absence of, DOMS (Fridén et al., 1983a), preservation of maximal force (Sacco and Jones, 1992) and an attenuated increase in serum creatine kinase activity (Schwane and Armstrong, 1983; Clarkson et al., 1987) after exercise which causes changes in the untrained state. However, no clear mechanisms have been shown experimentally which protect muscle from subsequent injury. The origin of the training response has been suggested to lie in alterations in the energy metabolism and in structural adaptations (Ebbeling and Clarkson, 1989). On the other hand, training-induced alterations in the energy metabolism of muscle are observed after a much longer period of prior training compared with training-induced protection against exercise injuries.

By using a training protocol involving an eccentric cycle ergometer that lasted 8 weeks, Fridén et al. (1983a) found improved coordination and reorganization of the contractile apparatus of muscle fibres compared with the structural changes observed after a single bout of eccentric work in untrained individuals. According to Fridén (1983), there are three possible structural adaptation mechanisms that can take place in the myofibrillar apparatus after eccentric training. These are increases in sarcomere length and/or number of longitudinal sarcomeres, an increase in the synthesis of Z-band proteins, or strengthening of myofibrils by intermediate filaments (e.g. desmin).

In addition to adaptations after relatively long-term training, rapid adaptations, even after a single bout of various types of exercise, have also been reported in both humans (Clarkson et al., 1987; Newham et al., 1987) and animals (Schwane and Armstrong, 1983; Salminen et al., 1984). Salminen et al. (1984) observed the development of protection against exercise injuries in mouse skeletal muscles after 3 days of light running training. Schwane and Armstrong (1983) reported that only a single prior level or downhill running bout prevented injury in rat vastus intermedius muscle after downhill exercise. Komulainen and Vihko (1992) showed that swimming exercise does not cause muscle injuries, and that swimming training, even lasting several weeks, does not protect mice from exercise injuries caused by a single bout of submaximal running This observation suggests that energy metabolic adaptations may have no influence on training-induced protection, and that prior protective training must resemble the damaging exercise, i.e. includes an eccentric component. These protective effects have been shown to be partial (Triffletti et al., 1988) or almost complete (Schwane and Armstrong, 1983), and develop on a muscle-specific basis (Komulainen and Vihko, 1995).

The training-induced protection disappears after a certain detraining period in both animals (Salminen et al., 1984; Sacco and Jones, 1992; Komulainen and Vihko, 1995) and humans (Byrnes et al., 1985). In addition, the duration of the adaptation effect may vary considerably depending on the estimate of muscle damage (Nosaka et al., 1991). To characterize the duration of adaptation of histological and functional indices of muscle damage, Sacco and Jones (1992) used a repeated damaging eccentric exercise protocol for mouse tibialis anterior muscle. They observed preservation in muscle fibre morphology and maximal force if the exercise was repeated within 3 weeks, but by 12 weeks muscles again became susceptible to injuries. In a view of results by Komulainen and Vihko (1995), it seems that acquired protection also disappears at different rates depending on the muscle in question.

Acknowledgements
The authors would like to thank the LIKES-Foundation (Jyväskylä, Finland) for the opportunity to write this article.

References

Amelink GJ (1990) *Exercise-induced muscle damage.* Dissertation, Kripps Repro, Meppel.

Altland PD and Highman B (1961) Effects of exercise on serum enzyme values and tissues of rats. *Am J Physiol* 201: 393–395.

Appell H-J, Glöser S, Duarte JAR, Zelner A and Soares JMC (1993) Skeletal muscle damage during tourniquet-induced ischemia. *Eur J Appl Physiol* 67: 342–347.

Appell H-J, Soares JMC and Duarte JAR (1992) Exercise, muscle damage and fatigue. *Sport Med* 13: 108–115.

Armstrong RB (1986) Muscle damage and endurance events. *Sport Med* 3: 370–381.

Armstrong RB (1990) Initial events in exercise-induced muscular injury. *Med Sci Sports Exerc* 22: 429–435.

Armstrong RB and Phelps RO (1984) Muscle fiber type of the rat hindlimb. *Am J Anat* 171: 259–272.

Armstrong RB, Ogilvie RW and Schwane JA (1983) Eccentric exercise-induced injury to rat skeletal muscle. *J Appl Physiol* 54: 80–93.

Armstrong RB, Warren GL and Warren JA (1991) Mechanisms of exercise-induced muscle fibre injury. *Sport Med* 12: 184–207.

Armstrong RB, Duan C, Delp MD, Hayes DA, Glenn GM and Allen GD (1993) Elevations in rat soleus muscle $[Ca^{2+}]$ with passive stretch. *J Appl Physiol* 74: 2990–2997.

Asmussen E (1956) Observations on experimental muscle soreness. *Acta Rheumatol Scand* 2: 109–116.

Babior BM, Kipnes RS and Curnutte JT (1973) The production by leukocytes of superoxide, a potential bactericidal agent. *J Clin Invest* 52: 741–744.

Bär PR, Rodenburg AJB, Koot RW and Amelink HGJ (1994) Exercise-induced muscle damage: Recent developments. *BAM* 4: 5–16.

Balnave CD and Allen DG (1995) Intracellular calcium and force in single mouse muscle fibres following repeated contractions with stretch. *J Physiol London* 488: 26–36.

Baracos V, Greenberg RE and Goldberg AL (1986) Influence of calcium and other divalent cations on protein turnover in rat skeletal muscle. *Am J Physiol* 250: E702–E710.

Bobbert MF, Hollander AP and Huijing PA (1986) Factors in delayed onset soreness of man. *Med Sci Sports Exerc* 18: 75–81.

Byrnes WC, Clarkson PM, White JS, Hsieh SS, Frykman PN and Maughan RJ (1985) Delayed onset muscle soreness following repeated bouts of downhill running. *J Appl Physiol* 59: 710–715.

Carpenter S and Karpati G (1989) Segmental necrosis and its demarcation in experimental micropuncture injury of skeletal muscle fibers. *J Neuropathol Exp Neurol* 48: 154–170.

Clarkson PM and Tremblay I (1988) Exercise-induced muscle damage, repair, and adaptations in humans. *J Appl Physiol* 65: 1–6.

Clarkson PM, Byrnes WC, Gillison E and Harper E (1987) Adaptation to exercise-induced muscle damage. *Clin Sci* 73: 383–386.

Clarkson PM, Nosaka K and Braun B (1992) Muscle function after exercise-induced muscle damage and rapid adaptation. *Med Sci Sports Exerc* 24: 512–520.

Darr KC and Schultz E (1987) Exercise induced satellite cell activation in growing and mature muscle. *J Appl Physiol* 63: 1816–1821.

Duan C, Delp MD, Hayes DA, Delp PD and Armstrong RB (1990a) Rat skeletal muscle mitochondrial $[Ca^{2+}]$ and injury from downhill walking. *J Appl Physiol* 68: 1241–1251.

Duan C, Hayes DA and Armstrong RB (1990b) Effects of Ca^{2+} and verapamil on muscle injury immediately after exercise. *Med Sci Sports Exerc* 22: S132.

Duncan CJ (1987) Role of calcium in triggering rapid ultrastructural damage in muscle: A study with chemically skinned fibres. *J Cell Sci* 87: 581–594.

Ebbeling CB and Clarkson PM (1989) Exercise-induced muscle damage and adaptation. *Sport Med* 7: 207–234.

Evans WJ (1987) Exercise-induced skeletal muscle damage. *Physician Sportsmed* 15: 89–100.

Evans WJ and Cannon JG (1991) The Metabolic Effects of Exercise-Induced Muscle Damage. *In*: J Holloszy (ed.): *Exercise and Sport Science Reviews*, Vol. 19, Williams & Wilkins, Baltimore, MA, pp 99–125.

Fielding RA, Manfredi TJ, Ding W, Fiatarone M, Evans WJ and Cannon JG (1993) Acute phase response in exercise III. Neutrophil and IL-1β accumulation in skeletal muscle. *Am J Physiol* 265: R166–R172.

Florini JR, Ewton DZ and Magri KA (1991) Hormones, growth factors, and myogenic differentiation. *Annu Rev Physiol* 53: 201–206.

Forrester JV, McMenamin PG, Holthouse I, Lumsden L and Liversidge J (1994) Localization and characterization of major histocompatibility complex class II-positive cells in the posterior segment of the eye: Implications for induction of autoimmune uveoretinitis. *Invest Opthalmol Visual Sci* 35: 64–77.

Fridén J (1983) Exercise-induced muscle soreness. A qualitative and quantitative study of human muscle morphology and function. *New series* 105, Medical Dissertations, Umeå University, Umeå, p. 19.

Fridén J (1984) Changes in human skeletal muscle induced by long-term eccentric exercise. *Cell Tissue Res* 236: 365–372.

Fridén J, Sjöström M and Ekblom B (1981) A morphological study of delayed muscle soreness. *Experientia* 37: 506–507.

Fridén J, Seger J, Sjöström M and Ekblom B (1983a) Adaptive response in human skeletal muscle subjected to prolonged eccentric training. *Int J Sport Med* 4: 177–183.

Fridén J, Sjöström M and Ekblom B (1983b) Myofibrillar damage following intense eccentric exercise in man. *Int J Sport Med* 4: 170–176.

Fridén J, Sfakianos PN and Hargens AR (1986) Delayed muscle soreness and intramuscular fluid pressure: comparison between eccentric and concentric load. *J Appl Physiol* 61: 2175–2179.

Fridén J, Sfakianos PN, Hargens AR and Akeson WH (1988) Residual muscular swelling after repetitive eccentric contraction. *J Orthopaed Res* 6: 493–498.

Fridén J, Lieber RL and Thornell LE (1991) Subtle indications of muscle damage following eccentric contraction. *Acta Physiol Scand* 142: 523–524.

Gibala MJ, McDougall JD, Tarnopolsky MA, Stauber WT and Elorriaga A (1995) Changes in human skeletal muscle ultrastructure and force production after acute resistance exercise. *J Appl Physiol* 78: 702–708.

Gillis JM (1985) Relaxation of vertebrate skeletal muscle. A synthesis of biochemical and physiological approaches. *Biochim Biophys Acta* 811: 97–145.

Gollnick PD and King DW (1969) Effect of exercise and training on mitochondria of rat skeletal muscle. *Am J Physiol* 216: 1502–1509.

Heuff G, van der Ende MB and Boutkan H (1993) Macrophage populations in different stages of induced hepatic metastases in rats: An immunohistochemical analysis. *Scand J Immunol* 38: 10–16.

Hikida RS, Staron RS, Hagerman FC, Sherman WM and Costill DL (1983) Muscle fiber necrosis associated with human marathon runners. *J Neurol Sci* 59: 185–203.

Horowits R (1992) Passive force generation and titin isoforms in mammalian skeletal muscle. *Biophys J* 61: 392–398.

Hortobágyi T and Denahan T (1989) Variability in creatine kinase: Methodological, exercise, and clinically related factors. *Int J Sport Med* 10: 69–80.

Hurley JV (1983) *Acute Inflammation.* Churchill Livingstone, New York, pp 1–117.

Jackson MJ and Edwards RHT (1986) Biochemical mechanisms underlying skeletal muscle damage. *In:* G Benzi, L Packer and N Silibrandi (eds): *Biochemical Aspects of Physical Exercise.* Elsevier Science, Amsterdam, pp 329–335.

Jackson MJ, Jones DA and Edwards RHT (1984) Experimental skeletal muscle damage: the nature of the calcium-activated degenerative processes. *Eur J Clin Invest* 14: 369–374.

Jacobs SCJM, Wokke JHJ, Bär PR and Bootsma AL (1995) Satellite cell activation after muscle damage in young and adult rats. *Anat Rec* 242: 329–336.

Jennische E, Amundson B and Haljamäe H (1979) Metabolic responses in feline "red" and "white" skeletal muscle to shock and ischemia. *Acta Physiol Scand* 106: 39–45.

Jones DA, Newham DJ, Round JM and Tolfree SEJ (1986) Experimental human muscle damage: morphological changes in relation to other indices of damage. *J Physiol* 375: 435–448.

Kasper CE (1996) Sarcolemmal disruption in reloaded atrophic skeletal muscle. *J Appl Physiol* 79: 607–614.

Komulainen J and Vihko V (1992) Swimming exercise and skeletal muscle damage in mice. *Med Sci Res* 20: 413–415.

Komulainen J and Vihko V (1994) Exercise-induced necrotic muscle damage and enzyme release in the four days following prolonged submaximal running in rats. *Pflügers Arch* 428: 346–351.

Komulainen J and Vihko V (1995) Training-induced protection and effect of terminated training on exercise-induced damage and water content in mouse skeletal muscles. *Int J Sport Med* 16: 293–297.

Komulainen J, Pitkänen R and Vihko V (1993) Muscle water content and exercise-induced damage in mice after submaximal running. *Med Sci Res* 21: 111–113.

Komulainen J, Kytölä J And Vihko V (1994) Running-induced muscle injury and myocellular enzyme release in rats. *J Appl Physiol* 77: 2299–2304.

Komulainen J, Takala T and Vihko V (1995) Does increased serum creatine kinase activity reflect exercise-induced muscle damage in rats? *Int J Sport Med* 16: 150–154.

Komulainen J, Hesselink M, Kuipers H and Vihko V (1996) Forced lengthening contractions and muscle fiber injury. *Med Sci Sports Exerc* 28: S188.

Kuipers H (1994) Exercise-induced muscle damage. *Int J Sport Med* 15: 132–135.

Kuipers H, Drukker J, Frederik PM, Geurten P and van Kranenburg G (1983) Muscle degeneration after exercise in rats. *Int J Sport Med* 4: 45–51.

Lieber RL and Fridén J (1988) Selective damage of fast glycolytic muscle fibers with eccentric contraction of the rabbit tibialis anterior. *Acta Physiol Scand* 133: 587–588.

Lieber RL and Fridén J (1993) Muscle damage is not a function of muscle force but active muscle strain. *J Appl Physiol* 74: 520–526.

Lieber RL, Woodburn TM and Fridén J (1991) Muscle damage induced by eccentric contractions of 25% strain. *J Appl Physiol* 70: 2498–2507.

Lieber RL, Schmitz MC, Mishra DK and Fridén J (1994) Contractile and cellular remodeling in rabbit skeletal muscle after cyclic eccentric contractions. *J Appl Physiol* 77: 1926–1934.

Lieber RL, Thornell L-E and Fridén J (1996) Muscle cytoskeletal disruption occurs within the first 15 min of cyclic eccentric contraction. *J Appl Physiol* 80: 278–284.

Lindena J, Kupper W and Trautschold I (1984) Enzyme activities in thoratic duct lymph and plasma of anaesthetized, conscious resting and exercising dogs. *Eur J Appl Physiol* 52: 188–195.

Lynn R and Morgan DL (1994) Decline running produces more sarcomeres in rat vastus intermedius muscle fibers than does incline running. *J Appl Physiol* 77: 1439–1444.

Mäkitie J and Teräväinen H (1977) Ultrastructure of striated muscle of the rat after temporary ischemia. *Acta Neuropathol* 37: 237–245.

Mair J, Koller A, Artner-Dworzak E, Haid C, Wicke K, Judmaier W and Puschendorf B (1992) Effects of exercise on plasma myosin heavy chain fragments and MRI of skeletal muscle. *J Appl Physiol* 72: 656–663.

McCully KK and Faulkner JA (1986) Characteristics of lengthening contractions associated with injury to skeletal muscle fibers. *J Appl Physiol* 61: 293–299.

McMahon TA (1984) *Muscles, Reflexes, and Locomotion*. Princeton University Press, Princeton, NJ.

McNeil PL and Khakee R (1992) Disruptions of muscle fiber plasma membranes. Role in exercise-induced damage. *Am J Pathol* 140: 1097–1109.

Mcpherson PCD, Schork MA and Faulkner JA (1996) Contraction-induced injury to single fiber segments from fast and slow muscles of rats by single stretches. *Am J Physiol* 271: C1438–C1446.

Movat HZ, Cybulsky MI, Colditz IG, Chan MKW and Dinarello CA (1987) Acute inflammation in gram-negative infection: endotoxin, interleukin-1, tumor necrosis factor and neutrophils. *Fed Proc* 46: 97–104.

Myllylä R, Salminen A, Peltonen L, Takala TES and Vihko V (1986) Collagen metabolism of mouse skeletal muscle during repair of exercise injuries. *Pflügers Arch* 407: 647–670.

Nathan CF (1987) Secretory products of macrophages. *J Clin Invest* 79: 319–326.

Newham DJ and Jones DA (1985) Intra-muscular pressure in the painful human biceps. *Clin Sci* 69: 27P.

Newham DJ, McPhail G, Mills KR and Edwards RHT (1983) Ultrastructural changes after concentric and eccentric contractions of human muscle. *J Neurol Sci* 61: 109–122.

Newham DJ, Jones DA and Clarkson PM (1987) Repeated high-force eccentric exercise: effects on muscle pain and damage. *J Appl Physiol* 63: 1381–1386.

Noakes T (1987) Effect of exercise on serum enzyme activities in humans. *Sport Med* 4: 245–267.

Nosaka K and Clarkson PM (1996) Variability in serum creatine kinase response after eccentric exercise of elbow flexors. *Int J Sport Med* 17: 120–127.

Nosaka K, Clarkson PM, McGuiggin ME and Byrne JM (1991) Time course of muscle adaptation after high force eccentric exercise. *Eur J Appl Physiol* 63: 70–76.

Nurenberg P, Giddings CJ, Stray-Gundersen J, Fleckenstein JL, Gonyea WJ and Peshock RM (1992) MR Imaging-guided muscle biopsy for correlation of increased signal intensity with ultrastructural change and delayed-onset muscle soreness after exercise. *Radiology* 184: 865–869.

Ogilvie RW, Armstrong RB, Baird KE and Bottoms CL (1988) Lesions in the rat soleus muscle following eccentrically biased exercise. *Am J Anat* 182: 335–346.

Peeze Binkhorst FM, Kuipers H, Heymans J, Frederik PM, Slaaf DW, Tangelder G-J and Reneman RS (1989) Exercise-induced focal skeletal muscle fiber degeneration and capillary morphology. *J Appl Physiol* 66: 2857–2865.

Peeze Binkhorst FM, Slaaf DW, Kuipers H, Tangelder G-J and Reneman RS (1990) Exercise-induced swelling of rat soleus muscle: its relationship with intramuscular pressure. *J Appl Physiol* 69: 67–73.

Peter JB, Barnard RJ, Edgerton VR, Gillespie CA and Stempel KE (1972) Metabolic profiles of three fiber types of skeletal muscle in guinea pigs and rabbits. *Biochemistry* 11: 2627–2634.

Pette D and Staron RS (1993) The molecular diversity of mammalian muscle fibers. *News Physiol Sci* 8: 153–157.

Reddy MK, Etlinger JD, Rabinowitz M, Fischman DA and Zak R (1975) Removal of Z-lines and alfa-actinin from isolated myofibrils by calcium-activated neutral protease. *J Biol Chem* 250: 4278–4284.

Ruth GR and van Vleet JF (1974) Experimentally induced selenium-vitamin E deficiency in growing swine: selective destruction of type I skeletal muscle fibers. *Am J Vet Res* 35: 237–244.

Sacco P and Jones DA (1992) The protective effect of damaging eccentric exercise against repeated bouts of exercise in the mouse tibialis anterior muscle. *Exp Physiol* 77: 757–760.

Salminen A (1985) Lysosomal changes in skeletal muscles during the repair of exercise injuries in muscle fibers. *Acta Physiol Scand* Suppl. 124 (539).

Salminen A and Kihlström M (1985) Lysosomal changes in mouse skeletal muscle during repair of exercise injuries. *Muscle Nerve* 8: 269–279.

Salminen A and Vihko V (1983) The susceptibility of mouse skeletal muscles to exercise injuries. *Muscle Nerve* 6: 596–601.

Salminen A, Hongisto K and Vihko V (1984) Lysosomal changes related to exercise injuries and training-induced protection in mouse skeletal muscle. *Acta Physiol Scand* 120: 15–19.

Schultz E (1989) Satellite cell behavior during skeletal muscle growth and regeneration. *Med Sci Sports Exerc* 21: S181–S186.

Schumann H-J (1972) Überlastungsnekrosen der Skelettmuskulatur nach experimentellem Laufzwang. *Zbl Allg Pathol* 116: 181–190.

Schwane JA and Armstrong RB (1983) Effect of training on skeletal muscle injury from downhill running in rats. *J Appl Physiol* 55: 969–975.

Smith LL (1991) Acute inflammation: the underlying mechanism in delayed onset muscle soreness? *Med Sci Sports Exerc* 23: 542–551.

Staron RS and Johnson P (1993) Myosin polymorphism and differential expression in adult human skeletal muscle. *Comp Biochem Physiol* 106B: 463–475.

Stauber WT, Fritz VK, Vogelbach DW and Dahlman B (1988) Characterization of muscles injured by force lengthening. I. Cellular infiltrates. *Med Sci Sports Exerc* 20: 345–353.

Stauber WT, Clarkson PM, Fritz VK and Evans WJ (1990) Extracellular matrix disruption and pain after eccentric muscle action. *J Appl Physiol* 69: 868–874.

Tidball JG (1995) Inflammatory cell response to acute muscle injury. *Med Sci Sports Exerc* 27: 1022–1032.

Tokuyasu KT, Dutton AH and Singer SJ (1982) Immunoelectron microscopic studies of desmin (skeletin) localization and intermediate filament organization in chicken skeletal muscle. *J Cell Biol* 96: 1727–1735.

Triffletti P, Litchfield PE, Clarkson PM and Byrnes WC (1988) Creatine kinase and muscle soreness after repeated isometric exercise. *Med Sci Sports Exerc* 209: 242–248.

van der Meulen JH (1991) *Exercise-induced muscle damage: morphological, biochemical and functional aspects.* Dissertation RL, University of Limburg, Datawyse, Maastricht.

van der Meulen JH, Kuipers H and Drukker J (1991) Relationship between exercise-induced muscle damage and enzyme release in rats. *J Appl Physiol* 71: 999–1004.

Vihko V and Arstila AU (1974) Ultrastructural mitochondrial changes in mouse skeletal muscle after forced exhaustive running exercise. *IRCS Med Sci* 2: 1144.

Vihko V and Salminen A (1986) Propagation and repair of exercise-induced skeletal fiber injury. *In:* G Benzi, L Packer and N Silibrandi (eds): *Biochemical Aspects of Physical Exercise.* Elsevier Science Publishers, Amsterdam, pp 337–346.

Vihko V, Rantamäki J and Salminen A (1978a) Exhaustive physical exercise and acid hydrolase activity in mouse skeletal muscle: a histological study. *Histochemistry* 57: 237–249.

Vihko V, Salminen A and Rantamäki J (1978b) Acid hydrolase activity in red and white skeletal muscle of mice during a two-week period following exhausting exercise. *Pflügers Arch* 378: 99–106.

Vihko V, Salminen A and Rantamäki J (1979) Exhaustive exercise, endurance training, and acid hydrolase activity in skeletal muscle. *J Appl Physiol* 47: 43–50.

Warren GL, Hayes DA, Lowe DA, Williams JH and Armstrong RB (1994a) Eccentric contraction-induced injury in normal and hindlimb-suspended mouse soleus and EDL muscles. *J Appl Physiol* 77: 1421–1430.

Warren GL, Lowe DA, Hayes DA, Farmer MA and Armstrong RB (1994b) Cell membrane damage in exercise-induced muscle fiber injury. *Med Sci Sports Exerc* 26: S124.

Woledge RC, Curtin NA and Homsher E (1985) *Energetic Aspects of Muscle Contraction.* Monographics of the Physiological Society, no. 41.

Wrogemann K and Pena SDJ (1976) Mitochondrial calcium overload: a general mechanism for cell necrosis in muscle diseases. *Lancet* 1: 672–674.

Oxidative Stress in Skeletal Muscle
A.Z. Reznick et al. (eds)
© 1998 Birkhäuser Verlag Basel/Switzerland

Free radical mechanisms in exercise-related muscle damage

M.J. Jackson

Muscle Research Centre, Department of Medicine, University of Liverpool, Liverpool L69 3GA, UK

Summary. Skeletal muscle contains a variety of potential sites for the generation of free radical species together with a multi-faceted defence system to prevent the deleterious effects of these substances. Most work in this area has concentrated on the role of mitochondria in the generation of free radical species and hence muscle fibre type composition may be an important determinant of the potential for free radical generation. Oxidative skeletal muscle contains substantial numbers of mitochondria and is subjected to large changes in oxygen flux during exercise. The capacity of muscle to deal with increased radical production also appears to be enhanced in oxidative fibres with this tissue containing relatively high concentrations and activity of a number of different antioxidant materials and enzymes.
Most data now indicate that there is increased free radical activity in muscle during exhaustive aerobic exercise where the muscle is contracting in a primarily concentric or isometric manner. However, this type of exercise does not normally lead to significant muscle damage implying that the well-developed muscle antioxidant system is usually capable of preventing cell damage due to this increased free radical activity.
Where exercise is of a type likely to cause muscle damage (i.e. during eccentric muscle activity) there is much less convincing evidence for increased free radical activity or for a primary protective role of antioxidants. There is, however, evidence for involvement of the cell-mediated immune system in the secondary damage which is a characteristic of substantial eccentric contractile activity, and this may generate oxygen radicals contributing to the secondary tissue damage.

Introduction

A great deal has been written about the beneficial effect of regular physical exercise and the positive adaptive changes that occur in muscle and other tissues (i.e. the "training "effect). In recent years scientists have turned their attention to the tissue damage that sometimes occurs after excessive or unaccustomed exercise. This damage occurs most frequently in skeletal muscle although it has also been described in bones, erythrocytes, gastrointestinal tissue etc. Muscle is unique in its ability to modify the rate of respiration rapidly during exercise and several workers have suggested that increased oxygen free radical production deriving from the large increase in oxygen metabolism by muscle underlies the susceptibility of this tissue to exercise-induced damage (see Sen et al., 1994, for a review).

This topic has been the subject of considerable research effort in recent years although there is little consensus concerning the exercise protocols where oxygen radical production is particularly relevant, the role of free radicals in exercise-induced muscle damage or the need for antioxidant supplementation of exercising subjects. This chapter attempts to examine current knowledge in these areas, to draw conclusions where they appear warranted and to indicate areas where further studies are required.

Potential sources of free radical generation in skeletal muscle

There are a number of potential sites for the generation of free radical species within exercising muscle tissue and further possible mechanisms by which free radical species may be increased secondary to damage induced by other mechanisms.

Primary sources of free radicals

Mitochondrial generation of free radicals

Oxidative metabolism in mammals involves the reduction of molecular oxygen in mitochondria. When oxygen is not limiting, this highly regulated system provides a means for the continuous generation of "high-energy" phosphates (ATP) for muscular contraction. The ability of muscle to undertake coordinated increases in mitochondrial oxygen consumption is substantial. As part of the process for delivery of energy supplies molecular oxygen generally undergoes four-electron reduction catalyzed by cytochrome oxidase. This process has been claimed to account for 95–98% of the total oxygen consumption of tissues, but the remainder (i.e. 2–5% of the total) may undergo one electron reduction with the production of the superoxide radical (O_2^{-}) Further, one-electron reduction of superoxide produces hydrogen peroxide (H_2O_2)which has been observed in studies of isolated mitochondria (Loschen et al., 1974; Boveris and Chance, 1973).

The site of this apparent "loss" of electrons to oxygen has been proposed as coenzyme Q (Sjodin et al., 1990). It is envisaged that the quinones that make up coenzyme Q are reduced to semiquinones by electrons from NADH; these lipophilic compounds are able to diffuse readily so coming into contact with oxygen, with the subsequent generation of superoxide radicals (Loschen et al., 1974). This hypothesis agrees well with what is known about the radical species that are visible through electron spin resonance (ESR) studies of exercising muscle (Davies et al., 1982; Jackson et al., 1985). The only free radical signal observed in intact normal muscle has a g value of about 2.004 and has been claimed to derive mainly from the mitochondria of cells. The actual nature of the radical species has been disputed (Swartz, 1972) although it is likely that it derives primarily from semiquinones (Chetverikov et al., 1964).

It is therefore apparent that an increase in aerobic metabolism during exercise could theoretically lead to an increased production of superoxide radicals as a result of the potentially greatly increased electron flux through the mitochondrial electron transport chain. However, it should also be noted that the mitochondrion has well-developed systems for protection against oxygen radical-mediated damage, with the presence of a specific mitochondrial superoxide dismutase (SOD) to prevent local superoxide-mediated degeneration of biomolecules.

Xanthine oxidase

The possible role of xanthine oxidase in the generation of free radical species has been promoted by McCord and co-workers (McCord, 1985). These workers have proposed that the process of ischemia may lead to formation of xanthine oxidase from xanthine dehydrogenase (via activation of a calcium-dependent protease) and also to a breakdown of ATP with the formation of AMP via the adenylate kinase reaction. This is then further metabolized to hypoxanthine, which is a substrate for both xanthine dehydrogenase and xanthine oxidase. Xanthine oxidase uses molecular oxygen as an electron acceptor with the formation of xanthine (eventually uric acid) and the superoxide radical.

For such a process to provide an important source of free radical species in exercising muscle, a number of criteria must be achieved. First, the enzyme (xanthine dehydrogenase/oxidase) must be present in skeletal muscle; second a failure of calcium homoeostasis must occur (to stimulate the calcium-activated protease) in exercising muscle; and finally the substrate hypoxanthine must be produced in substantial amounts by exercising muscle.

Most human tissues have only very low activities of xanthine dehydrogenase/oxidase in comparison to other species (Al-Khalidi and Cheglassian, 1965; Wigner and Harkness, 1989), but this enzyme has been localized to the capillary epithelium of human muscle, providing a potential source for superoxide production in close proximity to skeletal muscle tissue. The second prerequisite, that exercising muscle show a failure of calcium homoeostasis, also appears to be true. A number of studies have demonstrated that calcium metabolism is deranged during excessive contractile activity of skeletal muscles (Claremont et al., 1984; McArdle et al., 1992) and that contractile activity-induced damage of skeletal muscle can be reduced or prevented by modification of muscle calcium levels (Jones et al., 1984) or inhibition of calcium-mediated pathological processes (Jackson et al., 1984). Evidence that calcium-activated proteases are activated in skeletal muscle post-exercise is indirect, but may be inferred from the characteristic Z-line "streaming" which is observed on ultrastructural examination (Fridén et al., 1983).

The production of hypoxanthine by muscle during exercise has received considerable attention. Most forms of exercise have been associated with some changes in purine metabolism (Sutton et al., 1980), changes in hypoxanthine are particularly evident in subjects undertaking ischaemic exercise and hypoxanthine is rapidly released from exercising muscle to the serum (Sutton et al., 1980; Hellsten-Westing et al., 1991) and hence could provide a substrate for xanthine oxidase within the capillary endothelium.

It is therefore clear that xanthine oxidase within muscle, or the closely associated capillary endothelium, could play a role in superoxide radical generation during or after exercise. The requirement for both hypoxanthine (as a substrate), and calcium activation of a protease to allow the reaction to occur, suggests that this is most likely to occur in metabolically compromised muscle where the rate of ATP breakdown is greater than its generation and cellular ion homoeostasis is lost. Such a situation is more likely to occur in very high-intensity short-duration exercise than in chronic forms of exercise at a moderate or low intensity. Finally it should be noted that, if the site of generation of the xanthine oxidase-derived radicals is within the endothelium, then the protease-mediated conversion of the enzyme and activation of the other calcium-mediated generative processes should occur in that tissue. To the author's knowledge this has not been demonstrated.

Prostanoid metabolism
Prostaglandins are released from various cell types in response to stimuli and appear to be released from skeletal muscle subjected to various stresses, including excessive contractile activity (Rodemann et al., 1981; Smith et al., 1983; McArdle et al., 1991). Many of the intermediates in prostaglandin metabolism are free radical species and the active oxygen metabolites are produced

during prostanoid biosynthesis (Halliwell and Gutteridge, 1989). Arachidonic acid (the precursor of prostaglandins) can also be converted to active metabolites by lipoxygenase enzymes. This process also involves the production of free radical intermediates, but such enzymes do not appear to have been demonstrated in skeletal muscle. However, products of lipoxygenase metabolism have been suggested as mediators of some forms of damage to skeletal muscle (Jackson et al., 1987).

The potential role of prostanoids as a source of oxidative stress in exercise is therefore not yet clear. At the present time the increase in prostanoid production during muscle contractile activity does not have any known physiological role, although it is unlikely to be surreptitious. Further examination of this area is necessary to evaluate whether this represents a beneficial or deleterious effect of excessive muscular contraction.

NAD(P)H oxidase
A further site of free radical production is the NAD(P)H oxidase which is known to occur in neutrophils and various other cell types (Winterbourn, 1990) and has been proposed to be an important source in muscle (Duncan, 1991). However, it is unclear whether this enzyme system is present in skeletal muscle and whether it would be influenced by contractile activity. Hence, in the absence of this crucial information, this potential source remains speculative.

Secondary sources of free radicals

A number of other sources of radicals within muscle are likely to be important after the onset of damage initiated by other mechanisms. This secondary generation of radicals may be important in propagation or exacerbation of damaging processes or may merely be a part of the body's adaptive responses to ensure that efficient preparation of the damaged tissue allows regeneration to occur.

Radical generation by phagocytic white cells
It is clear that substantial injury to muscle fibres is followed by invasion of the area by macrophages and other phagocytic cells from the blood and interstitium (Armstrong, 1986). These infiltrating cells appear to be essential to prepare the tissue to allow fast, effective regeneration to occur. As part of the phagocytic process they release substantial amounts of oxygen radicals (Fantone and Ward, 1985) to aid in the degeneration of necrotic areas, and also to contribute to damage of surrounding viable tissue. It is relevant to note that this increase in free radical generation is non-specific and will occur in all tissue damaged *in vivo* regardless of the mechanism by which the cellular damage occurs. Thus direct trauma to muscle during exercise can cause damage that will eventually lead to a secondary increase in intramuscular free radical generation from phagocytic cells, but this does not equate to an "exercise-induced" increase in free radical generation although measurements of indicators of free radical activity may be abnormal. Similarly supplementation with antioxidants in this situation could not influence the extent of the initial muscle damage, although they might theoretically modify the "scavenging" role

of the phagocytic cells. Such changes can be inferred from the work of Cannon and co-workers (1990).

Generation of radicals secondary to muscle calcium accumulation

It was previously mentioned that activation of a calcium-dependent protease appeared necessary to form xanthine oxidase in ischaemic tissue (McCord, 1985). Many workers have suggested that a failure of muscle calcium homoeostasis is a key step in the degenerative process in exercise-induced muscle damage (Jones et al., 1984; Gollnick et al., 1989; McArdle et al., 1992). It is possible that this failure of calcium homoeostasis occurs secondary to an increase in free radical activity as has been proposed for other tissues but a number of studies argue against this for skeletal muscle (Phoenix et al., 1989, 1991; Jackson et al., 1991; McArdle and Jackson, 1994).

If it is assumed that a failure of muscle calcium homoeostasis is a primary event in exercise-induced muscle damage this could lead to increased free radical activity. Thus, a rise in intramuscular calcium will activate endogenous phospholipase and proteolytic enzymes leading to release of free fatty acids and disruption of intracellular membrane structures. Furthermore, in an attempt to "buffer" the rise in intracellular calcium, mitochondria will become overloaded with calcium, leading to an eventual failure of ATP production with an increase in superoxide production (for a recent review see McArdle and Jackson, 1994)

Increased oxidative stress induced secondary to intracellular calcium overload may therefore be caused by exercise. Again, simple measurements of indicators of free radical activity will only indicate a rise in oxidative reaction products with no evidence for whether they are primary or secondary; they should therefore be evaluated with caution.

Table 1. Characteristics of different types of contractile activity

	Eccentric	Isometric	Concentric
Movement of muscle	Lengthening	Static	Shortening
Mechanical forces	High	Low	Low
Metabolic cost	Low	High	High
Oxygen consumption	Low	High	High
Tendency to induce:			
damage	High	Low	Low
fatigue	High	Low	Low
pain	High	Low	Low
Free radical generation:			
Superoxide production by mitochondria	Low	High	High
Tendency to release hypoxanthine	Low	High	High
Likelihood of secondary neutrophil accumulation	High	Low	Low

Free radical formation caused by disruption of iron-containing proteins

The potential of "delocalised", "loosely-bound" or "free" iron to catalyze free radical reactions is well known (Halliwell and Gutteridge, 1989). Endurance running and other sports with a high mechanical impact may theoretically cause destruction of erythrocytes, with release of iron, and are therefore a potential source of "catalytic" iron. Damage to muscle tissue is also known to release relatively large amounts of the iron-containing protein, myoglobin, into the circulation. Rice-Evans and co-workers (Rice-Evans,1990) have demonstrated the potential of this and other haem proteins to catalyze the production of oxygen radical species; this is an alternative mechanism for possible secondary production of further oxidants after the initiation of cellular damage.

It is relevant to speculate on the likelihood of different types of muscular exercise generating free radicals by some of the above mechanisms. High intensity exercise (usually involving primarily concentric or isometric contractions) is more likely to produce radicals from mitochondria and release hypoxanthine than eccentric exercise, although the tendency to induce damage does not follow this pattern (Tab. 1). Conversely, the increased tendency of eccentric contractions to induce damage implies an increased risk of secondary accumulations of phagocytic cells after this type of exercise, with the consequent implications for subsequent free radical production (see Tab. 1 for a summary).

The endogenous antioxidant protection of skeletal muscle

Skeletal muscle contains a number of different systems for prevention of free radical formation, scavenging of free radicals and removal of potentially damaging products of free radical activity. These include the key antioxidant enzymes superoxide dismutase, catalase and glutathione peroxidase, and the chain-breaking antioxidants vitamin C, glutathione, vitamin E, ubiquinols and some carotenoids (Sen and Hänninen, 1994).

The activity or content of these protective substances appears to be somewhat dependent on the predominant fibre type of the muscle under study. Thus, highly oxidative rat soleus muscle contain a two- to threefold fold higher concentration of glutathione in comparison with fast glycolytic extensor digitorum longus muscles or gastrocnemius muscles of mixed fibre type composition (Jackson et al., 1991). In addition, it has been reported that slow oxidative type I fibres contain increased activities of catalase (Jenkins and Tengie, 1981), SOD (Higuchi et al., 1985) and glutathione peroxidase (Salminen and Vihko, 1983) than fast glycolytic type II fibres. These changes appear to reflect the increased mitochondrial content of oxidative skeletal muscle fibre and the consequent increased likelihood of generation of increased amounts of mitochondrial free radicals during exercise (see previous section).

Other proteins that may protect against oxidative stress (such as the heat shock proteins) are also found in skeletal muscle and their expression is increased in response to oxidative stress (Salo et al., 1991). Again, the endogenous expression of some of these proteins is increased in oxidative compared with glycolytic muscle fibres (Locke et al., 1991). This appears to be an adaptive response to chronic oxidative stress in these fibres.

Role of free radicals in exercise-induced muscle damage to skeletal muscle

A considerable amount of work has been undertaken in this area and it is relevant to address the problem under various headings.

Are free radicals produced in excess during exercise?

The unequivocal demonstration of increased free radical activity in complex biological tissues is difficult and usually only accepted if a variety of indicators provides supportive evidence. This evidence can be in the form of measurements of indirect indicators of free radical activity (products of lipid peroxidation, DNA oxidation, protein oxidation), direct detection of free radicals (ESR techniques), or prevention of the putative free radical-mediated effect by supplementation with relatively specific antioxidants.

Initial suggestions that free radical processes, such as lipid peroxidation, were elevated during exercise came from studies of whole-body exercise in humans (Dillard et al., 1978) and rats (Brady et al., 1979; Gee and Tappel, 1981). These were rapidly followed by studies of the products of free radical reactions within the tissues of exercising animals (Davies et al., 1982). These data indicated that exercise to exhaustion in rats resulted in decreased mitochondrial respiratory control, loss of sarcoplasmic reticulum integrity, increased lipid peroxidation, and increased free radical generation as shown by ESR studies. These are perhaps the most widely quoted data in support of a role for free radical species in exercise-induced damage to skeletal muscle (and other tissues). It is notable that the exercise regime used was an endurance protocol in which the muscles were primarily contracting in a concentric manner.

Similar ESR studies undertaken by the author's group have also demonstrated an increased "stable" free radical signal in response to excess contractile activity of muscle (Jackson et al., 1985), although interpretation of these results was somewhat different from that of Davies et al. (1982). In particular, the author's group studied the possibility that the increased ESR-visible free radical signal occurs after an exercise-induced accumulation of calcium (Jones et al., 1984) within the muscle cells (Johnson et al., 1988) and is therefore a secondary consequence of alternative damaging processes (McArdle and Jackson, 1994).

Of particular relevance in this area are the possible effects on free radical production produced by the way in which the muscle is used. Muscle may contract in a concentric manner (where the active muscle is allowed to shorten), an eccentric manner (where the muscle is lengthened) or an isometric manner (where the muscle remains at a fixed length). Eccentric contractions are considerably more damaging to muscle (Newham et al., 1983, 1986a, 1986b), but these different types of exercise have been studied relatively infrequently in the evaluation of their influence on free radical production (Jackson, 1994).

Is any production of excess free radicals during exercise damaging or beneficial to tissues?

It is generally assumed that excess free radical production is damaging to tissues, and the described association between excessive free radical production during exercise and tissue damage (Davies et al., 1982; Jackson et al., 1985) has supported this. However, it should be noted that, at the current stage, this has only been shown to be an association rather than a cause-and-effect relationship (Jackson, 1994), and further work is required in this area.

There has been a small amount of work examining the possibility that beneficial effects may derive from free radicals produced during exercise. Salo et al. (1991) reported that oxygen radicals stimulate the production of stress proteins in exercising muscle and that these proteins may play a role in mitochondrial biogenesis and the training response to exercise. Free radicals may therefore be playing an important second messenger role in this situation.

Is the tissue antioxidant capacity modified by exercise?

Exercise training is recognized to be an efficient way of reducing the susceptibility of muscles to exercise-induced muscle damage, and several studies have investigated the possibility that this may be associated with an increase in the tissue's defences against free radicals. Exercise training in rats appears to be associated with an increase in the activity of muscle SOD (Higuchi et al., 1985), and modifications in the concentration of antioxidants and in the activity of antioxidant enzymes has also been reported in humans (Robertson et al., 1991).

Does antioxidant supplementation reduce free radical activity during exercise?

There is considerable evidence that supplementation with specific antioxidants may reduce indicators of increased exercise-induced free radical activity in muscle or the circulation (Dillard et al., 1978; Davies et al., 1982; Meydani et al., 1994; Jackson et al., 1985; Warren et al., 1992; Reznick et al., 1992).

Does antioxidant supplementation during exercise have beneficial effects on tissues?

This is perhaps the most important question requiring an answer. Vitamin E is the antioxidant that has received most attention in this area. Davies et al. (1982) originally found that vitamin E-depleted animals had a reduced exercise endurance. These effects were confirmed by further studies from the same group (Packer, 1984), and supported by reports of the exacerbating effect of vitamin E deficiency on other models of exercise-induced muscle damage (Jackson et al., 1983; Amelink, 1990).

The author's group has specifically examined effects of vitamin E on damage processes in isolated skeletal muscle. These studies have demonstrated that vitamin E has protective effects against contractile activity-induced (Jackson et al., 1983; McArdle et al., 1993) and calcium ionophore-induced (Phoenix et al., 1989, 1990, 1991) damage to skeletal muscle *in vitro*, but the mechanisms by which this protective effect occurs do not appear to be as clearcut as some workers have proposed. Although the protective effects are apparent in animals fed diets rich in polyunsaturated fatty acids but not animals fed a diet rich in saturated fatty acids (S. O'Farrell and M.J. Jackson, unpublished data) – in general agreement with the concept that the excess vitamin E is preventing free radical-mediated peroxidation of membrane polyunsaturated fatty acids – the protective effects also appear to be mimicked by phytol, isophytol, and a number of other lipophilic, non-antioxidant substances with long hydrocarbon side chains (Phoenix et al., 1989, 1991). It is therefore clear that further work is required in this area to clarify the nature of the protection offered by vitamin E.

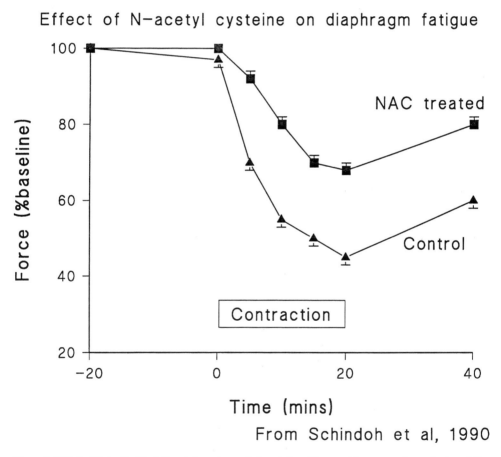

Figure 1. Effect of the antioxidant *N*-acetylcysteine on fatigue induced by repetitive contractions of strips of diaphragm *in vitro*. (Data derived from Schindoh et al., 1990).

All of the preceding studies were undertaken in exercise models in which the predominant form of muscle activity was not specified or in which it was entirely isometric. Eccentric exercise has been infrequently studied from the point of view of free radical processes, but where damage to skeletal muscle specifically induced by eccentric contractions has been studied, conflicting data have been reported. In a detailed study of damage to mouse extensor digitorum longus muscle induced by eccentric contraction, Zerba et al. (1990) found that treatment of animals with polyethylene glycol–SOD significantly reduced the amount of injury that was present 3 days after exercise in mice of various ages. However, in a study of animals that undertook lengthening contractions during downhill running, Warren et al. (1992) could show no protective effect of vitamin E supplementation. Nevertheless, studies of human subjects undertaking eccentric exercise have reported changes in blood parameters indicative of increased free radical activity (Packer and Viguie, 1989). Cannon and co-workers have also examined subjects undertaking downhill running. They found no protective effects of vitamin E supplementation against muscle damage (Cannon et al., 1990), although the supplements did appear to reduce oxidative stress (Meydani et al., 1994) these authors suggested that their data support a role for oxidants in the delayed-onset muscle damage.

Other antioxidants have been studied only infrequently as potential inhibitors of the deleterious effects of exercise, but the data that have been presented are inconclusive about the possible protective effects of these substances (Gerster, 1989; Bendich, 1991; Sastre et al., 1992). However, Jakeman and Maxwell (1993) have studied the effects of vitamin C supplementation on eccentric exercise in humans and reported a protective effect of this substance on the post-exercise muscle fatigue compared with untreated controls or a vitamin E-treated group. Comparable data concerning beneficial effects of the antioxidant, N-acetylcysteine on muscle fatigue have also been reported in animal studies by Schindoh et al. (1990) (Fig. 1) and in humans by Reid et al. (1994), although the mechanisms for these protective effects have not been elucidated.

Conclusions

It is clear from the preceding sections that there has been a considerable amount published regarding free radical activity and exercise, but there is little consensus about the importance of this in exercise-induced muscle damage. Most data indicate that there is increased free radical activity in muscle during exhaustive aerobic exercise where the muscle is contracting in a primarily concentric manner. However, this type of exercise does not normally lead to significant muscle damage, implying that the well-developed muscle antioxidant system is capable of preventing cell damage caused by this increased free radical activity.

Where exercise is of a type likely to cause muscle damage (i.e. during eccentric muscle activity) there is much less convincing evidence for increased free radical activity or for a primary protective role of antioxidants. There is, however, evidence for involvement of the cell-mediated immune system in the secondary damage which may follow eccentric contractile activity, this may generate oxygen radicals contributing to the secondary tissue damage.

References

Al-Khalidi UAS and Cheglassian TH (1965) The species distribution of xanthine oxidate. *Biochem J* 97: 318–320.

Amelink GJ (1990) *Exercise Induced Muscle Damage*. Utrecht, Netherlands: Univ. of Utrecht; PhD thesis.

Bendich A (1991) Exercise and free radicals: Effects of antioxidant vitamins. *Med Sport Sci* 32: 59–78.

Boveris A and Chance B (1973) The mitochondrial generation of hydrogen peroxide: general properties and effects of hyperbaric oxygen. *Biochem J* 134: 707–716.

Brady PS, Brady LJ and Ulrey DE (1979) Selenium, vitamin E and the response to swimming stress in the rat. *J Nutr* 109: 1103–1109.

Cannon JG, Orencole SF, Fielding RA, Meydani M, Meydani SN, Fiatarone MA, Blumberg JB and Evans WJ (1990) Acute phase response in exercise: Interaction of age and vitamin E on neutrophils and muscle enzyme release. *Am J Physiol* 259: R1214–R1219.

Chetverikov A, Kalmanson A, Kharitonenkov I and Blumenfield L (1964) Issledovanie metodom EPR svohodnykh radikalov v biologicheskikh ob ektakh, voznikailischikh vo vremiia protakaniia fermentativnykh reaktsii. *Biofizika* 9: 18.

Claremont D, Jackson MJ and Jones DA (1984) Accumulation of calcium in experimentally damaged mouse muscles. *J Physiol* 353: 57P.

Davies KJA, Quintanilha AT, Brooks GA and Packer L (1982) Free radicals and tissue damage produced by exercise. *Biochem Biophys Res Commun* 107: 1198–1205.

Dillard CJ, Litov RE, Savin WM and Tappel AL (1978) Effects of exercise, vitamin E and ozone on pulmonary function and lipid peroxidation. *J Appl Physiol* 45: 927–932.

Duncan CJ (1991) *In*: CJ Duncan (ed.): *Calcium, Oxygen Radicals and Cellular Damage*. Cambridge University Press, Cambridge, pp 97–113.

Fantone JC and Ward PA (1985) Polymorphonuclear leukocyte-mediated cell and tissue injury: oxygen metabolites and their relation to human disease. *Human Pathol* 16: 973–978.

Fridén J, Sjöstrom M and Ekbloom B (1983) Myofibrillar damage following intense exercise in man. *Int J Sport Med* 4: 170–176.

Gee DL and Tappel A L (1981) The effect of exhaustive exercise on expired pentane as a measure of *in vivo* lipid peroxidaion in the rat. *Life Sci* 28: 2425–2429.

Gerster H (1989) The role of vitamin C in athletic performance. *J Am Coll Nutr* 8: 636–643.

Gollnick PD, Hodgson D R and Byrd S K (1989) Exercise-induced muscle damage: A possible link to failures in calcium regulation. *In*: G Benzi (ed.): *Advances in Myochemistry* 2. John Libbey Euotext, London, pp 339–350.

Halliwell B and Gutteridge JMC (1989) *Free Radicals in Biology and Medicine*. Clarendon, Oxford.

Hellsten-Westling Y, Sollevi A and Sjodin B (1991) Plasma accumulation of hypoxanthine, uric acid and creatine kinase following exhausting runs of different durations in man. *Eur J Appl Physiol* 62: 380–384.

Higuchi M, Cartier LJ, Chen M and Holloszy JO (1985) Superoxide dismutase and catalase in skeletal muscle: Adaptive response to exercise. *J Gerontol* 40: 281–286.

Jackson MJ Exercise and oxygen radical production by muscle (1994) *In*: CK Sen, L Packer and O Hanninan (eds): *Exercise and Oxygen Toxicity*. Elsevier, London, 49–57.

Jackson MJ, Jones DA and Edwards RHT (1983) Vitamin E and skeletal muscle. *In*: R Porter and J Whelan (eds): *Biology of vitamin E*. Pitman, London, pp 224–239 (Ciba Foundation Symposium Series No. 101).

Jackson MJ, Jones DA and Edwards RHT (1984) Experimental muscle damage: The nature of the calcium-activated degenerative processes. *Eur J Clin Invest* 14: 369–374.

Jackson MJ, Edwards RHT and Symons MCR (1985) Electron spin resonance studies of intact mammalian skeletal muscle. *Biochim Biophys Acta* 847: 185–190.

Jackson MJ, Wagenmakers AJM and Edwards RHT (1987) The effect of inhibitors of arachidonic acid metabolism on efflux of intracellular enzymes from skeletal muscle following experimental damage. *Biochem J* 241: 403–407.

Jackson MJ, McArdle A and Edwards RHT (1991) Free radicals, calcium and damage in dystrophic and normal skeletal muscle. *In*: CJ Duncan (ed.): *Calcium, Free Radicals and Tissue Damage*. Cambridge Univeristy Press, Cambridge, pp 139–145.

Jakeman P and Maxwell S (1993) Effect of antioxidant vitamin supplementation on muscle function after eccentric exercise. *Eur J Appl Physiol* 67: 426–430.

Jenkins RR and Tengie J (1981) Catalase activity in skeletal muscle of varying fibre types. *Experientia* 37: 67–68.

Johnson KM, Sutcliffe LH, Edwards RHT and Jackson MJ (1988) Calcium ionophore enhances the electron spin resonance signal from isolated skeletal muscle. *Biochim Biophys Acta* 964: 285–299.

Jones DA, Jackson MJ, McPhail G and Edwards RHT (1984) Experimental muscle damage: The importance of external calcium. *Clin Sci* 66: 317–322.

Locke M, Noble EG and Atkinson BG (1991) Inducible isoform of HSP70 is constitutively expressed in a muscle fibre type specific pattern. *Am J Physiol* 261: C774–C779.

Loschen G, Azzi A, Richter C and Flohe L (1974) Superoxide radicals as precursors of mitochondrial hydrogen peroxide. *FEBS Lett* 42: 68–72.

McArdle A and Jackson MJ (1994) Intracellular mechanisms involved in damage to skeletal muscle. *Basic Appl Myol* 4: 43–50.

McArdle A, Edwards RHT and Jackson MJ (1991) Effects of contractile activity on indicators of muscle damage in the dystrophin-deficient *mdx* mouse. *Clin Sci* 80: 367–371.

McArdle A, Edwards RHT and Jackson MJ (1992) Accumulation of calcium by normal and dystrophin-deficient muscle during contractile activity *in vitro*. *Clin Sci* 82: 455–459.

McArdle A, Edwards RHT and Jackson MJ (1993) Calcium homeostasis during contractile activity of vitamin E-deficient skeletal muscle. *Proc Nutr Soc* 52: 83A.

McCord JM (1985) Oxygen derived free radicals in post-ischaemic tissue injury. *N Engl J Med* 312: 159–163.

Meydani M, Evans WJ, Handelman G, Biddle L, Fielding RA, Meydani SN, Burrill J, Fiatarone MA, Blumberg IB and Cannon JG (1994) Protective effect of vitamin E on exercise-induced oxidative damage in young and older adults. *Am J Physiol* 264: R992–R998.

Newham DJ, Mills KR, Quigley BM and Edwards RHT (1983) Pain and fatigue after concentric and eccentric muscle contractions. *Clin Sci* 64: 55–62.

Newham DJ, Jones DA and Edwards RHT (1986a) Plasma creatine kinase changes after eccentric and concentric contractions. *Muscle Nerve* 9: 59–63.

Newham DJ, Jones DA, Tolfree SEJ and Edwards RHT (1986b) Skeletal muscle damage: A study of isotope uptake, enzyme efflux and pain after stepping. *Eur J Appl Physiol* 55: 106–112.

Packer L and Viguie C (1989) Human exercise: Oxidative stress and antioxidant therapy. *In*: G Benzi (ed.): *Advances in Myochemistry 2*. John Libbey Eurotext, London, pp 1–17.

Phoenix J, Edwards RHT and Jackson MJ (1989) Inhibition of calcium-induced cytosolic enzyme efflux form skeletal muscle by vitamin E and related compounds. *Biochem J* 257: 207–213.

Phoenix J, Edwards RHT and Jackson MJ (1990) Effects of calcium ionophore on vitamin E deficient rat muscle. *Brit J Nutr* 64: 245–256.

Phoenix J, Edwards RHT and Jackson MJ (1991) The effect of vitamin E analoges and long hydrocarbon chain compounds on calcium-induced muscle damage: A novel role for α-tocopherol. *Biochim Biophys Acta* 1097: 212–218.

Reid MB, Stokic DS, Kech SM, Khawli FA and Leis AA (1994) *N*-Acetylcysteine inhibits muscle fatigue in humans. *J Clin Invest* 94: 2468–2474.

Reznick AZ, Witt E, Matsumoto M and Packer L (1992) Vitamin E inhibits protein oxidation of skeletal muscle of resting and exercised rats. *Bioch Biophys Res Commun* 189: 801–806.

Rice-Evans C (1990) Erythrocytes, oxygen radicals and cellular pathology. *In*: DK Dass and WB Essman (eds): *Oxygen Radicals: Systemic Events and Disease Processes*. Karger, Basle, pp 1–30.

Robertson JD, Maughan RJ, Duthie GG and Morrice PC (1991) Increased blood antioxidant systems of runners in response to training load. *Clin Sci* 80: 611–618.

Rodemann MP, Waxman L and Goldberg AL (1981) The stimulation of protein degradation by Ca^{2+} is mediated by postaglandin E_2 and does not require the calcium activated protease. *J Biol Chem* 257: 8716–8723.

Salminen A and Vihko V (1983) Lipid peroxidation in exercise myopathy. *Exerc Mol Pathol* 38: 380–388.

Salo DC, Donavan CM and Davies KJA (1991) HSP70 and other possible heat shock or oxidative stress proteins are induced in skeletal muscle, heart and liver during exercise. *Free Radical Biol Med* 11: 239–246.

Sastre J, Asensi M, Gasco E, Pallardo FV, Ferrero JA, Furakawa T and Vina J (1992) Exhaustive physical exercise causes oxidation of glutathione status in blood: Prevention by antioxidant administration. *Am J Physiol* 263: R992–R995.

Schindoh C, Dioharco A, Thomas A, Manubray P and Supinski G (1990) Effect of *N*-acetyl cysteine on diaphragm fatigue. *J Appl Physiol* 68: 2107–2113.

Sen CK and Hänninen O (1994) Physiological antioxidants *In*: CK Sen, L Packer and O Hänninen (eds): *Exercise and Oxygen Toxicity*. Elsevier, Amsterdam pp 89–126.

Sen CK, Packer L and Hänninen O (1994) Exercise and oxygen toxicity. Elsevier, Amsterdam.

Sjodin B, Hellsten-Westing Y and Apple FS (1990) Biochemical mechanisms for oxygen free radical formation during exercise. *Sport Med* 10: 236–254.

Sutton JR, Toews CJ, Ward JR and Fox IH (1980) Purine metabolism during strenuous muscular exercise in man. *Metabolism* 29: 254–260.

Swartz HM (1972) Cells and tissues. *In*: HM Swartz, JR Bolton and DC Borg (eds): *Biological Applications of Electron Spin Resonance*. John Wiley & Sons, New York, pp 155.

Smith RH, Palmer RM and Reeds PJ (1983) Protein synthesis in rabbit forelimb muscles. *Biochem J* 214: 142–161.

Wigner M and Harkness RA (1989) Distribution of xanthine dehydrogenase and oxidase activities in human and rabbit tissue. *Biochim Biophys Acta* 991: 79–84.

Warren JA, Jenkins RR, Packer L, Witt EH and Armstrong PB (1992) Elevated muscle vitamin E does not attenuate eccentric exercise-induced muscle injury. *J Appl Physiol* 72: 2168–2175.

Winterbourn CC (1990) Neutrophil oxidants: Production and reactions *In*: DK Das and WB Essman (eds): *Oxygen Radicals: Systemic Events and Disease Processes*. Karger, Basel pp 31–70.

Zerba E, Komorowski TE and Faulkner JA (1990) Free radical injury to skeletal muscles of young adult and old mice. *Am J Physiol* 258: C429–C435.

Oxidative Stress in Skeletal Muscle
A.Z. Reznick et al. (eds)
© 1998 Birkhäuser Verlag Basel/Switzerland

The effects of exercise, ageing and caloric restriction on protein oxidation and DNA damage in skeletal muscle

Z. Radák[1] and S. Goto[2]

[1]*Laboratory of Exercise Physiology, University of Physical Education Budapest, H-1123 Budapest, Hungary*
[2]*Department of Biochemistry, School of Pharmaceutical Sciences, Toho University, Funabashi, Japan*

Summary. Constant generation of free radicals is an accompanying process of aerobic metabolism. The interaction of metal ion, hydrogen peroxide and amino acid residues of proteins results in a variety of structural modifications on amino acid residues and inactivation of proteins. Moderately modified proteins are sensitive to proteolytic degradation, while severely oxidized cross-linked proteins are resistant to proteolysis. Physiological and pathological conditions which alter the delicate balance of radical generation and antioxidant defence, including proteolytic systems, change the rate of accumulation of oxidized proteins. Moreover, changes in the rate of protein synthesis and degradation could influence the accumulation of oxidatively modified proteins. Oxidative modifications of proteins in some conditions may be a part of the controlling process in the cells.
Ageing and muscular diseases can increase the rate of accumulation of oxidized proteins, while caloric restriction and regular physical exercise of moderate intensity might have a beneficial effect on this accumulation. Mitochondria are suggested to be especially exposed to free radical insult by virtue of their being the dominant site of free radical generation. Mitochondrial DNA (mtDNA) is in the front line of the attack by radical species and due to the poor repair mechanisms and fast turnover rate of the DNA, the damage cannot be completely repaired. The replication of the mutant DNA should have serious consequences, since mtDNA encodes essential components of the electron transport chain. There is an increase in DNA damage adducts with age and in diseases with accelerated ageing. Regular exercise and caloric restriction could have retarding effects on the accumulation of DNA and protein damage. Skeletal and cardiac muscles as stable postmitotic tissues tend to harbour oxidatively modified proteins and DNA adducts.
The available data suggest that caloric restriction and moderate physical training via a variety of adaptive processes might help delay age related increases in oxidation of amino acid residues and DNA.

Introduction

Proteins are the most abundant constituent of cells, so it is not surprising that they are the object of the massive insult of endogenous and exogenous free radical species. A historical study of Dakin (1906) reported that oxidation of amino acid residues results in a variety of chemical modifications. The oxidation of amino acid residues is a chain reaction catalysed by various agents. It is widely accepted that metal-catalysed oxidation of proteins takes place *in vivo*, and in the last 15 years it became evident that many physiological and pathophysiological processes are associated with metal-catalysed oxidation (reviewed by Stadtman, 1986, 1990; Stadtman and Oliver, 1991; Goto et al., 1995; Dean et al., 1997).

Oxidative modifications of amino acid residues in proteins might lead to fragmentation of the proteins, accumulation of altered forms and increased susceptibility to proteolytic degradation (Davies, 1986; Stadtman, 1990). Some residues such as histidine and arginine are especially targeted and modified by radicals, because they are often positioned close to metal-binding sites that are supposed to be essential for protein oxidation (Stadtman, 1992, Stadtman and Berlett, 1991). Indeed, oxidative modification of proteins involves the interaction of metal ion, hydrogen peroxide and amino acid residues (Fig. 1). Hydrogen peroxide might be present in cells as a

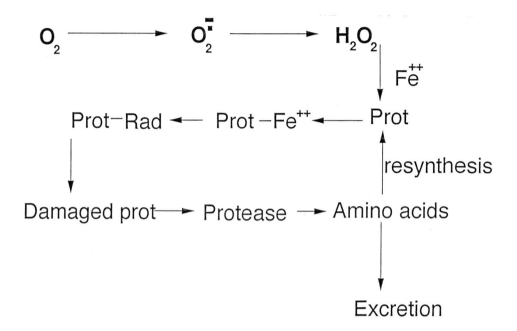

Figure 1. Suggested mechanism of metal-catalysed oxidation: Prot, protein; Prot–Rad, protein radical.

result of either superoxide dismutation by superoxide dismutase (SOD) or as enzymatic process. It is believed that, as a result of its moderate reactivity and long half-life, as well as its non-polarity, hydrogen peroxide can easily cross biological membranes (Yu, 1994). Transition metals, especially iron present in limited forms proteins and enzymes (haem rings) can be mobilized from its physiological sites to bind to potential metal-binding sites of various proteins (Puppo and Halliwell, 1988). It is hypothesized that hydroxyl radicals generated at the sites readily react with amino acid residues close by, resulting in various kinds of oxidative modifications in the proteins (Stadtman, 1990; Dean et al., 1997). One of the modified end-products of the free radical reaction in amino acid residues is the reactive carbonyl derivatives (RCDs) which are readily measurable by its reaction with 2,4-dinitrophenylhydrazine (DNPH) (Levine et al., 1990). Radical-induced oxidation of the peptide side chains of arginyl, aspartyl, glutamyl, lysyl, prolyl and threonyl residues can generate RCDs. Several studies (Levine et al., 1994; Reznick and Packer, 1994; Cao and Cutler, 1995; Nakamura and Goto, 1996; Buss et al., 1997) improved the original method of RCD measurement (Levine et al., 1990).

Despite the fact that RCDs can also be formed by reaction with unsaturated alkenals generated by lipid peroxidation or by glycation, the determination of RCDs is used extensively and there are

Table 1. Levels of reactive carbonyl derivatives in skeletal and cardiac muscles

Muscle	Condition	Reactive carbonyl derivatives (RCD)	Reference
Quadriceps	Exhaustive exercise	▲	Reznick et al. (1992a)
Hindlimb	8 weeks of running	▲	Witt et al. (1992)
Quadriceps	Immobilization	▲	Fares et al. (1996)
Pectoralis major	Dystrophy	▲	Murphy and Kehler (1989)
Quadriceps	4 weeks of running sea level	▶	Radák et al. (1997a)
	4 weeks of running at high altitude	▲	
Hindlimb muscle Mitochondria		▼	
Microsomes	4 weeks of swimming	▶	Radák et al. (1997c)
Cytoplasm		▶	
Soleus	4 weeks of caloric restriction	▼	Z. Radák et al. (unpublished)
	4 weeks of high altitude	▼	
Hindlimb	Mn deficiency	▲	Astier et al. (1996)
Cardiac muscle	Ischemia–reperfusion	▲	Reznick et al. (1992b)
Cardiac muscle	Ageing	▲	Sohal et al. (1994)

a number of reports that suggest a link between its accumulation and a variety of pathophysiological conditions such as ageing (Starke-Reed and Oliver,1989; Goto et al., 1995), rheumatoid arthritis (Chapman et al., 1989), Alzheimer's disease (Smith et al., 1991), smoking (Reznick et al., 1992, 1997), ischemia–reperfusion (Oliver et al., 1990), muscular dystrophy (Murphy and Kehler, 1989), exercise (Reznick et al., 1992; Witt et al., 1992; Radák et al., 1997a, 1998). The present review focuses on the literature relating to accumulation of RCDs in skeletal muscle (Tab. 1).

Protein turnover and protein oxidation

Metabolic turnover of macromolecules is of fundamental importance to the maintenance of life in all organisms. Skeletal muscles represent 30–40% of the human body mass, so control of protein turnover of the muscle is particular important. Protein turnover is a cyclical process involving synthesis and degradation of proteins. This process is economical because amino acids derived from degradation are recycled for new synthesis of proteins. Reports published by the laboratories of Stadtman and Davies revealed that free radical-induced oxidation of proteins alters the susceptibility to proteolytic degradation (Fagan et al., 1986; Davies and Goldberg, 1987; Stadtman, 1990). Indeed, the accumulation of oxidatively modified proteins does not necessarily correlate well with the extent of oxidative stress, because oxidatively modified proteins are more readily degraded (Levine et al., 1981; Rivett, 1985; Davies et al., 1987) compared with normal proteins. The massive oxidative stress and/or defect of proteolytic system, when oxidized proteins cannot

go through digestion, leads to the development of cross-linking and further aggregation of damaged proteins (Davies et al., 1987; Pacifici et al., 1993). On the other hand, proteolytic enzymes
contribute significantly to antioxidant defences, because the preferential degradation of oxidized
proteins is a preventive step against the accumulation of physiologically inactive "junk" (Rivett,
1985; Davies, 1986; Davies and Delsignore, 1987; Grune et al., 1995). Moreover, ATP-dependent
and -independent proteolytic systems involving proteasomes can reduce massive accumulation of
abnormal proteins efficiently (Grune et al., 1995). The extent of oxidative modification has a vital
role in the degradation, because moderately modified hydrophobic proteins are more readily
degraded and severely modified proteins, such as cross-linked ones, are resistant to degradation
(Pacifici et al., 1993). Extreme examples are β-amyloid proteins and paired helical filaments
found primarily in the brain of patients with Alzheimer's disease or with other types of amyloid.
These altered proteins are highly resistant to proteolytic degradation as a result of their very
compact secondary and tertiary structures.

Some physiological (ageing, fasting, exercise, altitude exposure) and pathophysiological (cancer, immobilization, muscular dystrophy) conditions significantly alter the rate of protein turnover
and some data indicate that the accumulation of oxidatively modified proteins is also altered. An
early report of Omaye and Tappel (1974) suggested that free radical-induced damage plays a role
in the pathogenesis of muscular dystrophy. The indirect signals of increased free radical
production such as elevated activities of antioxidant enzymes and level of lipid peroxidation and
protein oxidation, were reported in samples from dystrophic skeletal muscle (Omaye and Tappel,
1974; Mizuno, 1985; Murphy and Kehler, 1989). An interesting observation was made by
Murphy and Kehler (1989) as they measured the RCDs in pectoralis and soleus muscles of
dystrophic chickens of different ages. It was concluded that the difference in the accumulation of
RCDs in the muscle of dystrophic birds was the result of the decrease in RCDs in normal, but
not dystrophic, animals, and the elevated RCDs in dystrophic chicken was not an artefact of
physiological changes.

Immobilization is also associated with the increases in accumulation of RCDs. Carmeli et al.
(1993) reported that 4 weeks of immobilization, which resulted in a marked reduction of muscle
weight (about 30% decrease) of aged rats, increased the accumulation of RCDs by 400% of the
initial value and the treatment of growth hormone significantly attenuated the increase. Fares et al.
(1996) come to similar conclusions after immobilization of hindlimb muscle. Whether the
immobilization induces oxidative modification of proteins by direct physical stress in the muscle
is not clear because it also increases lipid peroxidation and protein oxidation in the brain as well
(Liu et al., 1996). High-altitude exposure suppresses both the rate of protein synthesis and free
radical formation (Rennie et al., 1983; Radák et al., 1994; Radák et al., 1997). A short-term (4-
week) exposure to an altitude of 4000 m in a hypobaric chamber leads to decreases in the
accumulation of RCDs in rat skeletal muscle (Z. Radák et al., unpublished data).

The authors hypothesized that increase in protein synthesis also influences the level of RCD
accumulation. The right wing of quails was loaded in order to increase the hypertrophy of
anterior latissimus dorsi muscle (Z. Radák et al., unpublished data). The left wing of the birds
served as an intra-animal control muscle. The stretch-induced hypertrophy of the muscle was
accompanied by changes in RCD content. The accumulation of RCDs apparently decreased on

day 16 of the loading compared with 5 h and 3 days. This observation suggests that the accumulation of RCDs is significantly altered in stretch-induced hypertrophy.

At present it cannot be excluded that free radical-induced damage of cellular proteins plays a controlling role in protein turnover via some poorly understood mechanisms, such as the possible induction of proteolysis by damaged proteins. It has been suggested by Toledo et al. (1995) that free radical-mediated post-translation damage of proteins plays a role in DNA repair and affects cell differentiation.

Ageing, caloric restriction, exercise and protein oxidation

Ageing, caloric restriction and exercise can influence the rate of free radical formation, efficiency of antioxidant defence, protein synthesis, degradation and level of RCD accumulation. There are a number of reports that the ageing process is associated with the accumulation of RCDs in different tissues (reviewed by Stadtman, 1992; Orr and Sohal, 1996), although some investigators found few age-related changes (Goto and Nakamura, 1997). However, the skeletal muscle is not a well-studied tissue with respect to protein damage and ageing. In the gastrocnemius muscle, the activity of the enzymes of mitochondrial electron transport chain decreases with age and caloric restriction retards this decline (Desai et al., 1996). This change in the function of electron transport chain is most probably associated with a parallel increase in accumulation of RCDs with age in cardiac muscle, and in the retarding effect of caloric restriction on accumulation of RCDs (Sohal et al., 1994). It should be mentioned that caloric restriction can increase protein turnover rate so as to reduce the amount of RCDs (Ishigami and Goto, 1990).

Physical exercise increases oxygen uptake manyfolds and could therefore result in enhanced free radical formation (Davies et al., 1982). However, it has to be remembered that the effects of single bout exercise and regular moderate training are very different. The regular moderate training can have beneficial effects on a variety of biological processes in different organs and, probably as a result of this adaptive process, it extends the lifespan of laboratory animals (Holloszy, 1993) and possibly of humans too (Brown, 1992; Sarna et al., 1993). On the other hand, single bout exercise may result in oxidative damage that could be the first step of the adaptive process in the long run (Davies et al., 1982). Reznick et al. (1992) published the first paper measuring the oxidative modifications of proteins after a single bout of running exercise in rats. The exercise induced a moderate increase in RCDs and vitamin E supplementation significantly reduced the level of RCDs in both control and supplemented rats. It was concluded that the increase in RCDs was the result of an inadequate response of the antioxidant system and the red fibres are better equipped with antioxidant defence. The study of Rajguru et al. (1994) reports that 10–15 min of swimming in rats leads to an increase in sulfhydryl levels in the microsomes of the skeletal muscle. It was suggested that elevation of the sulfhydryl level is part of the regenerative process, because reduced glutathione is involved in the removal of lipid peroxides, and the increased amount of sulfhydryl compounds in the blood works as a detoxifying agent. These results might indicate that even a single bout of exercise could initiate as adaptive body response.

The first study on the effects of physical training on accumulation of RCDs was conducted by Witt et al. (1992). An increase in RCD accumulation in hindlimbs was measured after an 8-week period of running training with moderate intensity. It was also demonstrated that the administration of vitamin E decreased the accumulation of RCDs.

Training at altitudes of 4000 m increased the RCD content in the red and white parts of quadriceps muscle of rats, whereas no significant changes were observed in group trained at sea level (Radák et al., 1997a). Therefore, it has been suggested that the combined effects of high altitude and training resulted in the accumulation of RCDs and training at sea level did not increase the oxidative modification of proteins.

The author's recent study supports the view that regular training does not increase RCD accumulation. The skeletal muscle was separated into mitochondrial, microsomal and cytoplasmic fractions (Radák et al., 1997b). Four weeks of swimming reduced the RCD content in cell fractions of trained rats compared with those of the untrained one. This result is supported by the findings that exercise increases the rate of both protein synthesis (Nair, 1995; Welle et al., 1995) and proteolytic degradation (Evans, 1992; Biolo et al., 1995), resulting in a better life-maintaining process through more frequent renewal of oxidatively modified proteins by newly synthesized ones in the cells. Taking the observations related to regular exercise together, it has been suggested that exercise via a variety of pathways (reduced generation of reactive oxygen species, increased antioxidant defence, faster protein synthesis and degradation) probably decreases or at least maintains the level of oxidatively modified proteins that would otherwise tend to increase. This could account for the improved cell functions associated with exercise-induced adaptation. In the author's study cited above, it was found that the magnitude of RCD accumulation was three- to fourfold larger in mitochondria than in other cell fractions measured. Hence, mitochondria play an exceptional role in free radical generation, which deserves a detailed review of mitochondria including their protein components.

Mitochondria: The "powerhouse" of free radical formation and oxidative damage

Mitochondria are designed structurally and functionally for respiration and oxidative phosphorylation. In the last step of oxidative phosphorylation, electrons from hydrogen atoms are transferred through the membrane-bound electron carriers to O_2 forming water. Protons are transported out from the mitochondria and the energy generated by the proton gradients drives ATP synthesis. During this vital physiological process, 2–4% of the O_2 consumed by mitochondria is supposed to be transformed into reactive oxygen species (Chance et al., 1979). Hence, mitochondria are not only the powerhouse of ATP synthesis but also have the adverse effects of free radical generation. Therefore, proteins that are located in the mitochondria, or encoded by mitochondrial DNA (mtDNA), are in the front line of an attack by free radicals.

Human mtDNA has 16 569 base-pairs, encoding about 10% of mitochondrial proteins and involving seven subunits of the NADH–ubiquinone reductase system, cytochrome b apoprotein of ubiquinone–cytochrome c reductase, three subunits of cytochrome c oxidase, and two subunits of ATP synthetase (Anderson et al., 1981; Chomyn et al., 1986). Only a relatively small amount (5–15%) of the whole cell protein content is located in the mitochondria. Although a larger part of mitochondrial proteins are encoded by nuclear DNA (nDNA), the expression of the entire mitochondrial genome is vital for the maintenance of respiration and oxidative phosphorylation and, therefore, mutations of mtDNA should have a serious outcome. Indeed, free radical-induced damage of mtDNA leads to a defect in the function of encoded proteins, resulting in a decreased ratio of state 3:state 4, and as age-related decrease of complexes I, II and IV (Torri et al., 1992).

In the cardiac muscle of rats, the mtDNA has an apparent turnover rate that is about five times faster than that of the nDNA (Gross et al., 1969). However, this rapid turnover could result in increased replications of unrepaired mutation and the concomitant shift of mtDNA genotype. Hence, the replication of mutant mtDNA could result in an increase in the inactivation of individual genes, the expression of mutant gene products having altered activities, or inappropriate gene expression. Clayton et al. (1974) reported that mitochondria are not equipped with the repair mechanism of pyrimidine dimer, which further jeopardizes the normal replication, because some byproducts of free radical attack have mutagenic potential (Shibutani et al., 1991; Shinegawa et al., 1994). The 8-hydroxydeoxyguanine (8-OHdG), for instance, induces G–C to T–A transversion during DNA replication (Kuchino et al., 1987; Cheng et al., 1992). It has been shown that cells have a capacity for removing 8-OHdG (Tchou et al., 1991) as well as nucleotide excision repair activity for modified nucleotides (Bessho et al., 1993). It is, however, unclear whether the activity of this enzyme changes with ageing or exercise.

Damage of mtDNA and diseases in skeletal muscle

Metals play an important role also in DNA damage, because free radicals derived by Fenton reactions, especially hydroxyl radical, easily modify the structure of DNA (Fig. 2). It was demonstrated that some oxidative stresses, e.g. ischemia–reperfusion lead to the mobilization of iron and copper from the functional sites, which catalyse the Fenton reaction (Chevion et al., 1993). It was also reported that overload of copper and iron results in massive DNA damage (Bacon and Britton, 1990). Skeletal muscle does not possess the high repair capacity that occurs in more mitotically active tissues, but it is able to respond easily to the increased oxidative demand by proliferation of mitochondria. The metabolic stress resulting from the defect of mitochondrial respiration could lead to massive increase in proliferation of mitochondria, development of ragged red fibres (with a large number of abnormal mitochondria) as a result of a metabolic feedback loop, in which cells attempt to compensate for the reduced oxidative potential. However, the proliferation of mitochondria with mutant mtDNA would not resolve the metabolic stress, but would rather increase it. The damage of mtDNA in skeletal muscle can therefore have serious consequences in a variety of diseases and ageing. Some diseases such as Kearn–Sayre syndrome, Parkinson's disease, mitochondrial myopathy, myotonic dystrophy, myoclonic epilepsy and neurogenic muscle weakness, could be regarded as a manifestation of accelerated ageing and accumulation of DNA damage is enhanced in the skeletal muscle of patients with these diseases.

As postmitotic tissues tend to harbour DNA modifications, numerous studies have been conducted on skeletal muscle to classify the types of damage. About 20 types of DNA modifications have been identified including transition of A to G at a variety of nucleotide positions (4317, 3243, 8344, 10006, 12246) in patients with myoclonus epilepsy and ragged red fibres or even in the skeletal muscle of healthy people (Shoffer et al., 1990; Tanaka et al., 1990; Munscher et al., 1993; Zhang et. al. 1993). The analysis of these point mutations is difficult because the

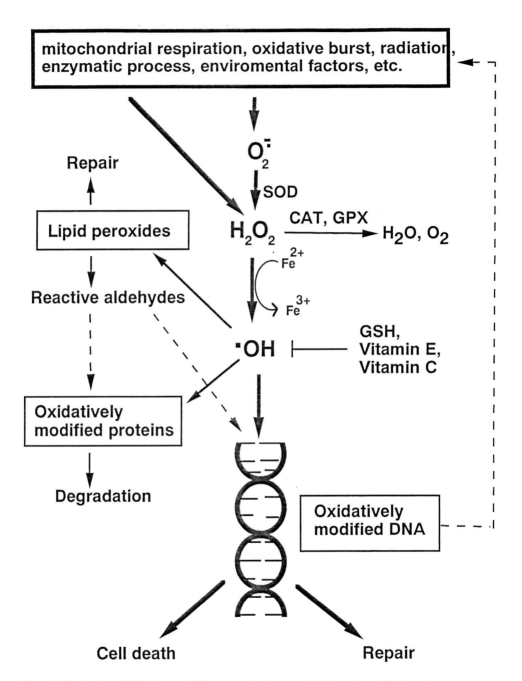

Figure 2. Mechanism of DNA damage: SOD, superoxide dismutase; CAT, catalase; GPX, glutathione peroxidase; GSH, glutathione.

techniques available for tracing mutations (polymer chain reaction) are not sensitive enough to detect the transition points.

Patients with mitochondrial myopathy harbour massive amounts of specific deletions (Wallace, 1992). Deletion of DNA could be further separated according to the localization, which involves non-coding or coding regions spanning numerous genes (Cann and Wilson, 1983; Schon et al., 1989). Single and multiple mtDNA deletions have been detected with increasing age (Zhang et al., 1992). Duplication of DNA is regarded as a possible intermediate step in deletion. The duplication of mutant mtDNA of muscle origin was present in other tissues of patients who had mitochondrial myopathy, indicating that mtDNA mutation can arise in early embryogenesis (Poulton et al., 1989) because mtDNA is maternally inherited (Giles et al., 1980). It should be kept in mind that single-stranded breaks of DNA and base modifications such as 8-OHdG represent different kinds of DNA modification; however, both can be different end-products of common chemical processes (Toyokuni and Sagripanti, 1996).

Repair

As DNA, especially mtDNA, is the subject of continuous insult of free radical species, a repair mechanism is present for reducing the accumulation of damaged nucleotides. The efficiency of the repair varies for different DNA adducts. Some components of the repair process are activated by DNA-damaging agents, such as radiation for c-Abl protein tyrosine kinase and some require DNA damage for being activated, such as double-stranded breaks or other DNA lesions for DNA-dependent protein kinase (Kharbanda et al., 1997). Functional interactions are present in these different repair processes to overcome the DNA damage (Kharbanda et al., 1997). As a result of the nature of the repairing process, the damage is repaired in a selective order. The O_6-methylguanine is repaired rapidly by an alkyl transferase that removes the methyl group, whereas the 7-methylguanine is repaired with moderate efficiency because it does not jeopardize the replication process (Park and Ames, 1988). The 8-OHdG is one of the markers of DNA oxidative stress most often used, and it was suggested that its formation results from mitochondria-derived free radicals rather than from alkylating agents (Park and Ames, 1988). Numerous base-excision proteins are involved in the repair of 8-oxoguanine through removal of 8-oxodG from DNA by Fpg protein (Grollman and Moriya, 1993). The adenine DNA glycosylase also contributes actively to the repair. Recent studies have suggested that oxidized free guanine nucleotides are incorporated into DNA in place of thymine nucleotides during replication, thus inducing mutation. The efficiency of the repair process could be altered by a variety of physiological and pathological processes such as exercise (Inoue et al., 1993), caloric restriction (Kaneko et al., 1997) and ageing (Moriwaki et al., 1996).

The effect of ageing, caloric restriction and exercise on DNA damage

In 1956, Harman proposed a free radical theory of ageing in which free radical-induced oxidative modifications in various cellular components contribute to the ageing process (Harman, 1956).

More recently, Linnane et al. (1989) named mitochondria as significant contributors to ageing by harbouring mutations of mtDNA due to which there is a substantial decline in functional, physiological cell capacity, loss of performance and appearance of age-related muscle weakness during lifespan. In non-replicating tissues, such as the brain, cardiac and skeletal muscles, it is particularly easy to follow the age-related accumulation of oxidatively damaged metabolites and adducts. It was reported by Trounce et al. (1989) that at respiration state III (activated) in human skeletal muscle, both substrate and enzyme activity (pyruvate + malate, glutamate + malate, succinate, monoamine oxidase, succinate–cytochrome c reductase and cytochrome oxidase) tend to fall with age, probably as a result of enhancement of mtDNA mutations. An increase in free radical production has also been reported with advancing age (Sohal et al., 1994); pro-oxidation of the cell redox cycle, and the ratios of GSH/GSSG (reduced and oxidized glutathione) and NADH/NAD$^+$ could be behind the increased accumulation of damaged DNA adducts (Noy et al., 1985). Furthermore, there is an increase in the availability of loosely bound transition metals, which readily react with hydrogen peroxide to generate hydroxyl radicals that damage proteins and DNA (Sohal and Weindruch, 1996). Moreover, the physiological function of ubiquinone, which serves not only as an electron and proton carrier but also as an antioxidant against the damage of mtDNA, decreases with age (Erstner and Dallner, 1995). A significant role for free radicals in the ageing process is supported by the life-extending effect of the over-expression of antioxidant enzymes (Cu,Zn-superoxide dismutase and catalase) in flies (Orr and Sohal, 1994) and the elevated level of antioxidant activity in long-lived mutants of nematodes (Larsen, 1993).

The accumulation of damaged DNA adducts could be the result of either the massive stress, which exceeds the ability of the defence systems (radical removal and repair systems) to deal with them or the failure of the defence systems under constant stress. The decline in antioxidant defence systems includes enzymes and other biological components. It has been suggested that decreases in the efficiency of overall, gene-specific and strand-specific repair decreases with age and this could be accounted for by the accumulation of damaged markers of DNA such as 7-methylguanine or 8-OHdG. The accumulation of 8-OHdG, the product of hydroxyl radical assault on the deoxyguanosine in DNA, was reported in ageing cardiac and skeletal muscles (Hayakawa et al., 1992; Hayakawa et al., 1991) and in diseased skeletal muscle (Richter, 1992). Numerous data indicate that there is an increase in the accumulation of oxidative DNA damage with age in various organs especially skeletal muscle (Fraga et al., 1990; Ozawa et al., 1995; Kaneko et al., 1996; Ozawa, 1995). The rearrangement of mtDNA in skeletal muscle of healthy people increases with ageing, and localization of mutations is heterogeneous and individual (Melov et al., 1995), and occurs in parallel with the decline in mitochondrial metabolic rate. Hayakawa et al. (1992) reported that the 8-OHdG content of mtDNA in the diaphragm increased 25-fold from the age of 55 to 85, the correlation between 8-OHdG and age being exponential.

Caloric restriction is the only known "natural" effective method that increases the lifespan of rodents and other animals (Yu, 1994b; Weindruch, 1996). It has been reported that caloric restriction reduces body fat, temperature, metabolic rate and oxidative damage of lipids, proteins and DNA, and increases the activity of antioxidant enzymes and DNA repair (Sohal and Weindruch, 1996). Recently, Kaneko et al. (1997) reported that caloric restriction apparently reduced age-associated increases in DNA damage in different organs involving cardiac muscle. However, the mechanism by which caloric restriction increases lifespan is not fully understood since it is generally believed that there is an inverse relationship between the metabolic rate and longevity in mammals (Porter and Brand, 1993), whereas caloric restriction in rats simultaneously increases lifespan and the metabolic rate by stimulating higher physical activity compared with rats fed *ad libitum* (Duffy et al., 1989). Moreover, regular physical exercise itself increased the mean and/or maximum life span in rats (Holloszy and Smith1987). Similar data were also reported in humans (Sarna et al., 1993).

Physical exercise significantly increases oxygen consumption as a result of the metabolic process and the mechanical work of skeletal muscle. Hence, an increase in markers of DNA damage in the blood or urine after exercise might be regarded as the result of muscle-derived metabolites or muscle work-induced metabolites. The study of Hartman et al. (1994) showed the DNA damage induced by treadmill running in white blood cells, measured by immunological comet assay. The damage peaked at 24 h after the running and recovered to normal levels after 72 h of exercise. Sen et al. (1994) confirmed that exercise could induce DNA damage in white blood cells. One year later, Hartman and co-workers (1995), using a similar protocol, reported that the administration of vitamin E significantly reduces the DNA damage in the blood induced by exercise. Moreover, administration of lipoic acid to patients affected by chronic progressive external ophthalmoplegia and muscle mitochondrial DNA deletion had a beneficial effect on the metabolic process during and after physical exercise (Barbiroli et al., 1995). In another patient who had severe defects with succinate–cytochrome c reductase in the skeletal muscle resulting from mtDNA deletion, the administration of menadione (vitamin K_3) and ascorbate significantly increased the efficiency of the recovery after the exercise (Eleff et al., 1984) and this was detected by NMR. A study conducted by Inoue et al. (1993) on DNA damage indicated that regular exercise tends to decrease the level of nuclear 8-OHdG. The plasma concentrations of hypoxanthine, xanthine and uric acid rose significantly, whereas the urine excretion of some metabolites decreased after the exercise probably as a result of increased repair of the oxidative DNA damage. Recent work by Pilger et al. (1997) also indicates that regular exercise does not induce DNA damage, measured by 8-OHdG in urine samples, in trained habitual long-distance runners compared with the control group.

Sandri et al. (1995) examined the effect of spontaneous wheel-running followed by 2 days of rest on untrained mice, and found an appearance of DNA fragmentation and polyubiquitin accumulation in the muscle; the latter was possibly an indication of impaired proteolysis. It was suggested that DNA fragmentation and apoptosis might be associated with the early stage of exercise that induces muscle damage which might be a first step in exercise-induced adaptation (Davies et al., 1982; Evans, 1992).

Conclusion

Free radical-induced oxidation of proteins is a normal physiological process which leads to a variety of end-products. The accumulation of oxidized proteins is altered by conditions that increase free radical formation and/or modify the rate of protein turnover. The oxidative modification of proteins may have a physiological, controlling roles. Diseases and ageing increase the level of accumulation, whereas caloric restriction and regular exercise might decrease it in skeletal and cardiac muscles. Mitochondria are one of the prominent sites of free radical formation and radical-induced damage. The accumulation of mutated mitochondrial genomes is more enhanced in static postmitotic cells such as those in skeletal and cardiac muscles. In particular mtDNA is subject to insult and is modified by free radicals; as a consequence of a weak mitochondrial repair mechanism and fast turnover rate, a significant part of the damage could not be repaired. The outcome of increased damage of mtDNA is a defect of mitochondrial respiration because the key components of the electron transport chain are encoded by mtDNA. The agglomeration of DNA damage is altered by a variety of diseases and ageing. Caloric restriction and regular exercise might decrease the accumulation of DNA damage either by reducing the rate of free radical formation and inducing the activity of the antioxidant system or by enhancing the efficiency of DNA repair.

References

Anderson S, Bankier AT, Barell BG, De Brujin MHL, Coulson AR, Drouin J, Eperson IC, Nierlich DP, Roe BA, Sanger F et al. (1981) Sequence of the human mitochondrial genome. *Nature* 290: 457−465.

Astier C, Rock E, Lab C, Gueux E, Mazur A and Rayssiguier Y (1996) Functional alterations in sarcoplasmic reticulum membranes of magnesium-deficient rat skeletal muscle as consequences of free radical-mediated process. *Free Radical Biol Med* 20: 667−74.

Barbiroli B, Medori M, Tritschler HJ, Klopstock T, Seibel P, Reichmann H, Iotti S, Lodi R and Zaniol P (1995) Lipoic (thioctic) acid increases brain energy availability and skeletal muscle performance as shown by *in vivo* 31P-MRS in a patient with mitochondrial cytopathy. *J Neurol* 242: 472−477.

Bacon BR and Britton RS (1990) The pathology of hepatic iron over-load: A free radical-mediated process. *Hepatology* 11: 127−137.

Bessho T, Tano K, Kasai T, Ohtsuka E and Nishimura SJ (1993) Evidence for two DNA repair enzymes for 8-hydroxyguannine (7,8-dihydro-8-oxoguanine) in human cells. *J Biol Chem* 268: 19416−19421.

Biolo G, Maggi SP, Williams BD, Tipton KD and Wolfe RR (1995) Increased rates of muscle protein turnover and amino acid transport after resistance exercise in humans. *Am J Physiol* 268: E514−E520.

Brown DR (1992) Physical activity, ageing, and psychological well-being: an overview of the research. *Can J Sport Sci* 17: 185−93.

Buss H, Chan TP, Sluis KB, Domigan NM and Winterbourn CC (1997) Protein carbonyl measurement by a sensitive ELISA method. *Free Radical Biol Med* 23: 361−366.

Cao G and Cutler RG (1995) Protein oxidation and aging. I. Difficulties in measuring reactive protein carbonyls in tissues using 2,4-dinitrophenylhydrazine. *Arch Biochem Biophys* 320: 106−114.

Carmeli E, Hochberg Z, Livne E, Lichtenstein I, Kestelboim C, Silbermann M and Reznick AZ (1993) Effect of growth hormone on gastrocnemius muscle of aged rats after immobilization: biochemistry and morphology. *J Appl Physiol* 75: 1529−1535.

Chance B, Sies H and Boveris A (1979) Hydroperoxide metabolism of mammalian organs. *Physiol Rev* 59: 527−605.

Chapman ML, Runbin BR and Gracy RW (1989) Increased carbonyl content of proteins in synovial fluid from patients with rheumatoid arthritis. *J Rheumatol* 16: 15−18.

Cheng KC, Cahill DS, Kasai H, Nishimura S and Loeb L (1992) 8-Hydroxyguanonine and abundant form of oxidative DNA damage, causes G to T and A to C substitutions. *J Biol Chem* 267: 166−172.

Chevion M, Jiang Y, Har-El R, Berenshtein E, Uretzky G and Kitrossky N (1993) Copper and iron are mobilized following myocardial ischemia: possible predictive criteria for tissue injury. *Proc Natl Acad Sci USA* 90: 1102−1106.

Chomyn A, Cleeter MWJ, Ragen CI, Riley M, Doolittle RF and Attardi G (1986) URF-6, last unidentified reading frame of human mtDNA, codes for NADH dehydrogenase subunit. *Science* 234: 614−618.

Clayton DA, Doda JN and Friedberg EC (1974) The absence of a prymidine dimer repair mechanism in mammalian mitochondria. *Proc Natl Acad Sci USA* 71: 2777−2781.

Dakin HD (1906) The oxidation of amino acids with the production of substances of biological importance. *J Biol Chem* 1: 171−176.

Davies KJA, Quintanilha AT, Brooks GA and Packer L (1982) Free radicals and tissue damage produced by exercise. *Biochem Biophys Res Commun* 107: 1198−1205.

Davies KJA (1986) Intracellular proteolytic system may function as secondary antioxidant defense: An hypothesis. *Free Radical Biol Med* 2: 155−173.

Davies KJA and Delsignore ME (1987) Proteins damage and degradation by oxygen radicals: III. Modification of denaturated protein. *J Biol Chem* 262: 9908–9913.

Davies KJA and Goldberg AL (1987) Oxygen radicals stimulate intracellular proteolysis and lipid peroxidation by independent mechanisms in erythrocytes. *J Biol Chem* 262: 8220–8226.

Davies KJA, Lin SW and Pacifici RE (1987) Proteins damage and degradation by oxygen radicals: IV. Degradation of denatureted proteins. *J Biol Chem* 262: 9914–9920.

Dean RT, Fu S, Stocker R and Davies MJ (1997) Biochemistry and pathology of radical-mediated protein oxidation. *Biochem J* 324: 1–18.

Desai VG, Weindruch R, Hart RW and Feuers RJ (1996) Influences of age and dietary restriction on gastrocnemius electron transport system activities in mice. *Arch Biochem Biophys* 333: 145–151.

Duffy PH, Feuers RJ, Leakey JA, Nakamura K, Turturro A and Hart RW (1989) Effect of chronic caloric restriction on physiological variables related to energy metabolism in the male Fisher 344 rat. *Mech Age Dev* 48: 117–133.

Eleff S, Kennaway NG, Buist NRM, Darley-Usmar VM, Capaldi RA, Bank WJ and Chance B (1984) ^{31}P NMR study of improvement in oxidative phosphorylation by vitamins K_3 and C in a patient with a defect in electron transport at complex III in skeletal muscle. *Proc Natl Acad Sci USA* 81: 3529–3533.

Erstner L and Dallner G (1995) Biochemical. physiological and medical aspects of ubiquinone function. *Biochim Biophys Acta* 1271: 195–204.

Evans WJ (1992) Exercise, nutrition and aging. *J Nutr* 122(suppl): 796–861.

Fagan JM, Waxman L and Goldberg AL (1986) Red blood cells contain a pathway for the degradation of oxidant-damaged hemoglobin that does not require ATP or ubiqitin. *J Biol Chem* 261;5705–5713.

Fares FA, Gruener N, Carmeli E and Reznick AZ (1996) Growth hormone (GH) retardation of muscle damage due to immobilization in old rats. Possible intervention with a new long-acting recombinant GH. *Ann N Y Acad Sci* 786: 430–443.

Fraga CG, Shinegawa MK, Park JW Degan P and Ames BN (1990) Oxidative damage to DNA during aging: 8-hydroxy-2'-deoxyguanosine in rat organ DNA and urine. *Proc Natl Acad Sci USA* 87: 4533–4537.

Giles RE, Blanc H, Cann HM and Wallace DC (1980) Maternal inheritance of human mitochondrial DNA. *Proc Natl Acad Sci USA* 77: 6715–6719.

Goto S and Nakamura A (1997) Age-associated, oxidatively modified proteins: A critical evaluation. *Age* 20: 81–89.

Goto S, Hasegawa A, Nakamoto H, Nakamura A Takahashi R and Kurochkin IV (1995) Age-associated changes of oxidative modification and turnover of proteins. *In*: RG Cutler, L Packer, J Bertram and A Mori (eds): *Oxidative Stress and Aging*. Birkhäuser Verlag, Basel, pp 151–158.

Grollman AP and Motiya M (1993) Mutagenesis by 8-oxoguanine: an enemy within. *Trends Genet* 9: 246–249.

Gross NJ, Getz GS and Rabinowitz M (1969) Apparent turnover of mitochondrial deoxyribonucleic and mitochondrial phospholipids in the tissues of the rat. *J Biol Chem* 244: 1552–1562.

Grune T, Reinheckel T, Joshi M and Davies KJA (1995) Proteolysis in cultured liver epithelial cells during oxidative stress Role of the multicatalytic proteinase complex, proteasome *J Biol Chem* 270: 2344–2351.

Harman D (1956) Aging – A theory based on free radical and radiation chemistry. *J Gerontol* 11: 298–300.

Hartmann A, Plallert U, Raddatz K, Grunert-Fuchs M and Speit G (1994) Does physical activity induce DNA damage? *Mutagenesis* 9: 269–272.

Hartmann A, Niess AM, Grunet-Fuchs M Poch B and Speit G (1995) Vitamin E prevents exercise-induced DNA damage. *Mutat Res* 1346: 195–202.

Hayakawa M, Hattori K, Sugiyama S, Tanaka M and Ozawa T (1991) Age-associated accumulation of 80hydroxydeoxyguanosine in mitochondrial DNA of human diaphragm. *Biochem Biophys Res Commun* 179: 1023–1029.

Hayakawa M, Hattori K, Sugiyama S and Ozawa T (1992) Age-associated oxygen damage and mutation in mitochondrial DNA I human hearts. *Biochem Biophys Res Commun* 189: 979–985.

Holloszy JO (1993) Exercise increase average longevity of female rats despite of increased food intake and no growth retardation. *J Gerontol* 48: B97–100.

Holloszy JO and Smith EK (1987) Effect s of exercise on longevity of rats. *Fed Proc* 46: 1850–1853.

Inoue T, Mu Z, Sumikawa K, Adachi K and Okochi T (1993) Effect of physical exercise on the content of 8-hydroxydeoxyguanosine in nuclear DNA prepared from human lymphocytes. *Jpn J Cancer Res* 84: 720–725.

Ishigami A and Goto S (1990) Effect pf dietary restriction on the degradation of proteins in senescent mouse liver parachymal cells in culture. *Arch Biochem Biophys* 283: 362–366.

Kaneko H, Tahara S and Matsuo S (1996) Non-linear accumulation of 8-hydroxy-2'-deoxyguanosine, a marker of oxidized DNA damage, during aging. *Mutat Res* 316: 277–285.

Kaneko H, Tahara S and Matsuo M (1997) Retarding effect of dietary restriction on the accumulation of 8-hyroxy-2'-deoxyguanosine in organs of Fisher 344 rats during aging. *Free Radical Biol Med* 23: 76–81.

Kharbanda S, Pandey P, Jin S, Inoue S, Bharti A, Yuan ZM, Weichselbaum R, Weaver D and Kufe D (1997) Functional interaction between DNA-PK and c-Abl in response to DNA damage. *Nature* 386: 732–735.

Kuchino Y, Mori F, Kasai H, Inoue H, Iwai S, Miura K, Ohtsuka E and Nishimura S (1987) DNA temples containing 8-hydroxydeoxyguanosine are misread both the modified base and at adjacent residues. *Nature* 327: 77–79.

Larsen PL (1993) Aging and resistance to oxidative damage in Caenorhabditis elegans. *Proc Natl Acad Sci USA* 90: 8905–8909.

Levine RL, Oliver CN, Fulks RM and Stadtman ER (1981) Turnover of bacterial glutamine synthetase: oxidative inactivation precedes proteolysis. *Proc Natl Acad Sci USA* 78: 2120–2124.

Levine RL, Garland D, Oliver CN, Amici A, Climet I, Lenz A, Ahn B, Shalteil S and Stadtman ER (1990) Determination of carbonyl content of oxididatively modified proteins. *Methods Enzymol* 186: 464–478.

Levine RL, Williams JA, Stadtman ER and Shacter E (1994) Carbonyl assay for determination of oxidatively modified proteins. *Methods Enzymol* 37: 346–357.

Linnane A, Marzuki S, Ozawa T and Tanaka M (1989) Mitochondrial DNA mutations as an important contributor to ageing and degenerative diseases. *Lancet* i: 642–645.

Liu J, Wang X, Shigenaga MK, Yeo HC, Mori A and Ames BN (1996) Immobilization stress causes oxidative damage to lipid, protein and DNA in brain of rats. *FASEB J* 10: 1532–1538.

Melov S, Shoffner M, Kaufman A and Wallace DC (1995) Marked increase in the number and variety of mitochondrial DNA rearrangemants in aging human skeletal muscle. *Nucl Acid Res* 23: 4122–4126.

Mizuno Y (1985) Glucose-6-phosphate dehydrogenase, 6-phosphogluconate dehydrogenase and glyceraldehydes-3-phosphate dehydrogenase activities in early stages of developement in dystrophic chickens *J Neurol Sci* 68: 47–60.

Moriwaki SI, Ray S, Tarone RE, Kraemer KH and Grossman L (1996) The effects of aging on the processing of UV-damaged DNA in human cells; reduced DNA repair capacity and increased DNA mutability. *Mutat Res* 364: 117–123.

Muncher C, Rieger T, Muller-Hocker J and Radenbach B (1993) The point mutation of mitochondrial DNA characteristic for MERRF disease is found also in healthy people of different ages. *FEBS Lett* 317: 27–30.

Murphy ME and Kehrer JP (1989) Oxidation state of tissue thiol groups and content of protein carbonyl groups in chicken with inherited muscular dysrophy. *J Biochem* 260: 359–364.

Nair KS (1995) Muscle protein turnover: methodological issues and the effect of aging. *J Gerontol* 50: 107–112.

Nakamura A and Goto S (1996) Analysis of protein carbonyls with 2,4-dinitrophenylhydrazine and its antibodies by immunoblot in two-dimensional gel electrophoresis. *J Biochem* 119: 768–774.

Noy N, Schwartz H and Gafni A (1985) Age related changes in the redox status of rat muscle cell and their role in enzyme aging. *Mech Age Dev* 29: 63–69.

Oliver CN, Starke-Redd PE, Stadtman ER, Liu GJ, Carney JM and Floyd RA (1990) Oxidative damage to brain proteins, loss of glutamine synthetase activity, and production of free radicals during ischemia/reperfusion-induced injury to gerbil brain. *Proc Natl Acad Sci USA* 87: 5144–5147.

Omaye ST and Tappel AL (1974) Glutathione peroxidase, glutathione reductase, and thiobarbituric acid-reactive products in muscles of chickens and mice with genetic muscular dystrophy. *Life Sci* 15: 137–145.

Orr WC and Sohal RS (1994) Extension of life-span by overexpression of superoxide dismutase and catalase in drosophila melanogaster. *Science* 263: 1128–1130.

Ozawa T (1995) Mechanism of somatic mitochondrial DNA mutation associated with aging and diseases. *Biochim Biophys Acta* 1271: 177–189.

Ozawa T, Katsumata K, Hayakawa M, Yoneda M, Tanaka M and Sugiyama S (1995) Mitochondrial DNA mutation and survival rate. *Lancet* 345: 189.

Pacificiti RE, Kono Y and Davies KJA (1993) Hydrophobicity as the signal for selective degradation of hydroxyl radical modified hemoglobin by the multicatalytic proteinase complex, proteasome. *J Biol Chem* 268: 15405–15411.

Park YW and Ames BN (1988) Methylguanine adducts in DNA are normally present at high levels and increase on aging: Analysis by HPLC with electrochemical detection. *Proc Natl Acad Sci USA* 85: 7467–7470.

Pilger A, Germadnik D, Formanek D, Zwick H, Winkler N and Rudiger HW (1997) Habitual long-distance running does not enhance urinary excretion of 8-hydroxydeoxyguanosine. *Eur J Appl Physiol* 75: 467–469.

Porter RK and Brand DM (1993) Body mass dependence of H^+ leak in mitochondria and its relevance to metabolic rate. *Nature* 362: 628–630.

Poulton J, Deadman ME and Gradiner RM (1989) Duplication of mitochondrial DNA in mytochondrial myopathy. *Lancet* i: 236–240.

Puppo A and Halliwell B (1988) Formation of hydroxyl radicals in biological system. Does myoglobin stimulate hydroxyl radical formation from hydrogen peroxide? *Free Radical Res Commun* 4: 415–422.

Radák Z, Lee K, Choi W, Sunoo S, Kizaki T, Oh-ishi S, Suzuki K, Taniguchi N, Ohno H and Asano K (1994) Oxidative stress induced by intermittent exposure at a simulated altitude of 4000 m decreases mitochondrial superoxide dismutase content in soleus muscle of rats. *Eur J Appl Physiol* 69: 392–395.

Radák Z, Asano K, Lee KC, Ohno H, Nakamura A, Nakamoto H and Goto S (1997a) High altitude increases reactive carbonyl derivatives but not lipid peroxidation in skeletal muscle of rats. *Free Radical Biol Med* 22: 1109–1114.

Radák Z, Asano K, Nakamura A, Nakamoto H and Goto S (1998) Anabolic interval exercise increases accumlation of reactive carbonyl derivatives in lungs of rats. *Pflüger Arch Eur J Physiol* 435: 439–441.

Radák Z, Asano K, Fu J, Nakamura A, Nakamoto H and Goto S (1997c) Does physical exercise induce oxidative stress to mitochondrial, microsomal, and cytoplasmic fractions of liver and skeletal muscle of rats? *BioFactors* (suppl); *in press*.

Rajguru SU, Yeargans GS and Seidler NW (1994) Exercise causes oxidative damage to rat skeletal muscle microsomes while increasing cellular sulfhydryls. *Life Sci* 54: 149–157.

Rennie MJ, Babij P, Sutton JR, Tonkins W, Read WW, Ford R and Holliday D (1983) Effects of acute hypoxia on forearm leucine metabolism. *In*: JR Sutton, CS Houston and NL Jones (ed.): *Hypoxia, Exercise and Altitude*. A.R. Liss, New York, pp: 317–323.

Reznick AZ and Packer L (1994) Oxidative damage to proteins spectrophotometric method for carbonyl assay. *Methods Enzymol* 233: 357–363.

Reznick AZ, Witt E, Matsumoto M and Packer L (1992a) Vitamin E inhibits protein oxidation in skeletal muscle of resting and exercising rats. *Biochem Biophys Res Commun* 189: 801–806.

Reznick AZ, Kagan VE, Ramsey R, Tsuchiya M, Khwaja S, Serbinova EA and Packer L (1992b) Antiradical effects in L-propionyl carnitine protection of the heart against ischemia-reperfusion injury: the possible role of iron chelation. *Arch Biochem Biophys* 296: 394–401.

Reznick AZ, Cross CE, Hu ML, Suzuki YJKhwaja S, Safadi A, Motchnick PAPacker L and Halliwell B (1992c) Modification of plasma proteins by cigarete smoke as measured by protein carbonyl formation. *Biochem J* 286: 607–611.

Reznick AZ, Han D and Packer L (1997) Cigarette smoke induced oxidation of human plasma proteins, lipids, and antioxidants; selective protection by the biothiols dihydrolipoic acid and glutathione. *Redox Report* 3: 169–174.

Richter C (1992) Reactive oxygen and DNA damage in mitochondria. *Mutat Res* 275: 2420–255.

Rivett JA (1985) The effect of mixed-function oxidation of enzymes on their susceptibility to degradation by a nonlysosomal cystein proteinase. *Arch Biochem Biophys* 243: 624–632.

Sandri M, Carraro U, Podhorska OM, Rizzi C Arslan P, Monti D and Franceseschi C (1995) Apoptosis, DNA damage and ubiquitin expression in normal and mdx muscle fibers after exercise. *FEBS Lett* 373: 291–295.

Sarna S, Sahi T, Koskenvuo M and Kaprio J (1993) Increased life expectancy of world class male athletes. *Med Sci Sports Exerc* 25: 237–244.

Schon EA, Rizutto R, Moraes CT, Nakase H, Zeviani M and Diamuro S (1989) A direct repeat is a hot spot for large-scale deletion of human mitochondrial DNA. *Science* 244: 346–349.

Sen CK, Rankinen T, Vaisanen S and Rauramaa R (1994) Oxidative stress after human exercise: effect of N-acetylcysteine supplementation. *J Appl Physiol* 76: 2570–2577.

Shibutani S, Takeshita M and Grollman AP (1991) Isertion of specific bases during DNA synthesis past the oxidation-damaged base of 8-oxydG. *Nature* 349: 431–434.

Shinegawa MK, Hagen TM and Ames BN (1994) Oxidative damage and mitochondrial decay in aging. *Proc Natl Acad Sci USA* 91: 10771–10778.

Shoffer JM, Lott MT, Lezza AM, Seibel P, Balinger SW and Wallace DC (1990) Myoclonic epilepsy and ragged-red fibre disease (MERRF) is associated with a mitochondrial tRNA^{Lys} mutation. *Cell* 61: 931–937.

Sohal RS and Weindruch R (1996) Oxidative stress, caloric restriction, and aging. *Science* 273: 59–63.

Sohal RS, Ku HH, Agarmal S, Foster MJ and Lal H (1994) Oxidative damage, mitochondrila oxidant generation and antioxidant defenses during aging and in a response to food restriction in the mouse. *Mech Ageing Dev* 74: 121–133.

Smith CD, Carney JM, Starke-Reed PE, Oliver CN, Stadtman ER, Floyd RA and Markesbery WR (1991) Excess brain protein oxidation and enzyme dysfunction in normal aging and in Alzheimer diseases. *Proc Natl Acad Sci USA* 88: 10540–10543.

Stadtman ER (1986) Oxidation of proteins by mixed-function oxidation system: implication in protein turnover, aging and neutrophil function. *Trends Biochem Sci* 11: 11–12.

Stadtman ER (1990) Metal ion-catalyzed oxidation of proteins: Biochemical mechanism and biological consequences. *Free Radical Biol Med* 9: 315–325.

Stadtman ER (1992) Protein oxidation and aging. *Science* 257: 1220–1224.

Stadtman ER and Berlett BS (1991) Fenton chemistry. *J Biol Chem* 266: 17201–17211.

Stadtman ER and Oliver CN (1991) Metal-catalyzed oxidation of proteins. *J Biol Chem* 266: 2005–2008.

Starke-Reed PE and Oliver CN (1989) Protein oxidation and proteolysis during aging and oxidative stress. *Arch Biochem Biophys* 275: 559–567.

Sugden PH and Fuller SJ (1991) Regulation of protein turnover in skeletal and cardiac muscle. *Biochem J* 273: 21–37.

Tanaka M, Ino H, Ohno K, Hattori K, Sato W and Ozawa T (1990) Mitochondrial mutation in fatal Infantile cardiomyopathy. *Lancet* 336: 1452.

Tchou J, Kasai T, Shibutani S, Chung MH, Lavai J, Grollman AP and Nishimura S (1991) 8-oxoguanine (8-hydroxyguanine) DNA glycosylase and its substrate specificity. *Proc Natl Acad Sci USA* 88: 4690–4694.

Toledo I, Rangel P and Hansberg W (1995) Redox imbalance at the start of each morphogenic step neuraspora crassa conidiation. *Arch Biochem Biophys* 319: 519–524.

Torri K, Sugiyama S, Tanaka M, Tagaki K, Hanaki Y, Iida K, Matsuyama M, Hirabayashi N, Uno Y and Ozawa T (1992) Aging-associated deletions of human diaohragmatic mitochondrial DNA. *Am J Respir Cell Molec Biol* 6: 543–549.

Trounce I, Byrne E and Marzuki S (1989) Decline in skeletal muscle mitochondrial respiratory chain function: possible factor in ageing. *Lancet* i: 637–639.

Toyokuni S and Sagripanti JL (1996) Association between 8-hydroxy-2'-deoxyguanosine formation and DNA strand breaks mediated by copper and iron. *Free Radical Biol Med* 20: 859–864.

Wallace DC (1992) Diseases of mitochondrial DNA. *Annu Rev Biochem* 61: 1175–1212.

Welle S, Thornton C and Statt M (1995) Myofibrillar protein synthesis in young and old human subjects after three months of resistance training. *Am J Physiol* 268: E422–E427.

Weindruch R (1996) Caloric restriction and aging. *Sci Am* 274: 46–52.

Witt E, Reznick AZ, Viguie CA, Sarke-Reed P and Packer L (1992) Exercise, oidative damage and effects of anti-oxidant manipulation. *J Nutr* 122: 766–773.

Yu BP (1994a) Cellular defense against damage from reactive oxygen species. *Physiol Rev* 74: 139–163.

Yu BP (1994b) *Modulation of Aging Processes by Dietary Restriction*. CRC Press, Boca Raton, FL.

Zhang C, Baumer A, Maxwell RJ, Linname AW and Nagley P (1992) Multiple mitochondrial DNA deletions in an elderly human individual. *FEBS Lett* 297: 34–38.

Zhang C, Linnane AW and Nagley P (1993) Occurrence of a particular base subunits (3243, A to G) in mito-chondrial DNA of tissues of aging humans. *Biochem Biophys Res Commun* 195: 1104–1110.

Oxidative Stress in Skeletal Muscle
A.Z. Reznick et al. (eds)
© 1998 Birkhäuser Verlag Basel/Switzerland

Antioxidant enzyme response to exercise and training in the skeletal muscle

L.L. Ji

Department of Kinesiology and Nutritional Science, University of Wisconsin-Madison, 2000 Observatory Drive, Madison, WI 53706, USA

Introduction

Oxygen-derived free radicals and other reactive oxygen species (ROS) are byproducts of normal cell life (Halliwell and Gutteridge, 1989). With the exception of strict anaerobes and a few species of bacteria, most aerobic organisms are equipped with a host of enzymes that are directly or indirectly involved in the antioxidant defence against ROS. Enzymes that provide primary defences include superoxide dismutase (SOD), catalase (CAT), and glutathione peroxidase (GPX). Glutathione reductase (GR) and enzymes producing NADPH, such as glucose-6-phosphate dehydrogenase (G6PDH), malic enzyme, and isocitrate dehydrogenase, are important in keeping glutathione in its reduced form (GSH) so that adequate cellular oxidoreductive (redox) status is maintained. Glutathione sulfur-transferase (GST) is an important enzyme in metabolizing prooxidant xenobiotics in the liver. Secondary defences include a group of loosely defined enzymes which either repair cellular damage caused by ROS or remove the damaged molecules, such as phospholipase A_2 or specific proteases (Yu, 1994). The catalytic mechanisms and the regulation of various antioxidant enzymes have been reviewed extensively by previous authors (Chance et al., 1979; Yu, 1994). Thus, this chapter focuses on the three primary antioxidant enzymes, i.e. SOD, CAT, and GPX in the skeletal muscle.

Molecular structure, catalytic mechanisms and kinetics

Superoxide dismutase

SOD (EC 1.15.1.1), discovered by McCord and Fridovich (1969), represents a family of metalloenzymes which catalyze a common one-electron dismutation of $O_2^{-\cdot}$ to H_2O_2.

$$2O_2^{-\cdot} + 2H^+ \longrightarrow H_2O_2 + O_2$$

This is the only function of SOD known to date. The reaction occurs naturally with a rather slow rate ($t_{1/2} = 7$ s) and is also pH dependent. With 0.35 μM SOD present, the rate of $O_2^{-\cdot}$ dismutation is accelerated dramatically with a $t_{1/2}$ of 0.5 ms, regardless of pH (Chance et al., 1979).

There are three types of SOD isozymes, depending on the metal ion bound to its active site. Copper–zinc superoxide dismutase (CuZnSOD) is a highly stable enzyme found primarily in the cytosolic compartment of the eukaryotic cells such as yeast, plants and animals, but not generally in the prokaryotes such as bacteria and algae. CuZnSOD is a dimer ($M_r = 32\,000$) and sensitive to cyanide and H_2O_2 inhibition (Fridovich, 1995). It is interesting that, although Cu and Zn are both required for the synthesis of the enzyme, the copper ion plays the primary function of dismutation by alternative oxidation and reduction, whereas the Zn ion appears to have no catalytic role apart from stabilizing the enzyme (Halliwell and Gutteridge, 1989). Manganese superoxide dismutase (MnSOD) is a tetramer with a much larger molecular weight of $88\,000$. It is present in the mitochondrial matrix of eukaryotes and insensitive to cyanide and H_2O_2. However, it is not as stable as CuZnSOD and can be inhibited by sodium dodecyl sulfate (SDS) and chloroform/ethanol treatments (Ohno et al., 1994). This distinction of cyanide sensitivity has been used to measure the activity of the two types of SOD in tissue extracts without isolating the mitochondria and cytosol. In addition to CuZnSOD and MnSOD, bacteria contain a third type of SOD that requires iron as a prosthetic group. FeSOD and MnSOD show a high degree of homology, indicating common ancestry during evolution. Several characteristics of SOD kinetics are worth a mention. First, unlike most enzymes, SOD lacks a Michaelis constant (K_m), i.e. saturation kinetics are not exhibited. The enzyme is partially occupied by its substrate (O_2^{-}) and its catalytic activity increases with increasing O_2^{-} concentration over a wide range (Chance et al., 1979). Further, high levels of H_2O_2 have been shown to inhibit SOD *in vitro* (Blum and Fridovich, 1985), however, organic peroxides and $OH^{\cdot}$ have no effect on SOD (Pigeolet et al., 1990). As a result of the kinetic properties mentioned above, assays of SOD are usually based on indirect methods, involving inhibition of a reaction in which O_2^{-} is generated (Fridovich, 1985). Therefore, it is difficult to compare maximal *in vivo* activity between studies using different assay methods.

Catalase

The primary reaction that (EC 1.11.1.6) catalyzes is the decomposition of H_2O_2 to H_2O, although CAT has other biological functions (Chance et al., 1979). It shares this function with GPX, but the substrate specificity and affinity as well as the cellular location of the two antioxidant enzymes are different (see below):

$$2H_2O_2 \longrightarrow 2H_2O + O_2$$

CAT is a tetramer with a relatively large molecular weight of about $\sim240\,000$. Haem (Fe^{3+}) is a required ligand for binding to the enzyme's active site to promote its catalytic function. CAT resembles SOD in many kinetic properties such as the lack of an apparent K_m and V_{max}, and its activity increases enormously with an increase in H_2O_2 concentration. In the presence of H_2O_2, CAT is also capable of reducing a limited number of hydroperoxides (peroxidatic function), but not *t*-butyl-hydroperoxide, to their respective aldehydes (Chance et al., 1979). Azide and cyanide are both inhibitors of CAT. This inhibition is often used to separate CAT activity from GPX activity for enzyme assays in crude tissue extracts. $O_2^{-\cdot}$ and $OH^{\cdot}$ both inactivate CAT but organic peroxide has no effect (Pigeolet et al., 1990). CAT is also strongly inhibited by aminotriazole, a specific CAT inhibitor that can be fed to animals without causing gross metabolic defects (Halliwell and Gutteridge, 1989). Assays of CAT typically involve the addition of H_2O_2 and its removal is followed spectrophotometrically at 240 nm (Aebi, 1984). Caution is necessary in reports of catalytic activity because it is determined not only by the enzyme protein present in the assay medium, but also by the concentration of H_2O_2 used. Without defining the conditions of the assay, the comparison of activities among studies is often meaningless.

Glutathione peroxidase

GPX (EC 1.11.1.9) catalyzes the reduction of H_2O_2 and organic hydroperoxide to H_2O and alcohol, respectively, using GSH as the electron donor (Flohe, 1982):

$$2GSH + H_2O_2 \longrightarrow GS\text{-}SG + 2\,H_2O$$

$$\text{or} \quad 2GSH + ROOH \longrightarrow GS\text{-}SG + ROH$$

It should be clarified that GPX refers only to the Se-dependent enzyme. The so-called Se-independent GPX is part of the GST (EC 2.5.1.18) family, which removes some species of ROOH (Habig et al., 1974). The primary function of GST is, however, to conjugate GSH with a variety of xenobiotic substances, making them more water soluble and hence easier to metabolize. GPX is highly specific for its hydrogen donor GSH, but has low specificity for hydroperoxide, ranging from H_2O_2 to complex organic hydroperoxides and including long-chain fatty acid hydroperoxides and nucleotide-derived hydroperoxides. This characteristic of GPX makes it a versatile hydroperoxide remover in the cell, so it plays an important role in inhibiting lipid peroxidation and preventing damage to DNA and RNA (Flohe, 1982). It is also important to recognize that, although GPX and CAT have an overlap of substrate H_2O_2, GPX (at least in mammals) has a much greater affinity for H_2O_2 at low concentrations ($K_m = 1$ μM) than CAT ($K_m = 1$ mM) (Sies, 1985). GPX is susceptible to $O_2^{-\cdot}$ and hydroperoxide inhibition *in vitro* as a result of the oxidation of the selenocysteine residue at the enzyme's active site (Blum and Fridovich, 1985). However, a high concentration of H_2O_2 (10^{-1} M) and organic peroxide (0.05–0.3 mM) is required to exert 50% inhibition of GPX (Pigeolet et al., 1990). Both SOD and GSH prevent the inactivation of GPX by removing $O_2^{-\cdot}$ and reducing the sulfhydryls of the enzyme, respectively.

Cellular distribution and function

In mammals, the highest SOD activity is found in the liver, followed by kidney, brain, adrenal, and heart (Halliwell and Gutteridge, 1989). In the skeletal muscle, SOD activity is similar to that in the heart and, unlike some other antioxidant enzymes (see below), is relatively consistent among

Table 1. Activity of superoxide dismutase (SOD), Glutathione peroxidase (GPX), and catalase (CAT) in various tissues

Tissues	SOD			GPX			CAT
	Cu/Zn	Mn	Total	Cytosolic	Mito-chondrial	Total	
Liver	500	50	14400	550	430	85	670
Heart	65	21	2610	150	70	17	84
Soleus	ND	ND	1300	ND	ND	13	61
DVL	21	8	1360	23	17	2	18
SVL	ND	ND	887	ND	ND	0.9	15
Erythrocytes	NA	NA	8.8	NA	NA	25	10

Units of enzyme activity in rat tissues: Cu/Zn SOD and MnSOD, unit/mg protein; SOD total, unit/g wet weight; cytosolic and mitochondrial GPX, nmol/min·mg^{-1} protein; catalase, $K \times 10^{-2}$/g wet weight; activity for other enzymes, μmol/min·g^{-1} wet weight. Activity in human erythrocytes: units per g Hb. NA, not applicable; ND, not determined.

muscle fibre types (Ji et al., 1992a) (Tab. 1). The distribution of SOD activity in the cell varies from tissue to tissue, and also depends on species (Halliwell and Gutteridge, 1989). Thus, in rat liver, about 90% of SOD activity is found in the cytosol and 10% in the mitochondria, whereas in the myocardium and skeletal muscles, the contribution of MnSOD to total SOD activity is higher, ranging from 15% to 20% (Leeuwenburgh et al., 1997).

SOD is the first line of enzymatic defence against intracellular free radical production, terminating a free radical chain reaction caused by $O_2^{-\cdot}$. Although $O_2^{-\cdot}$ itself is not highly toxic, it can extract an electron from many biological compounds close to its generation sites, such as the mitochondrial inner membrane, causing a free radical chain reaction. Therefore, it is essential for the cell to keep superoxide anions in check. The superoxide theory of oxygen toxicity has been challenged over the years because it is argued that the product of SOD, H_2O_2, is also capable of generating OH· which would be more toxic (Halliwell and Gutteridge, 1989). However, $O_2^{-\cdot}$ can generate OH· in the cell by attacking the ion–sulfur protein (4Fe–4S) cluster thereby releasing Fe^{2+}, which sets the stage for OH· production via the Fenton reaction or the Haber–Weiss reaction. This "cooperation" of $O_2^{-\cdot}$ and H_2O_2 is an important mechanism for cellular, particular DNA, damage by $O_2^{-\cdot}$ (Fridovich, 1995).

Mammalian species also contain extracellular SOD (EC SOD) located in the plasma and interstitial fluid. It is a CuZnSOD; unlike its cytosolic counterpart, however, EC SOD is tetramer with a larger molecular weight and it can be bound to heparin. Its primary function is to remove $O_2^{-\cdot}$ generated outside the cell membrane as a result of irradiation, inflammation, and ischemia–reperfusion (Fridovich, 1995; Ohno et al., 1994).

The essential role of SOD is best illustrated by the phenotype of SOD mutants. *Escherichia coli* that cannot produce MnSOD (sodA) or FeSOD (sodB) are normal under anaerobic conditions compared with their wild-type counterparts, but they show higher levels of susceptibility to oxygen and prooxidant chemicals. Hypersensitivity towards O_2 was also observed in yeasts carrying both CuZnSOD and MnSOD mutants. In humans, mutagenesis of CuZnSOD in Lou Gehrig's disease results in apoptosis of spinal neurons (amyotrophic lateral sclerosis) (Fridovich, 1995). However, unless both forms of SOD are defective, serious oxidative stress is rarely seen.

CAT is located primarily in the organelle called the peroxisome (Aebi, 1984). However, mito-chondria and other intracellular organelles such as endoplasmic reticulum also contain some CAT activity. Heart and liver mitochondria in rats are well-known examples of this, but a recent study by Luhtala et al. (1994) also showed a high CAT activity in submitochondrial particles in skeletal muscle. There is a considerable debate about whether the detected CAT activity in these sources is a result of contamination caused by cell fractionation during assay preparation (Halliwell and Gutteridge, 1989). The activity of CAT in mammalian tissues follows the order of SOD, with liver being the highest and skeletal muscle the lowest (Tab. 1). Among the various muscle types, type 1 muscle (soleus) displays the highest CAT activity, followed by type 2a muscle (deep vastus lateralis). Type 2b muscle (superficial vastus lateralis) had the lowest CAT activity. The differ-ences in CAT activity presumably reflect the aerobic capacity of the muscles. These patterns are consistent with an early finding by Burge and Neil (1916–17) that CAT activity in skeletal muscles of wild animals was significantly higher than in those of domestic animals.

The primary function of CAT is to remove H_2O_2 produced in the peroxisomes by flavoprotein dehydrogenase in the β-oxidation of fatty acids, urate oxidase and enzymes in the metabolism of D-amino acids. When the H_2O_2 concentration is high, the catalytic function of CAT prevails, i.e. H_2O_2 will be decomposed to O_2 and H_2O. Alternatively, CAT may exhibit a peroxidase activity when the H_2O_2 concentration is low ($< 10^{-6}$ M) and in the presence of a suitable hydrogen donor, such as ethanol (Chance et al., 1979).

GPX is located in both the cytosol and mitochondrial matrix of the cell, with a distribution ratio of about 2:1 (Chance et al., 1979). This allows it to reach a number of cellular sources of hydroperoxide generation. The activity of GPX is high in the liver and erythrocytes, moderate in the brain, kidney, and heart, and low in skeletal muscle (Tab. 1). However, oxidative type 1 muscle (soleus) possesses a GPX activity close to the level in the heart (Tab. 1) (Leeuwenburgh et al., 1997). Selenium deficiency dramatically decreases tissue GPX activity in all body tissues in a dose-responsive fashion and this can be reversed by re-feeding animals with a diet containing adequate amounts of Se (Hill et al., 1987). There is evidence that mitochondria are more resistant to dietary Se deficiency. When postneonatal rats were fed a low Se diet for 8 weeks, cytosolic GPX activity in liver, heart, and muscle was decreased to 3–4% of control values, whereas mito-chondrial GPX was maintained at 9, 20, and 24%, respectively (Ji et al., 1988b, 1992b).

In mammalian cells the removal of the hydrogen and organic peroxides (e.g. lipid peroxide) is catalyzed by GPX, forming water and alcohol, respectively. By donating a pair of hydrogen ions, GSH is oxidized to glutathione disulfide (GSSG). Reduction of GSSG is catalyzed by GR, a flavin-containing enzyme, wherein NADPH is the reducing power. This reaction takes place to-gether with the GPX reaction, so providing a redox cycle for the regeneration of GSH from GSSG. However, because GR activity in most tissues is well below that of GPX, cells export GSSG under oxidative stress to maintain an intracellular redox ratio (Ishikawa and Sies, 1984). NADPH is supplied by the hexose monophosphate pathway, controlled by G6PDH or malic enzyme. In the skeletal muscle, isocitrate dehydrogenase is considered more important than G6PDH or malic

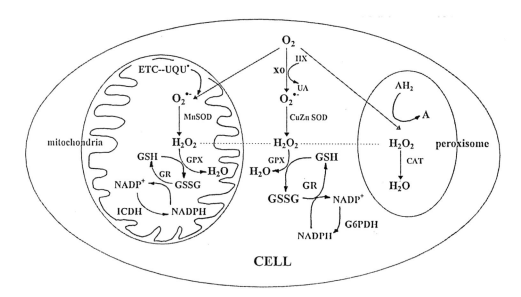

Figure 1. A schematic summary of the synergistic contribution of SOD, CAT and GPX to the removal of O_2^{-} and H_2O_2 from the cell. AH_2, a nonspecific hydrogen ion donor; CAT, catalase; ETC, electron transport chain; G6PDH, glucose-6-phosphate dehydrogenase; GSH, reduced glutathione; GSSG, oxidized glutathione; GPX, glutathione peroxidase; HX, hypoxanthine; ICDH, isocitrate dehydrogenase; UQU', semiubiquinone; SOD, superoxide dismutase; UA, uric acid; XO, xanthine oxidase. (Adapted from Chance et al., 1979).

enzyme in supplying NADPH (Reed, 1986). Figure 1 provides a schematic summary of the synergistic contribution of SOD, CAT and GPX to the removal of ROS.

Gene regulation of antioxidant enzymes

In prokaryotes such as bacteria, antioxidant enzymes are regulated via transcriptional pathways (Storz et al., 1990). The gene in *E. coli* that controls the coordinated expression of some 30 proteins is *oxyR*; these include CAT, alkyhydroperoxidase, and GR, which have known antioxidant functions (Harris, 1992). Another gene locus apart from *oxyR* is called *soxR*, and this controls the expression of nine proteins, including SOD. The genes *oxyR* and *soxR* seem to be induced by H_2O_2 and O_2^{-}, respectively (Harris, 1992).

In contrast to the prokaryotes, the regulatory mechanisms for cellular antioxidants in the eukaryotes are not well understood. The complexity of the gene expression of antioxidant enzymes in eukaryotic cells is caused, at the very least, by the following factors: (1) Cells are compartmentalized into a hydrophobic (e.g. membrane bilayers) and hydrophilic (e.g. cytosol) environment. In addition, cells are bathed in extracellular fluid which is connected to the plasma. As most ROS (except H_2O_2) have a limited diffusibility, antioxidant enzymes need to be located in compatible cell compartments. (2) Eukaryotes are equipped with low-molecular-weight antioxidants such as α-tocopherol, ascorbate, β-carotene, and GSH. These non-enzymatic antioxidants complement and/or serve as substrates for antioxidant enzymes. Thus, expression of antioxidant enzymes is influenced by the levels and localization of these compounds in the cell. (3) The product of SOD, i.e. H_2O_2, is also harmful and is the substrate for CAT and GPX. Therefore unless there is coordinated expression of all three antioxidant enzymes, cells will not reduce their dangerous exposure to ROS as a whole. In fact, over-or under-expression of one antioxidant enzyme may subject the cell to greater oxidative stress (Harris, 1992).

Gene regulation of antioxidant enzymes is influenced by many factors including nutrition, oxidative status of the cell, development, and aging. Extensive research has been conducted with regard to the roles of transition metals (Cu, Zn, and Mn) in SOD, of Fe in CAT, and of Se in GPX gene expression (Harris, 1992). In the past few years, there have been some major advances in the research of gene regulation of antioxidant enzymes in eukaryotes, although most of the work is based on *in vitro* systems and studies on skeletal muscle are scarce. In general, cells have three strategies for regulating antioxidant enzyme gene expression in response to changed oxygen tension and hence the production of ROS. First, nuclear proteins known to be activating transcription of antioxidant enzymes possess conserved cysteine residues in the DNA-binding domain, which are redox sensitive and can be covalently modified to affect their targeting specificity and binding potential. Second, DNA binding by regulatory heme proteins can be affected as a result of O_2 binding to the heme molecule, thus regulating transcription rate. Third, change in O_2 tension can change a variety of hormones, such as catecholamines and steroids, and metabolites, such as ATP and cAMP, which then affect DNA binding of nuclear proteins regulated by these factors (Cowan et al., 1993).

SOD

It is well known that decreased availability of Cu can result in a low CuZnSOD activity in the liver, lung, heart, and skeletal muscle (Betteger et al., 1978). Zn is not as essential as Cu in determining CuZnSOD expression, because Zn deficiency has only a minor effect on SOD activity. Manganese is essential for the expression of MnSOD. If the supply of Mn is restricted in the diet, there is an over-expression of CuZnSOD. Similarly, Cu deficiency can result in a higher level of MnSOD to make up the total cellular SOD activity (Halliwell and Gutteridge, 1989). Suzuki et al. (1993) demonstrated that, in rats which are enable to express CuZnSOD, there was a compensatory increase in MnSOD in the liver. Lai et al. (1994) showed that dietary Cu deficiency elicited tissue-specific responses of CuZnSOD and MnSOD in rats. Although CuZnSOD activity, protein content, and mRNA abundance in the liver and heart were all decreased with Cu deficiency, they remained relatively constant in the brain. Furthermore, MnSOD activity, protein, and mRNA levels were elevated as a result of Cu deficiency in the heart and brain, and MnSOD mRNA was elevated in the liver. Thus, transcriptional, post-transcriptional, and post-translational mechanisms are involved in the regulation of SOD in response to changes in transition metal status.

The two types of SOD have quite different characteristics in terms of protein turnover. Recombinant human SOD (r-hSOD) studies reveal that CuZnSOD has a half-life ($t_{1/2}$) of 6–10 min, whereas MnSOD has a much longer $t_{1/2}$ of 5–6 h (Gorecki et al., 1991). This longer half-life, along with its relatively homogeneous amino acid sequence across all species and tissues, makes the MnSOD a much more desirable enzyme for therapeutic use (see below). The relative abundance of CuZnSOD mRNA has clear tissue-specific differences, with the liver possessing the highest levels, followed by the heart, lung, and then skeletal muscle (Fig. 2). Within different muscle fibres, CuZnSOD mRNA levels were higher in the type 1 muscle (soleus) than in an intermediate muscle (plantaris) which, in turn, were higher than in type 2 muscle (deep and superficial vastus lateralis, and gastrocnemius). Consistent with the mRNA levels, type 1 muscle also displays a higher level of enzyme protein content compared with type 2 muscle fibres (Fig. 3). These data are also in agreement with the enzyme activity of CuZnSOD displayed in Table 1. In general, MnSOD activity, content and mRNA abundance follow the same order across tissues as displayed by CuZnSOD (Fig. 2). These findings resemble those reported by Lai et al. (1994) who showed that liver had the highest CuZnSOD mRNA levels followed by

Northern Blots of SOD in Various Tissues

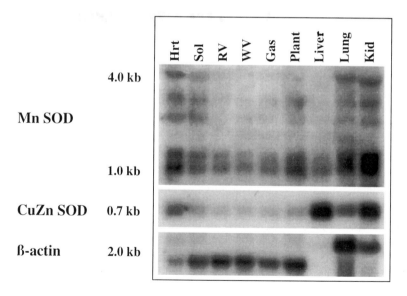

Figure 2. Northern blot of mRNA for CuZnSOD and MnSOD in the various rat tissues. Hrt, heart; Sol, soleus; RV, red vastus lateralis; WV, white vastus lateralis; Gas, gastrocnemius; Plan, plantaris; Kid, kidney. The cDNA probes for CuZnSOD and MnSOD were the kind gifts of Dr Ye-Shih Ho, Wayne State University, Detroit, Michigan. (Gore et al., 1997).

heart and brain. These data also agree with a recent report by Oh-Ishi et al. (1995) comparing mRNA levels, enzyme protein content, and activity of CuZnSOD and MnSOD in soleus versus extensor digitorum longus (type 2) muscle.

In the eukaryotic cells MnSOD is encoded by a nuclear gene. A large precursor enzyme form is synthesized in the cytosol and transported into the mitochondria via an energy-dependent process (Zhang, 1996). MnSOD has been shown to be inducible under oxidative stress and the upregulation of MnSOD gene expression is mediated, at least partially, by a transcriptional mechanism (Whisett et al., 1992). A number of potential inducers of MnSOD have been identified, including tumor necrosis factor α (TNFα), interleukin 1 (IL-1), and lipopolysaccharide (Visner et al., 1990). The MnSOD promoter contains both nuclear factor κB (NFκB) and activator protein-1 (AP-1) binding sites and the effects of TNFα and IL-1 on MnSOD gene expression were mediated in part by a thiol redox modulation of NFκB activation (Das et al., 1995).

Western Blots for Mn SOD and Cu-Zn SOD

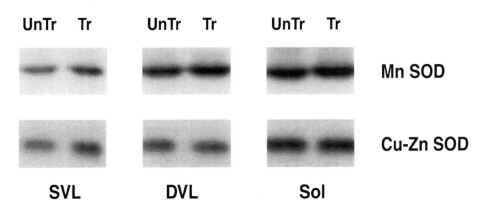

Figure 3. Western blot analysis for CuZnSOD and MnSOD in three types of rat skeletal muscle. SVL, superficial vastus lateralis; DVL, deep vastus lateralis; Sol, soleus; Tr, trained (treadmill running at 25 min/min, 10% grade for 2 h/day for 8 weeks); UnTr, untrained control. The CuZnSOD and MnSOD antiobodies were the kind gifts of Drs Keiichiro Suzuki and Naoyuki Taniguchi from the Department of Biochemistry, Osaka University, Japan. (Hollander et al., 1997).

CAT

Murine CAT gene regulation has been studied extensively (Holmes and Duley, 1975). A single gene (*Cs*) is located on chromosome 2 responsible for coding the primary structure of the enzyme. Once synthesized the polypeptide may be modified epigenetically in terms of sulfhydryl groups and carbohydrate or protein moieties. The normal (N) and epigenetically modified (E) polypeptidal subunits can produce five tetrameric isozymes, similar to the isozyme patterns of lactate dehydrogenase (Holmes and Masters, 1970).

There are relatively few data on gene regulation of CAT in skeletal muscle. As a peroxisomal enzyme, CAT is more sensitive to oxidative stress in the liver and lung. For example, administration of ciperofibrate, a peroxisome proliferator, increased hepatic CAT activity possibly as a result of enhanced β-oxidation and generation of H_2O_2 (Dhaunsi et al., 1994). In the lung of prenatal guinea-pigs, there is a coordinated elevation of mRNA levels for MnSOD, CAT and GPX which correlated with activities of these enzymes, suggesting a pre-translational control mechanism (Yuan et al., 1996). These changes presumably reflect a gene regulation in preparation for the increased oxygen tension at birth. In contrast, liver mRNA levels for SOD and CAT showed a significant increase only after birth.

GPX

The role of Se levels in tissue GPX activity has been well studied (Hill et al., 1987; Spallholz et al., 1981). Using an ELISA (enzyme-linked immunosorbent assay) method, Knight and Sunde (1987) showed that both GPX activity and protein decreased progressively in post-weaning rats fed a Se-deficient diet for 28 days. During the initial stage of Se-deficiency, there seemed to be an inactivation of GPX protein, indicated by its longer $t_{1/2}$ than that of GPX activity. Lang et al. (1987) showed that rats fed a Se-deficient diet significantly decreased muscle and liver GPX (by 80%), but increased total GSH in liver, muscle, and plasma. Ji et al. (1988b) found that, although Se-deficiency decreased GPX activity in the cytosol of liver and skeletal muscle to about 4% of controls, mitochondrial GPX activity of these tissues was maintained at 10–20% of the normal levels.

Recent studies focused on the gene regulation of GPX in the mammalian tissues have made some significant progress in understanding of this Se-dependent antioxidant enzyme. GPX is a homotetramer with each 22-kDa subunit bound to a selenium atom, existing as a selenocysteine. The expression of the GPX gene, *hgpx1*, occurs in a wide range of tissues controlled by a number of different mechanisms in mammalian tissues (Moscow et al., 1992). First, GPX expression is influenced by oxygen tension; thus, lung and erythrocytes have a higher GPX activity compared with most other tissues, as predicted. Second, metabolic rate seems to play a significant role in GPX expression because liver, kidney, and pancreas have high GPX activities (Flohe, 1982). Third, GPX appears to be developmentally regulated; GPX activity in the lung of prenatal rats increased dramatically several days before birth in anticipation of high oxygen tensions. Fourth, GPX activity in some tissues demonstrates sensitivity to hormones such as estrogen and testosterone. Finally, toxins and xenobiotics can induce GPX especially in the liver. The relative abundance of GPX mRNA shows clear tissue-specific differences, with the liver being the highest, followed by the heart and soleus (almost identical), type 2a (red vastus) and type 2b (white vastus) muscle (Fig. 4).

Myocardial GPX has been shown to be induced by high oxygen tension (Cowan et al., 1992). GPX activity and mRNA levels were significantly high in cultured myocytes at 20 versus 5.3 kPa (150 versus 40 mmHg). The differences reported for GPX activity induced by oxygen tension was found to be paralleled at the mRNA and transcriptional levels. Furthermore, two oxygen response elements (OREs) located at −1232 to −1213 and −282 to −275 in the 5'-flanking region of the human GPX gene have been identified (Cowan et al., 1993). In the liver, prooxidant xeno-biotics have a great impact on the expression of GPX. Thus, ethanol induces GPX mRNA in rat

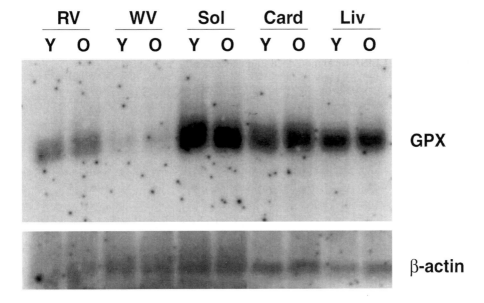

Figure 4. Northern blot of mRNA for GPX in the various rat tissues. Liv, liver; Card, cardiac muscle; Sol, soleus; RV, red vastus lateralis; WV, white vastus lateralis. Y, 4-month-old rats; O, 26.5-month-old rats. The cDNA probe for GPX was a kind gift of Dr Ye-Shih Ho, Wayne State University, Detroit, Michigan.

liver and this induction appears to be related to an increased lipid peroxidation (Nanji et al., 1995). To date, little work has been done on GPX gene regulation in the skeletal muscle.

Response to acute exercise

Strenuous aerobic exercise has been shown to be associated with increased ROS production in skeletal muscle, the liver (Davies et al., 1982; Jackson et al., 1985), and the heart (Kumar et al., 1992). The enzymes SOD, CAT, and GPX provide the first line of defence against the ROS generated during exercise, so it is expected that exercise may have a direct impact on these enzymes. Indeed, an acute bout of exercise has been shown to increase SOD activity in a number of tissues, including liver (Alessio and Goldfarb, 1988; Ji et al., 1988a,b, 1990), skeletal muscle (Quintanillha and Packer, 1983; Ji et al., 1990; Lawler et al., 1993), heart (Quintanillha and Packer, 1983; Ji and Mitchell, 1994), lung (Quintanillha et al., 1982), red blood cells (Lukaski et al., 1990; Mena et al., 1991), and platelets (Buczynski et al., 1991). With a few exceptions (Quintanillha and Packer, 1983), most of the studies also indicate that exercise increases CuZnSOD rather than MnSOD activity. This activation of SOD was initially proposed to be caused by increased O_2^- production during exercise based on *in vitro* SOD kinetics that partial occupancy of the enzyme by O_2^- can increase its activity (Ji, 1993). As we now know that CuZnSOD has a quick turn-over rate and a short $t_{1/2}$ in the range of minutes (see earlier), *de novo* synthesis of new enzyme protein cannot be ruled out in explaining the SOD responses to acute exercise. Recently, Radák et al. (1995) showed that enzyme activity and immunoreactive enzyme content of both CuZnSOD and MnSOD in rat soleus and tibialis muscles were significantly elevated after a single bout of exhaustive treadmill running lasting 60–70 min. Interestingly, CuZnSOD activity and content gradually returned to resting levels after 1–3 days, whereas MnSOD activity and protein content continued to increase during the post-exercise period. This finding indicates that the stimulating effects of exercise on CuZnSOD and MnSOD gene expression may be different in terms of threshold required and time course of induction.

Most of the previous studies have revealed no significant alteration in CAT activity with acute exercise (Meydani and Evans, 1993; Ji, 1995). However, there are exceptions when CAT activity was found to be increased significantly after an acute bout of exercise to exhaustion or at high intensity in rats (Ji and Fu, 1992; Ji et al., 1992a). Only the deep portion of the vastus lateralis muscle (DVL) showed this activation, whereas the superficial vastus lateralis (SVL), soleus, liver, and heart did not display any appreciable change with exercise. With a catalytic mechanism similar to SOD, one may expect CAT activity also to increase as a result of an increased H_2O_2 production during exercise. However, CAT is located primarily in the peroxisomes whereas the main source of H_2O_2 during short-term acute exercises is the mitochondria (Ji and Leichtweis, 1997). Furthermore, mitochondrial and cytosolic GPX are probably more effective in competing for the H_2O_2 produced in these two cell compartments because of its closer proximity to the source of ROS and comparatively lower K_m than CAT.

The effect of an acute bout of exercise on GPX activity in various tissues has been inconsistent as reported in the literature. Several studies showed no change in this enzyme in skeletal muscle after acute exercise (Vihko et al., 1978; Brady et al., 1979; Ji et al., 1990; Leeuwenburgh and Ji,

1995), whereas others reported significant elevation of GPX activity (Quintanillha, 1984; Ji and Fu, 1992; Ji et al., 1992a; Leeuwenburgh and Ji, 1996; Oh-ishi et al., 1997). Furthermore, heart (Quintanillha, 1984) and platelet (Buczynski et al., 1991) GPX activities have been shown to be elevated after exercise, but liver GPX seems to be unaffected in all studies reported (Ji, 1995). Although no clear explanation can be provided, the discrepancies may be related to the exercise intensity (i.e. percentage maximum oxygen consumption $VO_{2\,max}$) which varied in the aforementioned studies. Muscle fibre-specific responses of GPX have also been noticed, for example, GPX activity was found to increase as a function of treadmill speed in DVL and SVL, but not in soleus muscle (Ji et al., 1992a). However, Radák et al. (1995) reported an increased GPX activity one day after an acute bout of treadmill running to exhaustion in rat soleus, but not tibialis muscle. The mechanism responsible for the increased GPX activity with acute exercise is unknown.

One obvious approach to examining the effect of acute exercise on gene regulation of antioxidant enzymes is to measure their relative mRNA abundance in skeletal muscles. Oh-ishi et al. (1997) reported a significant down-regulation of mRNA levels for both CuZnSOD and MnSOD isozymes in soleus muscle of untrained rats, although in trained rats no down-regulation of exercise was observed. Recently the author's laboratory has investigated the effects of a single bout of prolonged exercise on mRNA abundance of muscle antioxidant enzymes and NF-κB activation in rats (Gore et al., 1997). The mRNA abundance of CuZnSOD, MnSOD, and CAT was not altered by exercise, but exercise decreased GPX mRNA levels by 21.6% and 60.8% ($p < 0.05$) in DVL and SVL, respectively. To gain some insight into the altered antioxidant enzyme gene regulation caused by exercise, NF-κB activation was measured using electrophoretic mobility shift gels with a consensus NF-κB binding site as the probe. Binding of NF-κB was significantly decreased ($p < 0.01$) in DVL with exercise but showed no change in SVL. These data demonstrate that an acute bout of exhaustive exercise may decrease the mRNA abundance of GPX, MnSOD, and CuZnSOD. The altered binding of the redox-sensitive transcriptional regulator NF-κB implicates an involvement of transcriptional control in antioxidant enzyme regulation during exercise.

Overall, antioxidant enzymes may be selectively activated during an acute bout of strenuous exercise. This activation may depend on the oxidative stress imposed on the specific tissue as well as the intrinsic antioxidant defence capacity. Skeletal muscle may be subjected to a greater level of oxidative stress during exercise than liver and heart as a result of increased oxygen consumption. It is estimated that muscle oxygen uptake can increase up to 100-fold in active contracting muscle cells compared with resting levels (Meydani and Evans, 1993); the muscle therefore needs greater antioxidant protection against potential oxidative damage occurring during and/or after exercise. To understand the mechanism for the observed rapid increases in antioxidant enzyme activity during acute exercise is still a great challenge. Although evidence has just started emerging that mammalian tissues can up-regulate gene expression of antioxidant enzymes in response to short-term oxidative stress, the signal transduction pathway is still elusive and the application of current knowledge to exercise condition is limited. Some significant progress has been made regarding SOD gene regulation influenced by acute and chronic exercise resulting, at least in part, from the availability of monoclonal antibody to muscle SOD isozymes and ELISA (Ohno et al., 1994). The study of muscle CAT and GPX is, however, hindered by the lack of specific antibodies to these enzymes, making it difficult to achieve accurate measurement of enzyme protein content. Thus, increased binding of substrates (ROS) to respective enzymes, causing allosteric and/or covalent modulation of the enzyme's catalytic activity, is still a viable explanation.

Adaptation to chronic training

Nature has made an almost perfect match between the antioxidant defence capacity of an organism and its living environment. Strict anaerobes have little, if any, antioxidant enzymes present in the body. Appearance of oxygen on the earth some two billion years ago provoked the development of the organism's antioxidant defence mechanism (Halliwell and Gutteridge, 1989). In mammals and birds, CAT activity in skeletal muscle was found to be greater in the wild species compared with their domestic counterparts (Burge and Neil, 1916–1917). It has long been recognized that living organisms are also capable of inducing antioxidant defence system by relatively rapid mechanisms to cope with oxidative stress. Both animals and certain plant species can induce SOD (mainly MnSOD) upon increased exposure to oxygen, paraquat (a known producer of O_2^{-}), and X-irradiation (Oberley et al., 1987). As an acute bout of strenuous physical exercise can cause a tremendous increase in oxygen consumption, and hence ROS production in skeletal muscle, repeated exercise exposure during chronic training is expected to induce muscle antioxidant systems in order to protect against potential oxidative damage. Training adaptation of SOD is not limited to skeletal muscle and several detailed reviews are available on this topic (Jenskins, 1993; Ohno et al., 1994; Ji, 1995).

The activity of SOD in skeletal muscle has been reported to increase significantly after training (Jenkins et al., 1983; Higuchi et al., 1985; Sen et al., 1992; Leeuwenburgh et al., 1994; Power et al., 1994; Oh-ishi et al., 1997; Leeuwenburgh et al., 1997). Most of the studies involved rats running on a motor-driven treadmill for 8–10 weeks. However, many studies failed to detect SOD training adaptation even though similar animal training models were used (Alessio and Goldfarb, 1988; Laughlin et al., 1990; Ji, 1993). Furthermore, a recent study by Tiidus et al. (1996) failed to find a training effect of SOD in human leg muscle after 8 weeks of bicycle training. The discrepancies may be explained by the following factors: (1) the different SOD isozymes studied; (2) the different SOD assays used; (3) the different training intensity and frequency used which imposed different oxidative stress on the muscle; and (4) different muscle fibre types tested. A careful study, conducted by Powers et al. (1994), investigated the influences of exercise intensity, duration, and muscle fibre type on training response of SOD in rats. Prominent training adaptation of SOD occurred in soleus muscle, whereas in red gastrocnemius muscle increased SOD activity was observed only at high training intensity. In soleus muscle, SOD activity increased as a function of exercise duration rather than intensity.

As SOD is present in both the cytosol (CuZnSOD) and mitochondria (MnSOD), effort has been made to identify which isozyme form is induced by training. Using cyanide as an inhibitor of CuZnSOD, Higuchi et al. (1985) demonstrated that MnSOD is primarily responsible for increased SOD activity with training. Although Ji et al. (1988b) showed no training effect on MnSOD in rat hindlimb muscle, enzyme activity was expressed per mitochondrial protein which is known to be increased after training. Thus, total muscle MnSOD was apparently elevated in the trained state. The above-mentioned methods for identifying SOD isozymes in response to training are subject to error as a result of the different preparation and assay procedures, so different results may be obtained (Leeuwenburgh et al., 1997). The availability of monoclonal antibody to SOD isozymes and ELISA has provided a powerful tool in the investigation of the gene regulation of SOD. Oh-ishi et al. (1997) have recently studied the interrelationship of SOD isozyme activity, protein content, and mRNA abundance in rat soleus muscle, after endurance training and in response to a single bout of exercise. Resting CuZnSOD activity was significantly increased with training but the enzyme protein content was not altered. There was no significant change in CuZnSOD mRNA level with training. In contrast, MnSOD showed an increased activity and protein content, whereas mRNA levels were not affected. Only CuZnSOD activity showed an elevation after an acute exercise bout. These data suggest that training induction of both types of SOD is caused by post-transcriptional mechanisms and that post-translational modulation may play a role in CuZnSOD. In general agreement with Oh-ishi et al. (1997), a recent study in the author's laboratory (Hollander et al., 1997) showed that MnSOD is inducible by endurance training, but the induction is muscle fibre specific. Significant increases in both MnSOD activity (+70%) and enzyme protein content (+26%) were found in rat DVL muscle after 10 weeks of treadmill running, with only marginal increases in MnSOD protein, but not its activity, in soleus muscle. CuZnSOD activity and protein content showed no effect in either type of muscle. Resting mRNA levels for both MnSOD and CuZnSOD were unaltered with training. These two studies both pointed to the importance of MnSOD training adaptation, which seems to support the notion that mitochondrial inner membrane is as a major source of ROS production during exercise (Ji and Leichtweis, 1997).

CAT activity has been shown to increase after training in skeletal muscle by some authors (Quintanillha, 1984; Jenkins, 1993; Oh-ishi et al., 1997), but most studies have reported no change in muscle catalase with training (see Jenkins, 1988; Meydani and Evans, 1993; Ji, 1995) and a few even reported a decrease (Laughlin et al., 1990; Leeuwenburgh et al., 1994). More consistent findings have been reported with respect to GPX adaptation to training and most studies have shown an increase in GPX activity after training (Ji et al., 1988a,b; Laughlin et al., 1990; Sen et al., 1992; Lawler et al., 1993; Leeuwenburgh et al., 1994, 1997; Powers et al., 1994a; Oh-ishi et al., 1996). However, a recent study by Tiidus et al. (1996) failed to confirm this training adaptation of GPX in human subjects undergoing a bicycle training program for 8 weeks. Adaptation of GPX also demonstrates a muscle fibre-specific pattern with type 2a muscle being the most responsive to training. Powers et al. (1994a) showed a 45% increase in GPX activity in red gastrocnemius muscle after endurance training in rats, and the level of increment appeared to be dependent upon running time rather than speed. Soleus and white gastrocnemius muscles revealed no training effect of GPX regardless of training intensity and duration. Consistent with the findings of Powers et al. (1994a), Leeuwenburgh et al. (1997) reported a 62% increase in GPX activity in DVL (type 2a) muscle in response to treadmill training for 2 h/day at moderate intensity (25 m/min, 10%). Soleus and myocardium showed no GPX training effect. A small but significant training induction of GPX (16%) was found when mixed vastus lateralis muscle in rats was studied (Sen et al., 1992). Although GPX is expressed as a uniform enzyme in the various cellular compartments, there is evidence that mitochondrial GPX undergoes a greater training adaptation than the cytosolic GPX in rat skeletal muscle (Ji et al., 1988b). As GPX has a wider range of substrate specificity with respect to peroxides and a lower K_m (1 µM) than CAT (1 mM), an increased GPX activity facilitates the removal of both H_2O_2 and lipid peroxides generated in the mitochondrial inner membrane (Nanji et al., 1995).

It is noteworthy that, as a striated muscle, diaphragm muscle also exhibits considerable training adaptation of antioxidant enzymes (Powers et al., 1994b). The level of upregulation was found to be considerably greater in the costal versus the crural diaphragm, possibly reflecting the different workload and metabolic rate of the two regions.

The physiological significance of a training induction of antioxidant enzymes is not totally clear or agreed upon at present. It is conceivable that an up-regulation of antioxidant enzymes offers greater protection to the various tissues during exercise. However, Higuchi et al. (1985) argued that the increase in mitochondrial MnSOD with training is relatively small compared with increases in other mitochondrial enzymes and therefore there is less antioxidant protection in the trained state. On the contrary, Davies et al. (1982) proposed that endurance training helps to decrease oxygen flux in each respiratory chain, thereby reducing oxidative damage. Ji et al. (1988a) showed that trained rats had a lesser degree of mitochondrial sulfhydryl oxidation during acute exercise and that exercise-induced inactivation of muscle enzymes was less pronounced in trained animals. These protective effects of training coincided with an enhanced antioxidant enzyme activity with training (Ji, 1988a,b). In humans, a higher antioxidant enzyme activity was reported to correlate with $VO_{2\,max}$ and trained athletes were shown to have greater SOD and CAT activities in skeletal muscle (Jenkins, 1993). To establish a clear benefit of training, future studies should be directed to the investigation of muscle functional improvement as a result of antioxidant enzyme adaptation.

Antioxidant enzymes in aging skeletal muscle

Aging has been proposed to be caused by the accumulative action of the deleterious oxygen free radicals throughout lifespan (Harman, 1956). It is well known that aging influences both free radical production and cellular antioxidant defence systems in the various organs and tissues (Ames et al., 1993; Meydani and Evans, 1993; Nohl, 1993; Yu, 1994). However, aging response of antioxidant enzymes seem to be highly tissue specific, and skeletal muscle especially displays some unique changes during aging. Although most tissues show a decline of cellular antioxidant defences (Matsuo, 1993), senescent skeletal muscle appears to increase antioxidant enzyme activities (Lammi-Keefe et al., 1984; Vertachy et al., 1989; Ji et al., 1990; Lawler et al., 1993; Leeuwenburgh et al., 1994; Luhtala et al., 1994). Lammi-Keefe et al. (1984) were among the first to observe that, contrary to the aging response of most tissues, several skeletal muscles showed an increased SOD activity with aging. Ji et al. (1990) showed that activities of all major antioxidant enzymes, such as SOD, CAT and GPX were significantly higher in DVL muscle of old versus young rats. In addition, the activities of GST, GR and G6PDH also increased with age. These changes occurred despite a general decline of muscle mitochondrial oxidative capacity. Leeuwenburgh et al. (1994) reported similar increases in the activities of all the antioxidant enzymes with aging in both DVL and soleus muscles in rats. In addition, γ-glutamyl transpeptidase activity was significantly elevated in DVL, indicating that aged muscle has greater potential for

GSH uptake. Lawler et al. (1993) found a significant increase in GPX, but not SOD, activity in the gastrocnemius and soleus muscle in old *versus* young rats. Luhtala et al. (1994) discovered that elevation of muscle antioxidant enzyme activities during aging was markedly affected by dietary restriction in Fischer 344 rats. The progressive increases in CAT and GPX activities from 11 to 34 months of age were prevented by a 30% reduction of food intake, whereas an age-related increase in MnSOD was also attenuated in hindlimb muscles. However, these age-related increases in antioxidant enzyme activities did not seem to prevent lipid oxidation in senescent muscles (Starnes et al., 1989; Ji et al., 1990; Leeuwenburgh et al., 1994). It has been proposed by Meydani and Evans (1993) that, even though antioxidant enzyme activities increase with age, these enzymes may become less efficient in protecting the muscle from oxidative damage. There is evidence that an acute bout of eccentric exercise may cause more severe oxidative injury to muscles in aged mice (Zerba et al., 1990) and men (Meydani and Evans, 1993) than their young counterparts.

The mechanisms responsible for the increased antioxidant enzyme activities caused by aging in skeletal muscle are still elusive. One possibility is that mitochondria from aged muscles produce greater levels of ROS (Nolh and Hegner, 1978), which may induce antioxidant enzyme gene expression. Aged muscle may be more susceptible to injury, which can also trigger an acute phase response that causes further ROS production (Meydani and Evans, 1993). The effect of aging on gene regulation of antioxidant enzymes has been investigated in a number of tissues, such as heart, kidney, brain, and liver. Decreased CuZnSOD and CAT activities were found to parallel lower levels of mRNA coding these enzymes (Semsei et al., 1989; Rao et al., 1990; Xia et al., 1995). However, litter data are available on skeletal muscle. Oh-ishi et al. (1995) reported that 24 month-old rats had higher activities of CuZnSOD, MnSOD (after normalization with citrate synthase activity), GPX, and CAT in soleus muscle compared with 4-month old rats. CuZnSOD protein content was also elevated in aged muscle, although no significant change was noticed in relative abundance of mRNA for both SOD isozymes. Preliminary studies in the author's laboratory revealed no significant age differences in mRNA levels for CuZnSOD, MnSOD, and GPX in DVL muscle despite prominent increases in their activities with old age (unpublished data). Only CAT mRNA showed a significant increase with age in soleus muscle (unpublished data), which coincided with a large (two- to three-fold) increase in catalase activity with age (Leeuwenburgh et al., 1994). Thus, the influence of aging on muscle antioxidant enzyme gene regulation may be a complicated matter involving both transcriptional and post-transcriptional mechanisms that require further investigation.

Antioxidant enzyme response to an acute bout of exercise seems to be no different in old versus young rats (Ji et al., 1990; Lawler et al., 1993); however, endurance training has been shown to increase some antioxidant enzyme activities in the senescent muscle. Ji et al. (1991) showed a training induction of muscle GPX in 27-month-old Fischer 344 rats, but SOD and CAT activities remained unchanged with training. Hammeren et al. (1993) also reported a training adaptation of GPX in various muscle fibres in old Fischer 344 rats. Leeuwenburgh et al. (1994) found that 10 weeks of endurance training increased GPX and SOD in DVL muscle of young rats, but had little effect on antioxidant enzyme activity in either DVL or soleus muscle in the old rats. There are two potential problems associated with training studies of antioxidant enzymes in senescent muscle: (1) the oxidative stress imposed on old animals could be limited by their exercise ability; and (2) the training threshold of senescent muscles could be raised as a result of an age-related increase in the antioxidant enzyme activity. Therefore, it is difficult to conclude whether training is more or less effective in the induction of antioxidant enzymes in senescent skeletal muscle.

Therapeutic use of antioxidant enzymes

The list of diseases that may have an etiologic background of free radical generation continues to grow and the significance of antioxidant enzymes in preventing pathologic processes is now well accepted (Ames et al., 1993; Yu, 1994). One classic example is that CuZnSOD mutation, leading to decreased cytosolic SOD activity, has been linked to amyotrophic lateral sclerosis (ALS), or Lou–Gehrig's disease (Fridovich, 1995). Glycation of CuZnSOD, causing its fragmentation and inactivation, is believed to be related to the etiology of cataract, diabetes, and adult progeria (Werner's syndrome) (Ohno et al., 1994). To date, most of the therapeutic uses of antioxidant enzymes are with SOD and CAT, with substantial applications concentrated on cardiac ischemia–reperfusion, pulmonary diseases, rheumatologic disorders, and diabetes (Greenward, 1990; Yu, 1994).

Supplementation of SOD is limited by its relatively large molecular weight and short half-life (see earlier). Thus, although SOD tablets are available even over the counter at local health product stores, a carefully conducted study revealed that 90% of the labeled ^{65}Zn ion of SOD apoenzyme administered to mice orally came out in the feces (Greenwald, 1990). To overcome these problems, several methods have been developed to enhance the bioavailability of SOD, including conjugation with polyethylene glycol (PEG), amidation, and SOD polymerization. The most recent approach is to attach a poly-(stylene-co-maleic acid)butyl ester to CuZnSOD to form an albumin-borne SOD derivative (SM-SOD) which has a much longer half-life of 6 h and is capable of accumulating in injured tissues with a lower pH (Inoue et a. 1989). This "smart SOD" has been shown to be effective in reducing oxidative damage under a number of pathologic conditions including ischemia–reperfusion, cold stress, and burn injury (Ohno et al., 1994). Radák et al. (1995) reported that intraperitoneal administration of SM-SOD in rats attenuated an elevation of plasma levels of xanthine oxidase and thiobarbituric acid-reactive substance (TBARS) that was associated with an acute bout of exhaustive exercise in the untreated control rats. Furthermore, CuZnSOD and MnSOD activities and immunoreactive enzyme contents in the skeletal muscles were significantly increased after exercise in the control rats, whereas SM-SOD-treated rats showed a decreased response. The same authors also reported decreased TBARS in the liver and kidney after exhaustive exercise in SM-SOD-treated rats, compared with untreated controls (Radák et al., 1996). These findings were taken as evidence that exogenous SM-SOD supplementation might be effective in reducing exercise-induced oxidative stress in various tissues.

Conclusions

Skeletal muscle plays a vital role in providing animals with mobility, which is essential for survival. A highly specialized skeletal muscle, the diaphragm, is vital for respiration (see Powers et al., this volume, for a detailed review of this muscle). To ensure a sufficient energy supply during muscular contraction, the muscles must use a high level of oxygen as the terminal acceptor of electrons provided via the intermediary metabolism of energy substrates. Some other metabolic pathways are also activated as a result of increased energy demand. These processes give rise to a variety of oxygen derivatives known as ROS. Generally speaking, the rate of ROS production is increased as exercise intensity increases.

It is ironic that skeletal muscle also has one of the lowest antioxidant defence capacities in the body. Apart from genetic reasons, this results in part from a relatively low rate of oxygen utilization in this organ, at least for laboratory animals and humans, the two sedentary subjects most commonly studied. However, muscle antioxidant enzymes demonstrate a remarkable adaptability in response to acute exercise and chronic training. This adaptation may be viewed as a restoration of antioxidant capacity to the levels reflecting their physically more active ancestors.

Among the three major antioxidant enzymes, MnSOD and GPX are located in the mitochondria, a major site of ROS generation, and demonstrate a more dynamic pattern of augmentation in response to exercise. CuZnSOD and cytosolic GPX may be induced by ROS that have generated from cytosolic enzyme systems or ROS that have diffused out from the mitochondria. CAT has a more definitive role in the peroxisomes which are sparse in skeletal muscle, so it is affected by exercise the least.

Gene regulation of antioxidant enzymes in skeletal muscle is still largely unknown. With more molecular and immunochemical methods available, our understanding of these highly regulatory enzymes is expanding constantly. The recent discoveries of NFκB and AP-1 binding sites on the SOD gene and of antioxidant responsive elements on the GPX gene have opened up new doors for research and therapeutic uses of these important antioxidants in healthy by and diseased states.

Acknowledgment
The author wishes to thank Dr Mitch Gore, Dr Christiaan Leeuwenburgh, Dr Steve Leichtweis, Russel Fiebig, and John Hollander for their contribution to the work presented in the current chapter. The author also want to acknowledge the grant funds received from NIH DK 42034 and AHA National Center 91-1602.

References

Aebi H (1984) Catalase. *In*: L Packer (ed.): *Methods in Enzymology*. Academic Press, Orlando, FL, Vol. 105, pp 121–125.
Alessio HM and Goldfarb AH (1988) Lipid peroxidation and scavenge enzymes during exercise: adaptive response to training. *J Appl Physiol* 64: 1333–1336.
Ames BN, Shigenaga MK and Hagen TM (1933) Oxidant, antioxidants and degenerative diseases of aging. *Proc Natl Acad Sci USA* 90: 7915–7922.
Bettger WJ, Fish TJ and O'Dell BL (1978) Effect of copper and zinc status of rats on erythrocyte stability and superoxide dismutase activity. *Proc Soc Exp Biol Med* 158: 279–282.

Blum J and Fridovich I (1985) Inactivation of glutathione peroxidase by superoxide radical. *Arch Biochem Biophys* 240: 500–508.

Brady PS, Brady LJ and Ullrey DE (1979) Selenium, vitamin E and the response to swimming stress in rats. *J Nutr* 109: 1103–1109.

Buczynski A, Kedziora J, Tkaczewski W and Wachowicz B (1991) Effect of submaxiaml physical exercise on antioxidative protection of human blood platelets. *Int J Sport Med* 12: 52–54.

Burge WE and Neil AJ (1916–17) Comparison of the amount of catalase in the muscle of large and of small animals. *Am J Physiol* 42: 373–377.

Chance B, Sies H and Boveris A (1979) Hydroperoxide metabolism in mammalian organs. *Physiol Rev* 59: 527–605.

Cowan DB, Weisel RD, Williams WG and Mickle DAG (1992) The regulation of glutathione peroxidase gene expression by oxygen tension in cultured human cardiomyocytes. *J Mol Cell Cardiol* 24: 423–433.

Cowan DB, Weisel RD, Williams WG and Mickle DAG (1993) Identification of oxygen responsive elements in the 5'-flanking region of the human glutathione peroxidase gene. *J Biol Chem* 268: 26904–26910.

Das KC, Lewis-Molock Y and White CW (1995) Thiol modulation of TNF and IL-1 induced Mn SOD gene expression and activation of NF-κB. *Mol Cell Biochem* 148: 45–57.

Davies KJA, Quantanilla AT, Brooks GA and Packer L (1982) Free radicals and tissue damage produced by exercise. *Biochem Biophys Res Commun* 107: 1198–1205.

Dhaunsi G, Singh I, Orak JK and Kingh AK (1994) Antioxidant enzymes in ciprofibrate-induced oxidative stress. *Carcinogenesis* 15: 1923–1930.

Flohe L (1982) Glutathione peroxidase brought into focus. *In*: W. Pryor (ed.): *Free Radicals in Biology and Medicine* 5. Academic Press, New York, pp 223–253.

Fridovich I (1985) Quantitation of superoxide dismutase. *In*: RA Greenwald (ed.): *Handbook of Methods for Oxygen Free Radical Research*. CRC Press, Boca Raton, FL, pp 213–215.

Fridovich I (1995) Superoxide radical and superoxide dismutases.*Annu Rev Biochem* 64: 97–112.

Gore M, Fiebig R, Hollander J and Ji LL (1997) Acute exercise alters mRNA abundance of antioxidant enzyme and nuclear factor B activation in skeletal muscle, heart and liver. *Med Sci Sports Exerc* 29: S229.

Gorecki M, Beck Y, Hartman JR Fischer M, Weiss L, Tochner Z, Slavin S and Nimrod A (1991) Recombinant human superoxide dismutases: production and potential therapeutical uses. *Free Radical Res Commun* 12–13: 401–410.

Greenwald RA (1990) Superoxide dismutase and catalase as therapeutic agents for human diseases: a critical review. *Free Radical Biol Med* 8: 201–209.

Habig WH, Pabst MJ and WB Jakoby WB (1974) Glutathione S-transferases. *J Biol Chem* 249: 7130–7139.

Halliwell B and Gutteridge JMC (1989) *Free Radicals in Biology and Medicine*. Clarendon Press, Oxford.

Hammeren J, Powers S, Lawler J, Criswell J, Lowenthal D and Pollock M (1993) Exercise training-induced alterations in skeletal muscle oxidative and antioxidant enzyme activity in senescent rats. *Int J Sport Med* 13: 412–416.

Harman D (1956) Aging: a theory based on free radical and radiation chemistry. *J Gerontol* 11: 298–300.

Harris ED (1992) Regulation of antioxidant enzymes. *FASEB J* 6: 2675–2683.

Higuchi M, Cartier L-J, Chen M and Holloszy JO (1985) Superoxide dismutase and catalase in skeletal muscle: Adaptive response to exercise. *J Gerontol* 40: 281–286.

Hill KE, Burk RF and Lane JM (1987) Effect of selenium depletion and repletion on plasma glutathione and glutathione-dependent enzymes in the rat. *J Nutr* 117: 99–104.

Hollander J, Gore M, Fiebig R, Bejma J and Ji LL (1997) Exercise training alters superoxide dismutase gene expression in rats. *FASEB J* 11: A584.

Holmes RS and Duley JA (1975) Biochemical and genetic studies of peroxisomal multiple enzyme systems: alpha-hydroxyacid oxidase and catalase. *In*: DL Markert (ed.): *Isozymes I. Molecular structure*. Academic Press, New York, pp 191–211.

Holmes RS and Master CJ (1970) Epigenetic interconversions of the multiple forms of mouse liver catalase. *FEBS Lett* 11: 45–48.

Inoue M, Ebashi I, Watanabe N and Morino Y (1989) Synthesis of a superoxide dismutase derivative that circulates bound to albumin and accumulates in tissues whose pH is decreased. *Biochemistry* 28: 6619–6624.

Ishikawa T and Sies H (1984) Cardiac transport of glutathione disulfide and S-conjugate. *J Biol Chem* 259: 3838–3843.

Jackson ML, Edwards RHT and Symons MCR (1985) Electron spin resonance studies of intact mammalian skeletal muscle. *Biochim Biophys Acta* 847: 185–190.

Jenkins RR (1983) *In*: HG Knuttgen, JA Vogel and J Poortmans (eds): *Biochemistry of Exercise*. Vol. 13, Campaign, Human Kinetics Publishers, pp 467–471.

Jenkins RR (1988) Free radical chemistry: relationship to exercise. *Sport Med* 5: 156–170.

Jenkins RR (1993) Exercise, Oxidative stress and antioxidant: A review. *Int J Sport Nutr* 3: 356–375.

Ji LL (1993) Antioxidant enzyme response to exercise and aging. *Med Sci Sports Exerc* 25: 225–231.

Ji LL (1995) Exercise and free radical generation: role of cellular antioxidant systems. *In*: J Holluszy (ed.): *Exercise and Sport Science Review*. Williams & Wilkins, Baltimore, pp 135–166.

Ji LL and Leichtweis S (1997) Exercise and oxidative stress (invited review). *Age* 20: 91–106.

Ji LL and Mitchell EW (1994) The effect of adrimycin on heart mitochondrial function in rested and exercised rats. *Biochem Pharmacol* 47: 877–885.

Ji LL and Fu RG (1992) Responses of glutathione system and antioxidant enzymes to exhaustive exercise and hydroperoxide. *J Appl Physiol* 72: 549–554.

Ji LL, Stratman FW and Lardy HA (1988a) Antioxidant enzyme systems in rat liver and skeletal muscle. *Arch Biochem Biophys* 263: 150–160.

Ji LL, Stratman FW and Lardy HA (1988b) Enzymatic downregulation with exercise in rat skeletal muscle. *Arch Biochem Biophys* 263: 137–149.

Ji LL, Dillon D and Wu E (1990) Alteration of antioxidant enzymes with aging in rat skeletal muscle and liver. *Am J Physiol* 258: R918–R923.

Ji LL, Wu E and Thomas DP (1991) Effect of exercise training on antioxidant and metabolic functions in senescent rat skeletal muscle. *Gerontology* 37: 317–325.

Ji LL, Fu RG and Mitchell EW (1992a) Glutathione and antioxidant enzymes in skeletal muscle: effects of fiber type and exercise intensity. *J Appl Physiol* 73: 1854–1859.

Ji LL, Stratman FW and Lardy HA (1992b) The impact of selenium deficiency on myocardial antioxidant enzyme systems and related biochemical properties. *J Am Coll Nutr* 11: 79–86.

Ji LL, Fu RG, Mitchell E, Griffiths M, Waldrop TG and Swartz HM (1994) Cardiac hypertrophy alters myocardial response to ischemia and reperfusion *in vivo*. *Acta Physiol Scand* 151: 279–290.

Knight SAB and Sunde RA (1987) The effect of progressive selenium deficiency on anti-glutathione peroxidase antibody reactive protein in rat liver. *J Nutr* 117: 732–738.

Kumar CT, Reddy VK, Plasad M, Thyagaraju K and Reddanna P (1992) Dietary supplementation of vitamin E protects heart tissue from exercise-induced oxidant stress. *Mol Cell Biochem* 111: 109–115.

Lai CC, Huang WH, Askari A, Wang Y, Savazyan N, Klevay L and Hhiu TH (1994) Differential regulation of superoxide dismutase in copper-diffident rat organs. *Free Radical Biol Med* 16: 613–620.

Lammi-Keefe CJ, Swan PB and Hegarty PVJ (1984) Copper–zinc and manganese superoxide dismutase activities in cardiac and skeletal muscles during aging in male rats. *Gerontology* 30: 153–158.

Lang JK, Gohil K, Packer L and Burk RF (1987) Selenium deficiency, endurance exercise capacity, and antioxidant status in rats. *J Appl Physiol* 63: 2532–2535.

Laughlin M H, Simpson T, Sexton WL, Brown OR, Smith JK and Korthuis RJ (1990) Skeletal muscle oxidative capacity, antioxidant enzymes, and exercise training. *J Appl Physiol* 68: 2337–2343.

Lawler JM, Powers SK, Visser T, Van Dijk H, Korthuis MJ and Ji LL (1993) Acute exercise and skeletal muscle antioxidant and metabolic enzymes: Effect of fiber type and age. *Am J Physiol* 265: R1344–R1350.

Leeuwenburgh C and LL Ji (1995) Glutathione depletion in rested and exercised mice: biochemical consequence and adaptation. *Arch Biochem Biophys* 316: 941–949.

Leeuwenburgh C and Ji LL (1996) Glutathione regulation during exercise in unfed and refed rats. *J Nutr* 126: 1833–1843.

Leeuwenburgh C, Fiebig R, Chandwaney R and Ji LL (1994) Aging and exercise training in skeletal muscle: Response of glutathione and antioxidant enzyme systems. *Am J Physiol* 267: R439–R495.

Leeuwenburgh C, Hollander J, Leichtweis S, Fiebig R, Gore M and Ji LL (1997) Adaptations of glutathione antioxidant system to endurance training are tissue and muscle fiber specific. *Am J Physiol* 272: R363–R369.

Lukaski H, Hoverson BS, Gallagher SK and Bolonchuck WW (1990) Physical training and copper, iron, and zinc status of swimmers. *Am J Clin Nutr* 51: 1093–1099.

Luhtala T, Roecher EB, Pugh T, Feuers RJ and Weindruch R (1994) Dietary restriction opposes age-related increases in rat skeletal muscle antioxidant enzyme activities. *J Gerontol* 49: B321–B328.

Matsuo M (1993) Age-related alterations in antioxidant defense. *In*: BP Yu (ed.): *Free Radicals in Aging*. CRC Press, Boca Raton, FL, pp 143–181.

Mc Cord IM and Fridovich I (1969) Superoxide dismutase: an enzymatic function for erythrocuprein (hemocuprein). *J Biol Chem* 244: 6049–6055.

Mena P, Maynar M, Gutierrez JM, Maynar J, Timon J and Campillo JE (1991) Erythrocyte free radical scavenger enzymes in bicycle professional racers. Adaptation to Training. *Int J Sport Med* 12: 563–566.

Meydani M and Evans WJ (1993) Free radicals, exercise, and aging. *In*: BP Yu (ed.): *Free Radicals in Aging*. CRC Press, Boca Raton, FL, pp 183–204.

Moscow JA, Morrow CS, He R, Mullenbach GT and Cowan KH (1992) Structure and function of the 5'-flanking sequence of the human cytosolic selenium-dependent glutathione peroxidase gene (*hgpx 1*). *J Biol Chem* 267: 5949–5958.

Nanji AA, Griniuviene B, Sadrzadeh SMH, Levitsky S and McCully JD (1995) Effect of type of dietary fat and ethanol on antioxidant enzyme mRNA induction in rat liver. *J Lipid Res* 36: 736–744.

Nohl H (1993) Involvement of free radicals in aging: a consequence or cause of senescence. *Brit Med Bull* 49: 653–667.

Nohl H and Hegner D (1978) Response of mitochondrial superoxide dismutase, catalase, and glutathione peroxidase activities to aging. *Mech Age Dev* 11: 145–151.

Oberley LW, St Clair DK, Autor AP and Oberley TD (1987) Increase in manganese superoxide dismutase activity in the mouse heart after X-irradiation. *Arch Biochem Biophys* 254: 69–80.

Oh-ishi S, Kizaki T, Yamashita H, Nagata N, Suzuki K, Taniguchi N and Ohno H (1995) Alteration of superoxide dismutase iso-enzyme activity, content, and mRNA expression with aging in rat skeletal muscle. *Mech Age Dev* 84: 65–76.

Oh-ishi S, Kizaki T, Nagasawa I, Izawa T, Komabayashi T, Nagata N, Suzuki K, Taniguchi N and Ohno H (1997) Effects of endurance training on superoxide dismutase activity, content, and mRNA expression in rat muscle. *Clin Exp Pharmacol Physiol* 24: 326–332.

Ohno H, Suzuki K, Fujii J, Yamashita H, Kizaki T, Ohishi S and Taniguchi N (1994) Superoxide dismutases in exercise and disease. *In*: CK Sen, L Packer and O Hänninen (eds): *Exercise and Oxygen Toxicity*. Elsevier Science, New York, pp 127–161.

Pigeolet E, Corbisier P, Houbion A, Lambert D, Michiels C, Raes M, Zachary M-D and Remacle J (1990) Glutathione peroxidase, superoxide dismutase, and catalase inactivation by peroxides and oxygen derived free radicals. *Mech Age Dev* 51: 283–297.

Powers SK, Criswell D, Lawler J, Ji LL, Martin D, Herb R and Dudley G (1994) Influence of exercise intensity and duration on antioxidant enzyme activity in skeletal muscle differing in fiber type. *Am J Physiol* 266: R375–R380.

Powers SK, Criswell D, Lawler J, Martin D, Ji LL and Dudley G (1994) Training-induced oxidative and antioxidant enzyme activity in the diaphragm: Influence of exercise intensity and duration. *Respir Physiol* 95: 226–237.

Quintanilha AT (1984) The effect of physical exercise and/or vitamin E on tissue oxidative metabolism. *Biochem Soc Trans* 12: 403–404.

Quintanilha AT and Packer L (1983) *In*: R Porter and J Whelan (eds): *Biology of Vitamin E*. Ciba Foundation Symposium no. 101. Pitman, London, pp 56–69.

Quintanilha AT, Packer L, Davies JMS, Racanelli T and Davies KJA (1982) Membrane effects of vitamin E deficiency: Bioenergetics and surface charge density studies of skeletal muscle and liver mitochondria. *Ann N Y Acad Sci* 399: 32–47.

Radák Z, Asano K, Inoue M, Kizaki T, Oh-Ishi S, Suzuki K, Taniguchi N and Ohno H (1995) Superoxide dismutase derivative reduces oxidative damage in skeletal muscle of rats during exhaustive exercise. *J Appl Physiol* 79: 129–135.

Radák Z, Asano K, Inoue M, Kizaki T, Ohishi S, Suzuki K, Taniguchi N and Ohno H (1996) Superoxide dismutase derivative prevents oxidative damage in liver and kidney of rats induced by exhausting exercise. *Eur J Appl Physiol Occup Physiol* 72: 189–194.

Rao G, Xia E and Richardson A (1990) Effect of age on the expression of antioxidant enzymes in male Fischer F344 rats. *Mech Age Dev* 53: 49–60.

Reed D (1986) Regulation of reductive processes by glutathione. *Biochem Pharmacol* 35: 7–13.

Remacle J, Lambert D, Raes M, Pigeolet E, Michiels C and Toussaint O (1992) Importance of various antioxidant enzymes for cell stability. *Biochem J* 286: 41–46.

Semsei I, Rao G and Richardson A (1989) Changes in the expression of superoxide dismutase and catalase as a function of age and dietary restriction. *Biochem Biophys Res Commun* 164: 620–625.

Sen CK and Packer L (1996) Antioxidant and redox regulation of gene transcription. *FASEB J* 10: 709–20.

Sen CK, Marin E, Kretzschmar M and Hänninen O (1992) Skeletal muscle and liver Glutathione homeostasis in response to training, exercise and immobilization. *J Appl Physiol* 73: 1265–1272.

Sen CK, Atalay M and Hänninen O (1994) Exercise-induced oxidative stress: glutathione supplementation and deficiency. *J Appl Physiol* 77: 2177–87.

Spallholz J, Martin JL and Ganther HE (1981) *Selenium in Biology and Medicine*. AVI Press Westport, CT.

Starnes JW, Cantu G, Farrar RP and Kehrer JP (1989) Skeletal muscle lipid peroxidation in exercise and food-restricted rats during aging. *J Appl Physiol* 67: 69–75.

Storz G, Tartaglia LA and Ames BN (1990) Transcriptional regulator of oxidative stress-inducible genes: Direct activation by oxidation. *Science* 248: 189–194.

Suzuki K, Miyazawa N, Nadata T, Seo HG, Usgiyama T and Ganiguchi N (1993) High copper and iron levels and expression of Mn-superoxide dismutase in mutant rats displaying hereditary hepatitis and hepatoma (LEC rats). *Carcinogenesis* 14: 1881–1884.

Tiidus PM, Pushkarenko J and Houston ME (1996) Lack of antioxidant adaptation to short-term aerobic training in human muscle. *Am J Physiol* 271: R832–R836.

Vertechy M, Cooper MB, Chirardi O and Ramacci MT (1989) Antioxidant enzyme activities in heart and skeletal muscle of rats of different ages. *Exp Gerontol* 24: 211–218.

Visner GA, Dougall WC, Wilson JM, Burr IM and Nick HS (1990) Regulation of manganese superoxide dismutase by lipopolysaccharide, interleukin-1, and tumor necrosis factor. *J Biol Chem* 265: 2856–2864.

Vihko V, Salminen A and Rantamaki J (1978) Oxidative lysosomal capacity in skeletal muscle of mice after endurance training. *Acta Physiol Scand* 104: 74–79.

Whisett JA, Clark JC, Wispe JR and Pryhuber GS (1992) *Am J Physiol* 262: 688–693.

Xia E, Rao G, Van Remmen H, Heydari AR and Richardson A (1995) Activities of antioxidant enzymes in various tissues of male Fischer 344 rats are altered by food restriction. *J Nutr* 125: 195–201.

Yu BP (1994) Cellular defenses against damage from reactive oxygen species. *Physiol Rev* 74: 139–162.

Yuan HT, Bingle CD and Kelly FJ (1996) Differential patterns of antioxidant enzyme mRNA expression in guinea pig lung and liver during development. *Biochim Biophys Acta* 1305: 163–171.

Zerba E, Komorowski TE and Faulkner JA (1990) Free radical injury to skeletal muscle of young, adult and old mice. *Am J Physiol* 258: C429–C435.

Zhang N (1996) Characterization of the 5'-flanking region of the human Mn SOD gene. *Biochem Biophys Res Commun* 220: 171–180.

Glutathione: A key role in skeletal muscle metabolism

C.K. Sen

Department of Molecular and Cell Biology, University of California at Berkeley, CA 94720, USA and Department of Physiology, University of Kuopio, 70211 Kuopio, Finland

Summary. Glutathione (GSH) is a low-molecular-weight thiol that is redox active and mostly present in mM concentrations in mammalian cells. High activity of GSH-dependent enzymes and remarkable GSH synthesizing ability of the skeletal muscle suggest that this tissue is a significant component of the complex inter-organ GSH homoeostasis. The hypothesis that skeletal muscle is a major player in whole body GSH metabolism has also been strongly supported by studies on hepatectomized rats. In addition to the above-mentioned functions of GSH, a role of this thiol in the regulation of muscle contraction has been proposed. Myoblast GSH status has also been shown to markedly regulate the inducible activation of the redox sensitive transcription factor NF-κB. Skeletal muscle GSH levels vary depending on the metabolic profile of the tissue. In healthy human skeletal muscle fibers, the level of reduced glutathione is higher in aerobic type I fibers than in anaerobic type II fibers. Another major determinant of skeletal muscle GSH status is the state of physical activity of the tissue. Endurance and sprint training enhances, whereas immobilization down-regulates, the skeletal muscle GSH level. Factors such as the central role of GSH in the antioxidant network, lowering of skeletal muscle GSH during exercise, and certain pathophysiological conditions have generated considerable interest in the search for effective pro-GSH nutritional supplements. Among the supplements that have been tested for their ability to serve as pro-GSH agents, α-lipoate and is N-acetyl-L-cysteine hold promise for human use.

Introduction

Glutathione (L-γ-glutamyl-L-cysteinylglycine) is a low-molecular-weight thiol that is redox active and mostly present in mM concentrations in mammalian cells. The most significant biological functions of endogenously synthesized and ubiquitously distributed reduced glutathione (GSH) include: (1) regulation of the cellular redox state, and thus redox-sensitive cell signaling; (2) defence against reactive species e.g. oxidants; (3) detoxification of xenobiotics, including drugs and pollutants; and (4) a reservoir of the amino acid cysteine for protein synthesis. The presence of GSH in cells is vital to ensure normal cell function. Decreased content of GSH in several cells and tissues has been associated with several health disorders, including myopathies. The primary objective of this work is to present a concise overview of the current understanding of skeletal muscle GSH metabolism, with reference to muscular function and damage.

Intracellular synthesis of GSH is a tightly regulated two-step process; both these processes are ATP dependent. γ-Glutamylcysteine synthetase (also referred to as glutamate–cysteine ligase) catalyzes the formation of the dipeptide γ-glutamylcysteine and subsequently the addition of glycine is catalyzed by glutathione synthase (Fig. 1). Substrates for such synthesis are provided both by direct amino acid transport and by γ-glutamyl transpeptidase (also known as glutamyl transferase) which couple the γ-glutamyl moiety to a suitable amino acid acceptor for transport into the cell. GSH is also generated intracellularly from its oxidized form glutathione disulfide (GSSG) by glutathione disulfide reductase (GSSG reductase) activity in the presence of NADPH. In this context it should be noted that NADH may be transhydrogenated by an energy-dependent mechanism to form NADPH. Thus, although GSSG reductase activity requires NADPH as a co-

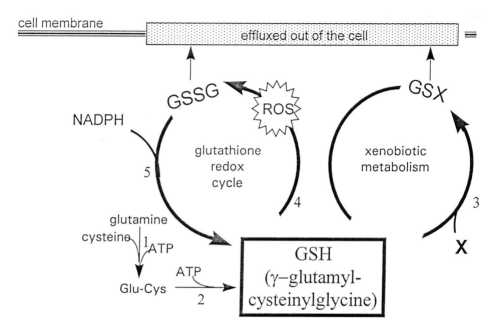

Figure 1. Glutathione-dependent metabolism of reactive oxygen species (ROS) and electrophilic xenobiotics (X). (1) γ-Glutamyl cysteinyl synthetase; (2) GSH synthase (activity of this inhibited by high levels of GSH in the cell); (3) GSH S-transferase; (4) GSH peroxidase; (5) GSSG reductase. GSX, adduct of X with GSH. The availability of reduced cysteine (Cys) inside the cell is a rate-limiting factor in GSH synthesis. Under conditions in which the generation of GSSG inside the cell exceeds the capacity of the cell to recycle it to GSH, there is efflux of GSSG from the cell. High levels of GSSG in the cell are cytotoxic.

factor, cellular NADH may also contribute to the reduction process. In cells challenged with the oxidant *tert*-butylhydroperoxide, almost half of the GSSG formed in the mitochondria has been observed to be reduced by NADPH regenerated from NADH, suggesting that NADH may markedly contribute to the recycling of GSSG to GSH. The activity of γ-glutamylcysteine synthetase, the first step in GSH synthesis, is regulated *in vitro* by feedback inhibition of GSH itself. This represents an important regulatory mechanism by which the maximum amount of *in vivo* GSH in the tissues is limited.

The antioxidant function of GSH may be implicated through two general mechanisms of reaction with reactive oxygen species: direct or spontaneous, and catalyzed by glutathione peroxidase (Fig. 1). As a major byproduct of such reactions GSSG is produced. Glutathione peroxidase is specific for its hydrogen donor, GSH, but may use a wide range of substrates extending from H_2O_2 to organic hydroperoxides. The cytosolic and membrane-bound monomer GSH phospholipid, hydroperoxide–glutathione peroxidase, and the distinct tetramer, plasma glutathione peroxidase, are able to reduce phospholipid hydroperoxides without the necessity for prior hydrolysis by phospholipase A_2. The protective action of phospholipid hydroperoxide–glutathione peroxidase against membrane-damaging lipid peroxidation has been directly demonstrated. GSH is also a major detoxicant. Glutathione S-transferases catalyze the reaction between the thiol-SH group of GSH and potential alkylating agents, thereby neutralizing their electrophilic sites and rendering them more water soluble (e.g. GSX in Fig. 1). Glutathione S-transferases represent a major group of phase II detoxification enzymes. The level of expression of glutathione S-transferases is a crucial determinant of the sensitivity of cells to a broad range of toxic chemicals. Several recent reviews have comprehensively addressed GSH metabolism (Taylor et al., 1996; Anderson, 1997; Droge et al., 1997; Sen, 1997).

In vitro studies carried out on skeletal muscle-derived myoblasts show high activity of GSH-dependent enzymes, suggesting that skeletal muscle may represent a significant component in complex inter-organ GSH homoeostasis (Sen et al., 1993). Based on estimated intracellular water space in these cells, the intracellular GSH level was calculated to be as high as 9 mM (Sen et al., 1995, 1997a). The hypothesis that skeletal muscle is a major player in whole-body GSH metabolism has been strongly supported by studies on hepatectomized rats (Kretzschmar et al., 1992). Surgical manipulation was carried out for total removal of the liver, the major GSH-synthesizing and -exporting organ of the body, while preserving the hepatic portal and vena caval vasculature. Three to six hours after removal of the liver, the plasma GSH level was decreased (about 50% of that in sham-treated rats) but stable. The GSH levels of brain and kidney were not changed in response to hepatectomy. With an increasing time period after hepatectomy, the heart and lung GSH levels were decreased. Removal of the liver resulted in only a marginal decrease of skeletal muscle GSH content 4 and 6 h after the surgery. These observations were indicative of a high GSH export capacity of extrahepatic tissues contributing about 50% of the total GSH influx into the circulation. This work also suggested that the skeletal musculature is probably an important source of GSH in the plasma. The presence of a high activity of GSH-synthesizing enzymes in skeletal muscle tissue (Sen et al., 1992; Marin et al., 1993) and the large mass of the tissue strongly suggest a key role of skeletal muscle in inter-organ glutathione homoeostasis (Kretzschmar and Muller, 1993).

Skeletal muscle GSH levels may vary depending on the muscle type. In male Sprague–Dawley rats total glutathione levels in different muscle types have been reported to be as follows: soleus > deep vastus lateralis > superficial vastus lateralis muscle (Ji et al., 1992). Soleus muscle, with marked oxidative metabolic capacity, appears also to have a remarkably well developed antioxidant defence system when compared with other skeletal muscles. In addition to having higher levels of total glutathione, activities of GSH-dependent antioxidant enzymes, such as GSH peroxidase, GSSG reductase, and the peroxide-decomposing enzyme catalase, were observed to be markedly higher in soleus muscle type (Ji et al., 1992). In human skeletal muscle, both GSSG reductase and GSH peroxidase activities have been observed to be relatively low when compared with those of other tissues. No difference was found among fibre types (I, IIA, and IIB) with regard to the reductase activity, but in contrast GSH peroxidase activity was significantly lower in type IIB fibres than in the other types. These results suggest that type IIB fibres may have a diminished ability to cope with hydroperoxides generated during oxidative stress, which, in turn, could lead to increased damage to membrane structures by lipid peroxidation or oxidation of sensitive intracellular thiol (-SH) containing enzymes by hydrogen peroxide (Austin et al., 1988).

The localization of GSH in skeletal muscle fibres of patients with inherited or acquired neuro-muscular diseases, and of those with no apparent disease of the neuromuscular system, has been studied histochemically (Meijer, 1991). In healthy human skeletal muscle fibres, the level of GSH is higher in aerobic type I fibres than in anaerobic type II fibres. This finding suggests that glutathione in these healthy fibres is maintained in the reduced state mainly by the activity of the decarboxylating and NADPH regenerating enzyme NADP-dependent isocitrate dehydrogenase. In diseased muscle fibres, there is generally a positive relationship between the activity of the NADPH producing enzymes glucose-6-phosphate dehydrogenase and 6- phosphogluconate dehydrogenase and the level of GSH. This positive relationship suggests that glutathione is maintained in these diseased fibres in the reduced state, mainly by the activity of both enzymes of the pentose phosphate pathway (Meijer, 1991).

Muscle contraction

As early as in 1970, a possible role of GSH in muscle contraction was observed (Kosower, 1970). Recently, a mechanically skinned fibre model was used to examine the role of oxidation–reduction in the control of Ca^{2+} release, and contraction in rat and toad skeletal muscle fibres under physiological conditions of myoplasmic $[Mg^{2+}]$ and $[ATP]$ and sarcoplasmic reticulum Ca^{2+} load. Among the reducing agents tested, i.e. dithiothreitol, GSH, and cysteine, only GSH had a marginal effect on the duration of peak contraction response in rat fibres. This was caused by an effect of GSH on the Ca^{2+} sensitivity of the contractile apparatus. The findings of this study suggest that oxidation of an intracellularly accessible site in the muscle fibre can interfere with excitation–contraction coupling, in a process made more sensitive by voltage sensor inacti-vation (Posterino and Lamb, 1996). A putative role of intracellular GSH to protect against oxida-tion of that site has been suggested.

Redox signaling

Oxidation–reduction (redox)-based regulation of signal transduction and gene expression is emerging as a fundamental regulatory mechanism in cell biology (Sen and Packer, 1996; Sen, 1998). Electron flow through side chain functional CH_2-SH groups of conserved cysteinyl resi-dues in proteins account for their redox-sensing properties. As in most intracellular proteins, thiol groups are strongly "buffered" against oxidation by the highly reduced environment inside the cell, only accessible protein thiol groups with high thiol–disulfide oxidation potentials are likely to be redox-sensitive. The list of redox-sensitive signal transduction pathways is steadily growing, and current information suggests that manipulation of the cell redox state may prove to be an important strategy for the management of AIDS and some forms of cancer. The endogenous thioredoxin and glutathione systems are of central importance in redox signaling (Muller et al., 1997). Recently, it was observed that skeletal muscle GSH plays a central role in the regulation of activation of the redox-sensitive transcription factor NF-κB (Sen et al., 1997a). Tumor necrosis factor α (TNFα), a cytokine product of monocytes and macrophages, is implicated in several

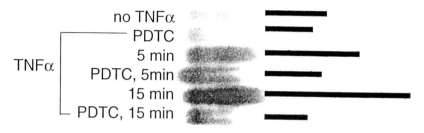

Figure 2. Inhibition of NF-κB activation induced by tumor necrosis factor α (TNFα; 50 ng/ml) by the thiol antioxidant pyrrolidine-dithiocarbamate (200 μM, 4 h; PDTC) in skeletal muscle-derived L6 myoblasts. PDTC treatment significantly increased L6 cell GSH content. (For more details, see Sen et al., 1997a).

skeletal muscle pathologies including muscle wasting of cachexia (see Buck and Chojkier, this volume). Muscle wasting and other conditions such as physical exercise and immobilization are also associated with disturbances in muscle glutathione status. Hence, the author's group sought to investigate the role of endogenous glutathione status in TNFα-induced activation of NF-κB in L6 cells derived from skeletal muscle. In GSH-deficient cells treated with buthionine sulfoximine, TNFα-induced activation of NF-κB was markedly potentiated, suggesting that such activation was sensitive to cellular GSH, but may have been independent of high levels of intracellular GSSG. Treatment of the cells with pyrrolidine-dithiocarbamate increased cell GSH levels and inhibited TNFα-induced activation of NF-κB (Fig. 2). Results from GSSG reductase-inhibited cells suggested that GSSG may participate in, but is not required for, this activation. The inhibitory effect of pyrrolidine-dithiocarbamate on NF-κB activation correlated with its effect on ICAM-1 (intercellular adhesion molecule 1) expression, suggesting that this GSH status-modifying agent not only influenced nuclear translocation of NF-κB proteins but also regulated κB-dependent transcription (Sen et al., 1997a).

Ischemia–reperfusion

Postischemic–reperfusion causes tissue damage to heart and skeletal muscles. In skeletal muscles, ischemia–reperfusion decreases tissue GSH levels (Duarte et al., 1997). In tibialis anterior muscle of the rat hindlimb, after 2 and 4 h of tourniquet ischemia and 1 and 5 h of reperfusion, the levels of GSSG/GSH have been compared with those in tibialis anterior muscles of nonischemic controls. In muscles subjected to 2 h of ischemia, the levels of GSH, GSSG, and the GSSG:GSH ratio did not differ significantly from those of nonischemic controls. After 4 h of ischemia without reperfusion, the GSH levels were slightly but significantly increased, compared with controls. However, after 1 h of reperfusion following 4 h of ischemia, the levels of GSH decreased by 50% compared with controls, and after 5 h of reperfusion the levels of GSH were still

50% lower than control levels. The GSSG:GSH ratio did not change over 1 and 5 h of reperfusion compared with control. This study showed that severe (4 h) but not moderate (2 h) ischemia depletes tissue GSH levels (Sirsjo et al., 1996b). Another study by this group showed that intermittent hypoxia or anoxia in muscle tissue, through hypoperfusion or ischemia, decreases intracellular GSH stores by leakage, weakening intracellular antioxidant defences and increasing the risk of oxidative reperfusion injury upon final normalization of tissue blood supply (Sirsjo et al., 1996a). A protective effect of hyperbaric oxygen treatment on ischemia-dependent derangements, and an improvement effect on recovery in postischemic skeletal muscle after 3 h of ischemia followed by reperfusion, have been observed (Haapaniemi et al., 1995).

Physical exercise

Information regarding the possible effect of physical exercise on the GSH status of human tissues, such as the skeletal muscles, heart or liver, is not currently available. However, several studies have investigated this in experimental animals. Physical exercise clearly influences GSH metabolism in rat skeletal muscle and liver (Sen and Hänninen, 1994). Exhaustive exercise decreases both liver and muscle glutathione (Lew et al., 1985). Consistently, the author's studies showed that exhaustive exercise decreases total glutathione content in the liver and active skeletal muscles, e.g. red gastrocnemius and mixed vastus lateralis. This GSH-decreasing effect was not, however, observed in the less active longissimus dorsi muscle (Sen et al., 1992).

Exercise-induced decrease of the total glutathione pool in the liver, red gastrocnemius muscle, mixed vastus lateralis muscle and heart of rats was also seen in another independent study carried out in the author's laboratory. This effect was not, however, seen in the lung (Sen et al., 1994). Duarte et al. (1993) confirmed that a single bout of exercise results in glutathione loss from skeletal muscle. Exercising resulted in a 50% decrease in total glutathione content in left soleus muscle, an effect that was interpreted as an index of oxidative stress. Recovery of muscle glutathione levels was slow in the post-exercise recovery period. This recovery was remarkably faster in mice supplemented with allopurinol, an inhibitor of the superoxide-producing enzyme xanthine oxidase. Such observations suggest that exercise-induced increase in xanthine oxidase-dependent superoxides causes oxidative stress to muscle tissues located in close proximity, and that this stress is manifested as a loss of tissue glutathione.

In another study, mice were subjected to strenuous running exercise and their soleus muscles were examined for changes in ultrastructure and content of GSH (Duarte et al., 1994). It was hypothesized that invading leukocytes contributed to oxidative stress and they were functionally inhibited in one experimental group by the administration of colchicine. Exercise led to an immediate decrease in muscle GSH content of about 60%, which slowly recovered over 96 h after exercise. With the administration of colchicine after exercise, GSH content was higher than in the muscles of the untreated exercise group 48 h after exercise, indicating an inhibition of the ability of leukocytes to produce oxidative stress. However, at 96 h after exercise, muscle GSH content was lower in the treated exercise group than in the untreated group. The morphological evaluation of the percentage of affected fibres showed that the invasion of leukocytes increased muscle fibre damage. These results suggest that invading leukocytes enhance the production of reactive species of oxygen which may have participated in inducing muscle damage. However, inhibition of leukocyte invasion did not permit their scavenger action of removing cell debris, which appeared to produce even more oxidative stress in the muscle (Duarte et al., 1994).

Exhaustive treadmill exercise induces GSH oxidation in plasma, skeletal muscle and liver of rats (Lew et al., 1985). This effect was confirmed in rat studies by the author's group, in which exhaustive treadmill exercise markedly increased the levels of GSSG in the liver, red gastrocnemius muscle, mixed vastus lateralis muscle, blood, and plasma (Sen et al., 1994). In muscle, metmyoglobin is oxidized by both H_2O_2 and other hydroperoxides to a species with a higher iron valency state and the spectral characteristics of ferryl-myoglobin. GSH reduces the latter species back to metmyoglobin with parallel oxidation to its disulfide (i.e. GSSG), but it cannot reduce metmyoglobin to ferrous myoglobin. Under aerobic conditions, the GSH-mediated reduction of ferryl-myoglobin is associated with O_2 consumption and amounts of GSSG formed are in excess of added peroxide.

After an oxidant challenge, GSH is transformed within the cell to GSSG. When this rate of oxidation is low, much of the GSSG so produced may be enzymatically reduced in the presence of GSSG reductase activity to GSH. However, in the event of a more severe oxidative stress, the rate of GSSG reduction cannot match the rate of its formation, which could result in the accumulation of intracellular GSSG, high concentrations of which may be cytotoxic.

As in erythrocytes and cardiac muscle cells, excess intracellular GSSG is pumped out from skeletal muscle cells (Sen et al., 1993). Such efflux of GSSG (see Fig. 1) from oxidatively stressed tissues may account for the exercise-induced decrease in muscle glutathione content. Skeletal muscle GSH level is also known to be depleted in critically ill patients (Hammarqvist et al., 1997). Critical illness is associated with decreased muscle GSH content and a decreased ratio between reduced and total glutathione, indicating a situation of oxidative stress in this tissue. This decrease may impair the muscle's defense against oxygen free radicals and influence amino acid transport, thus contributing to the loss of balance between protein synthesis and protein degradation that is characteristic of protein catabolism. Marked depletion of GSH in the muscle may lead to mitochondrial damage and initiate a vicious cycle leading to tissue degeneration (Martensson and Meister, 1989).

Exercise training

Several studies have consistently shown that exercise training may influence skeletal muscle GSH metabolism. Increased glutathione content and enhanced activity of GSH-dependent enzymes have been observed in exercise-trained skeletal muscle tissue (Sen, 1995). Ji's (1993) study with rats showed that a bout of exhaustive treadmill exercise increases the activities of GSH peroxidase and GSSG reductase in skeletal muscle. A tissue-specific response to training was reported. Endurance treadmill training decreased GSH peroxidase activity in the liver, but increased the activity of this enzyme in deep vastus lateralis muscle. Lew and Quintanilha (1991) observed that, for the same amount of submaximal exercise, endurance-trained rats had an improved ability to maintain tissue glutathione redox status as reflected by the GSSG:total glutathione ratio compared with their untrained counterparts. The endurance training program significantly increased the activities of GSH peroxidase, GSSG reductase, and glucose-6-phosphate dehydrogenase in skeletal muscle and heart tissues.

The effect of ageing and exercise training on rat skeletal muscle antioxidant enzyme activities have been tested (Ji et al., 1991). Superficial glycolytic and deep oxidative vastus lateralis muscles were collected from rats aged from 2.5 months (young) to 27.5 months (senescent). Old rats had significantly lower GSH peroxidase activity in deep vastus lateralis muscle. After progressive treadmill training, activity of the hydroperoxide-metabolizing enzyme in deep vastus lateralis muscle significantly increased to a level higher than that observed in young sedentary rats. It could be concluded that, although ageing can adversely affect the antioxidant enzyme capacity in skeletal muscle, regular exercise can preserve such protective function.

Much of the research of the author's group been directed towards the study of tissue GSH metabolism in response to exercise and training. In one study (Sen et al., 1992), treadmill training of rats increased skeletal muscle citrate synthase activity, indicating enhanced oxidative capacity. Hepatic total glutathione content was elevated in the trained rats. However, such an effect was not observed in any of the skeletal muscles studied (i.e. red gastrocnemius, mixed vastus lateralis, and longissimus dorsi). Leg muscle GSH peroxidase activity was higher in trained rats. Treadmill training decreased GSSG reductase activity in red gastrocnemius muscle. This effect may be related to the high intensity of training that may have increased flavoprotein turnover and breakdown in the muscle. Endurance training also increased the activity of γ-glutamyl transpeptidase in both leg muscles, the effect being more pronounced in red gastrocnemius. In trained leg muscles, activated γ-glutamyl transpeptidase may facilitate the import of substrates required for GSH generation. Decreased γ-glutamyl transpeptidase activity was observed in control leg muscles after exercise (Sen et al., 1992). This effect was not, however, observed in trained leg muscles, indicating that during exercise trained muscles have a more active substrate import system for GSH generation compared with untrained controls. γ-Glutamyl transpeptidase activity of the trained liver decreased (about 50%) after the exercise bout. This response might ensure that fewer γ-glutamyl compounds are re-trapped in the liver when the needs of the active peripheral tissues, e.g. skeletal muscle, are acute. The contention that exercise training strengthens GSH-dependent antioxidant defences in the skeletal muscle was clearly supported by another study where swim training of rats markedly increased the activities of GSH peroxidase and GSSG reductase (Venditti and Di Meo, 1996).

Antioxidant protection in skeletal muscle that is GSH dependent is influenced by the state of physical activity; endurance training enhances and chronic activity restriction diminishes such protection (Sen et al., 1992). Dogs are more naturally endowed aerobic runners than rats, which have been the experimental animal in most studies about the response of glutathione metabolism to endurance training and exercise. Beagle dogs, commonly used as a laboratory animal, possess a well-developed musculoskeletal system apparently suited for running. Thus, the influence of treadmill training was studied on beagle dogs. Treadmill training (5.5–6.8 km/h, 40 km/day, 5 days/week, 15% uphill grade, for 40 weeks) increased the oxidative capacity of red gastrocnemius, extensor carpi radialis, and triceps muscles of the leg. Training-induced changes in the components of GSH metabolism were most pronounced in red gastrocnemius muscle which is predominantly oxidative by composition. Hepatic and red gastrocnemius total glutathione levels were elevated in response to training. In all three leg muscles mentioned above, training elevated

GSH peroxidase activity. This effect was also most pronounced in red gastrocnemius muscle. The activities of GSSG reductase in extensor carpi radialis and triceps muscles were higher in the trained dogs. Trained animals with higher hepatic total glutathione reserves also had higher GSH S-transferase activity, indicating that the liver of the trained animals had a higher detoxicant status. Training effects were not observed in splenius muscle of the neck and trunk region. In a separate dog experiment (Sen et al., 1992), the effect of chronic activity restriction on red gastrocnemius muscle of beagles was studied. The knee and ankle joints of the right pelvic limb of each dog were immobilized for 11 weeks in a light fibreglass cast. The left leg was used as the paired control. Chronic physical inactivity did not influence the activity of GSH-dependent enzymes; however, the total glutathione level of red gastrocnemius muscle was remarkably decreased in the immobilized leg. Decreased total glutathione level and increased GSSG have also been observed to be associated with skeletal muscle atrophy (Kondo and Itokawa, 1994).

Powers et al. (1994b) investigated the effect of intensity and duration of exercise on responses of training-induced antioxidant enzymes. Rats were exercised at low, moderate, or high intensity at one of three exercise durations (30, 60 or 90 min/day). The costal and crural diaphragm, plantaris muscle, and parasternal intercostal muscles were studied (Powers et al., 1994b). Training effects observed were highly tissue specific. All training programs markedly increased GSH peroxidase activity in the costal diaphragm, but not in the crural diaphragm. Exercise intensity or duration did not have any major influence on the training-induced elevation of GSH peroxidase activity in the costal diaphragm. In the crural diaphragm, however, moderate and high-intensity exercise training decreased tissue GSH peroxidase activity when the daily exercise duration was as long as 90 min. None of the training programs influenced GSH peroxidase activity of the parasternal intercostal muscle, although remarkable effects were observed in plantaris muscle, in which daily exercise duration had a marked effect on the GSH peroxidase activity response. Daily exercise of longer duration triggered a more marked response. Results of another similar study support training effects as being highly tissue specific (Powers et al., 1994a). Although exercise training increased GSH peroxidase activity in red gastrocnemius muscle of rats, such effects were not consistently seen in soleus or even white gastrocnemius muscles. Similar to the results obtained from plantaris muscle, daily exercise duration had a marked effect on GSH peroxidase activity response. Criswell et al. (1993) tested the hypothesis that high-intensity interval training is superior to moderate-intensity continuous exercise training in the up-regulation of antioxidant enzyme activity in skeletal muscle. It was observed that the former was more effective than the latter in elevating GSH peroxidase activity in rat soleus muscle.

Studies investigating the influence of physical training on tissue antioxidant status have mostly tested the effect of endurance training, which enhances the oxidative capacity of tissues. Information about the the effect of sprint training, which relies primarily on nonoxidative metabolism, on antioxidant defences is scanty. Recently, Atalay et al. (1996) examined the effect of a sprint training regimen on rat skeletal muscle and heart GSH system. Soleus muscle, predominantly made up of slow-oxidative fibres, was studied as representative of slow-twitch muscle. Plantaris and extensor digitorum longus muscles, consisting mainly of glycolytic fibres, and the superficial white portion of quadriceps femoris muscle consisting mainly of fast-oxidative glycolytic fibres were studied as representative of fast-twitch muscles. Mixed gastrocnemius muscle was examined as an antagonist of extensor digitorum longus muscle. Activities of lactate dehydrogenase and citrate synthase were measured in the muscles to test the effects of training on glycolytic and oxidative metabolism, respectively. The efficacy and specificity of the 6-week sprint training protocol were attested by markedly increased anaerobic, but not aerobic, metabolic capacity in primarily mixed and fast-twitch fibre muscles. Endurance training consistently up-regulates GSH-dependent defences and other antioxidant enzymes, with the most marked effects in highly oxidative muscle. In contrast, sprint training enhanced antioxidant defences primarily in fast glycolytic muscle. Compared with the control group, GSH peroxidase activities in gastro-cnemius, extensor digitorum longus muscles, and heart increased after sprint training. The train-ing program also increased GSSG reductase activity in extensor digitorum longus muscle and heart. Sprint training did not influence glutathione levels or GSH-related enzymes in the oxidative soleus muscle. Thus, depending on the type of work program, GSH metabolism of specific tissues may be expected to respond to physical training.

The effect of intermittent sprint cycle training on the level of muscle antioxidant enzyme pro-tection has also recently been studied in humans (Hellsten et al., 1996). Resting muscle biopsies, obtained before and after 6 weeks of training and 3, 24, and 72 h after the final session of an additional 1 week of more frequent training, were analyzed for activities of the antioxidant enzymes GSH peroxidase, GSSG reductase, and superoxide dismutase. Seven weeks of training significantly increased the activity of the glutathione-related enzymes, but not that of superoxide dismutase; this suggests that intermittent sprint cycle training which induces an enhanced capacity for anaerobic energy generation also improves the level of GSH-dependent antioxidant protection in the muscle (Hellsten et al., 1996).

Antioxidant enzyme activities are related to skeletal muscle oxidative capacity (Laughlin et al., 1990). Treadmill training of rats increased citrate synthase activity of triceps brachii muscle, indi-cating an increased oxidative capacity. In both sedentary control and trained rats, activities of catalase, superoxide dismutase, and GSH peroxidase were directly correlated with the percentages of oxidative fibres in the skeletal muscle samples, suggesting that such fibres possess an elevated enzymatic antioxidant defence status.

Pro-glutathione supplements

Factors such as the central role of GSH in the antioxidant network, and enhanced oxidation of tissue GSH during exercise, have generated considerable interest in the search for effective pro-glutathione nutritional supplements. Enhancing tissue GSH reserves in mammals is a challenging task, particularly as GSH itself is not available to tissues when administered orally or injected intraperitoneally. To test the time-dependent distribution of intraperitoneally administered GSH, the thiol was administered (1 g/kg body weight) to male Wistar rats (Sen et al., 1994). Injection of GSH solution resulted in a rapid appearance ($\uparrow 10^2$ times after 0.5 h of administration) of glutathione in the plasma. After such a response, a rapid clearance of plasma total glutathione was observed. Twenty-four hours after of the injection, plasma total glutathione was restored to the pre-injection control level. Excess post-injection plasma GSH was rapidly oxidized, as detected by the presence of GSSG. Supplemented GSH was not available to tissues such as the liver, skeletal muscles, lung, kidney, and heart. However, after repeated injection of GSH for 3 consecutive days, blood and kidney total glutathione levels increased. No such effect was observed in the liver, red gastrocnemius muscle, mixed vastus lateralis muscle, heart, or lung (Sen et al., 1994).

Recently, α-lipoate has generated considerable clinical interest as a thiol-replenishing and re-dox-modulating agent (Sen, 1997, 1998; Sen et al., 1997b, c). A unique property of lipoate is that it is a "metabolic antioxidant", i.e. enzymatic systems in human cells treat it as a substrate for bioreduction. Thus, supplemented lipoate is promptly taken up by cells and reduced to dihydrolipoate (DHLA) at the expense of cellular reducing equivalents, e.g. NADH and NADPH. Dihydrolipoate is a powerful reducing agent which has many antioxidant properties and mediates the pro-glutathione effects of lipoate (Sen, 1997). The first study testing the efficacy of α-lipoate supplementation in exercise-induced oxidative stress has just been reported. Khanna et al. (1997) studied the effect of intragastric lipoate supplementation (150 mg/kg body weight for 8 weeks) on lipid peroxidation and glutathione-dependent antioxidant defenses in the liver, heart, kidney, and skeletal muscle of male Wistar rats. Lipoate supplementation significantly increased total gluta-thione levels in the liver and blood. This information is consistent with results from previously discussed cell experiments, and shows that lipoate supplementation may indeed increase gluta-thione levels of certain tissues *in vivo*. Lipoate supplementation did not, however, affect the total glutathione content of organs such as the kidney, heart, and skeletal muscles. Increases in the hepatic glutathione pool that were dependent on lipoate supplementation were associated with increased resistance to lipid peroxidation. This beneficial effect against oxidative lipid damage was also observed in the heart and red gastrocnemius skeletal muscle. Lower lipid peroxide levels in certain tissues of lipoate-fed rats suggest strengthening of the antioxidant network defence in these tissues (Khanna et al., 1997).

Another clinically relevant pro-GSH agent that has been most extensively studied is *N*-acetyl-L-cysteine (NAC; 2-mercaptopropionyl glycine). In addition to its reactive oxygen-detoxifying properties, NAC is believed to function as a cysteine delivery compound which may stimulate cellular GSH synthesis (Sen, 1997). After free NAC enters a cell, it is rapidly hydrolyzed to release cysteine. NAC, but not *N*-acetyl-D-cysteine or the oxidized disulfide form of NAC, is deacetylated in several tissues to release cysteine. Hydrolysis of oxidized NAC, must however, be preceded by cleavage of the disulfide bridge. NAC appears to be safe for human use because it has been used as a clinical mucolytic agent for many years. Reid and associates have shown that antioxidant enzymes are able to depress contractility of unfatigued diaphragm fibre bundles and inhibit development of acute fatigue. NAC has been tested for similar effects. Fibre bundles were removed from diaphragms and stimulated directly using supramaximal current intensity. Studies of unfatigued muscle showed that 10 mM NAC reduced peak twitch stress, shortened time to peak twitch stress, and shifted the stress–frequency curve down and to the right. Fibre bundles incubated in 0.1–10 mM NAC exhibited a dose-dependent decrease in relative stresses developed during 30-Hz contraction with no change in maximal tetanic (200 Hz) stress. NAC (10 mM) also inhibited acute fatigue. This effect has also been tested in humans. Healthy volunteers were studied on two occasions each: subjects were pre-treated with NAC 150 mg/kg or 5% dextrose in water by intravenous infusion. It was evident that pre-treatment with NAC can improve per-formance of human limb muscle during fatiguing exercise, suggesting that oxidative stress plays a causal role in the fatigue process and identifying antioxidant therapy as a novel intervention which may be useful clinically (Khawli and Reid, 1994; Reid et al., 1994).

Acknowledgement
Supported by grants from the Finnish Ministry of Education and the Juho Vainio Foundation, Helsinki.

References

Anderson ME (1997) Glutathione and glutathione delivery compounds. *Adv Pharmacol* 38: 65–78.

Atalay M, Seene T, Hänninen O and Sen CK (1996) Skeletal muscle and heart antioxidant defences in response to sprint training. *Acta Physiol Scand* 158: 129–134.

Austin L, Arthur H, de Niese M, Gurusinghe A and Baker MS (1988) Micromethods in single muscle fibers. 2. Determination of glutathione reductase and glutathione peroxidase. *Anal Biochem* 174: 575–579.

Criswell D, Powers S, Dodd S, Lawler J, Edwards W, Renshler K and Grinton S (1993) High intensity training-induced changes in skeletal muscle antioxidant enzyme activity. *Med Sci Sports Exerc* 25: 1135–1140.

Droge W, Gross A, Hack V, Kinscherf R, Schykowski M, Bockstette M, Mihm S and Galter D (1997) Role of cysteine and glutathione in HIV infection and cancer cachexia: therapeutic intervention with *N*-acetylcysteine. *Adv Pharmacol* 38: 581–600.

Duarte JA, Appell HJ, Carvalho F, Bastos ML and Soares JM (1993) Endothelium-derived oxidative stress may contribute to exercise-induced muscle damage. *Int J Sport Med* 14: 440–443.

Duarte JA, Carvalho F, Bastos ML, Soares JM and Appell HJ (1994) Do invading leukocytes contribute to the decrease in glutathione concentrations indicating oxidative stress in exercised muscle, or are they important for its recovery? *Eur J Appl Physiol* 68: 48–53.

Duarte JA, Gloser S, Remiao F, Carvalho F, Bastos ML, Soares JM and Appell HJ (1997) Administration of tourniquet. I. Are edema and oxidative stress related to each other and to the duration of ischemia in reperfused skeletal muscle? Arch. *Orthop Trauma Surg* 116: 97–100.

Haapaniemi T, Sirsjo A, Nylander G and Larsson J (1995) Hyperbaric oxygen treatment attenuates glutathione depletion and improves metabolic restitution in postischemic skeletal muscle. *Free Radic Res* 23: 91–101.

Hammarqvist F, Luo JL, Cotgreave IA, Andersson K and Wernerman J (1997) Skeletal muscle glutathione is depleted in critically ill patients. *Crit Care Med* 25: 78–84.

Hellsten Y, Apple FS and Sjodin B (1996) Effect of sprint cycle training on activities of antioxidant enzymes in human skeletal muscle. *J Appl Physiol* 81: 1484–1487.

Ji LL (1993) Antioxidant enzyme response to exercise and aging. *Med Sci Sports Exerc* 25: 225–231.

Ji LL, Fu R and Mitchell EW (1992) Glutathione and antioxidant enzymes in skeletal muscle: effects of fiber type and exercise intensity. *J Appl Physiol* 73: 1854–1859.

Ji LL, Wu E and Thomas DP (1991) Effect of exercise training on antioxidant and metabolic functions in senescent rat skeletal muscle. *Gerontology* 37: 317–325.

Khanna S, Atalay M, Laaksonen DE, Gul M, Roy S, Packer L, Hänninen O and Sen CK (1997) Tissue glutathione homeostasis in response to lipoate supplementation and exercise. *In: Oxygen Club of California*, Annual Meeting, Feb 27–March, 1, Santa Barbara.

Khawli FA and Reid MB (1994) *N*-Acetylcysteine depresses contractile function and inhibits fatigue of diaphragm *in vitro*. *J Appl Physiol* 77: 317–324.

Kondo H and Itokawa Y (1994) Oxidative stress in muscular atrophy. *In*: CK Sen, L Packer and O Hänninen (eds): *Oxidative Stress in Muscular Atrophy*. Elsevier Science BV, Amsterdam, pp 319–342.

Kosower EM (1970) A role for glutathione in muscle contraction. *Experientia* 26: 760–761.

Kretzschmar M and Muller D (1993) Aging, training and exercise. A review of effects on plasma glutathione and lipid peroxides. *Sport Med* 15: 196–209.

Kretzschmar M, Pfeifer U, Machnik G and Klinger W (1992) Glutathione homeostasis and turnover in the totally hepatectomized rat: evidence for a high glutathione export capacity of extrahepatic tissues. *Exp Toxicol Pathol* 44: 273–281.

Laughlin MH, Simpson T, Sexton WL, Brown OR, Smith JK and Korthuis RJ (1990) Skeletal muscle oxidative capacity, antioxidant enzymes, and exercise training. *J Appl Physiol* 68: 2337–43.

Lew H and Quintanilha A (1991) Effects of endurance training and exercise on tissue antioxidative capacity and acetaminophen detoxification. *Eur J Drug Metab Pharmacokin* 16: 59–68.

Lew H, Pyke S and Quintanilha A (1985) Changes in the glutathione status of plasma, liver and muscle following exhaustive exercise in rats. *FEBS Lett* 185: 262–266.

Marin E, Kretzschmar M, Arokoski J, Hänninen O and Klinger W (1993) Enzymes of glutathione synthesis in dog skeletal muscles and their response to training. *Acta Physiol Scand* 147: 369–373.

Martensson J and Meister A (1989) Mitochondrial damage in muscle occurs after marked depletion of glutathione and is prevented by giving glutathione monoester. *Proc Natl Acad Sci USA* 86: 471–475.

Meijer AE (1991) The histochemical localization of reduced glutathione in skeletal muscle under different pathophysiological conditions. *Acta Histochem* 90: 147–154.

Muller JM, Rupec RA and Baeuerle PA (1997) Study of gene regulation by NF-kappa B and AP-1 in response to reactive oxygen intermediates. *Methods* 11: 301–312.

Posterino GS and Lamb GD (1996) Effects of reducing agents and oxidants on excitation-contraction coupling in skeletal muscle fibres of rat and toad. *J Physiol (Lond)* 496: 809–825.

Powers SK, Criswell D, Lawler J, Ji LL, Martin D, Herb RA and Dudley G (1994a) Influence of exercise and fiber type on antioxidant enzyme activity in rat skeletal muscle. *Am J Physiol* 266: R375–R380.

Powers SK, Criswell D, Lawler J, Martin D, Ji LL, Herb RA and Dudley G (1994b) Regional training-induced alterations in diaphragmatic oxidative and antioxidant enzymes. *Resp Physiol* 95: 227–237.

Reid MB, Stokic DS, Koch SM, Khawli FA and Leis AA (1994) *N*-acetylcysteine inhibits muscle fatigue in humans. *J Clin Invest* 94: 2468–2474.

Sen CK (1995) Oxidants and antioxidants in exercise. *J Appl Physiol* 79: 675–686.

Sen CK (1997) Nutritional biochemistry of cellular glutathione. *J Nutr Biochem* 8: 660–672.

Sen CK (1998) Redox signaling and the emerging potential of thiol antioxidants. *Biochem Pharmacol* 55; *in press*.

Sen CK and Hänninen O (1994) Physiological antioxidants. *In*: CKSen, L Packer and O Hänninen (eds): *Physiological antioxidants*. Elsevier Science Publishers BV, Amsterdam, pp 89–126.

Sen CK and Packer L (1996) Antioxidant and redox regulation of gene transcription. *FASEB J* 10: 709–720.

Sen CK, Marin E, Kretzschmar M and Hänninen O (1992) Skeletal muscle and liver glutathione homeostasis in response to training, exercise, and immobilization. *J Appl Physiol* 73: 1265–1272.

Sen CK, Rahkila P and Hänninen O (1993) Glutathione metabolism in skeletal muscle derived cells of the L6 line. *Acta Physiol Scand* 148: 21–26.

Sen CK, Atalay M and Hänninen O (1994) Exercise-induced oxidative stress: glutathione supplementation and deficiency. *J Appl Physiol* 77: 2177–2187.

Sen CK, Hänninen O and Orlov SN (1995) Unidirectional sodium and potassium flux in myogenic L6 cells: mechanisms and volume-dependent regulation. *J Appl Physiol* 78: 272–281.

Sen CK, Khanna S, Reznick AZ, Roy S and Packer L (1997a). Glutathione regulation of tumor necrosis factor-alpha-induced NF-kappa B activation in skeletal muscle-derived L6 cells (In Process Citation). *Biochem Biophys Res Commun* 237: 645–649.

Sen CK, Roy S, Han D and Packer L (1997b). Regulation of cellular thiols in human lymphocytes by alpha-lipoic acid: A flow cytometric analysis. *Free Radical Biol Med* 22: 1241–1257.

Sen CK, Roy S and Packer L (1997c). Therapeutic potential of the antioxidant and redox properties of alpha-lipoic acid. *In*: L Montagnier, R Olivier and C Pasquier (eds): *Therapeutic Potential of the Antioxidant and Redox Properties of Alpha-Lipoic Acid*. Marcel Dekker, New York, pp 251–267.

Sirsjo A, Arstrand K, Kagedal B, Nylander G and Gidlof A (1996a). *In situ* microdialysis for monitoring of extracellular glutathione levels in normal, ischemic and post-ischemic skeletal muscle. *Free Radic Res* 25: 385–391.

Sirsjo A, Kagedal B, Arstrand K, Lewis DH, Nylander G and Gidlof A (1996b). Altered glutathione levels in ischemic and postischemic skeletal muscle: difference between severe and moderate ischemic insult. *J Trauma* 41: 123–128.

Taylor CG, Nagy LE and Bray TM (1996) Nutritional and hormonal regulation of glutathione homeostasis. *Curr Top Cell Regul* 34: 189–208.

Venditti P and Di Meo S (1996) Antioxidants, tissue damage, and endurance in trained and untrained young male rats. *Arch Biochem Biophys* 331: 63–68.

Vitamin E and its effect on skeletal muscle

M. Meydani[1], R. Fielding[2] and K.R. Martin[1]

[1]*Jean Mayer USDA Human Nutrition Research Center on Aging at Tufts University, 711 Washington Street, Boston, MA 02111, USA*
[2]*Department of Health Sciences, Sargent College of Allied Health Profession, Boston University, Boston, MA 02118, USA*

Summary. Recent evidence supports the contention that unaccustomed and strenuous exercise may lead to disruption of the delicate oxidant/antioxidant balance, resulting in oxidative stress within the human body. It is generally accepted that many instances of cellular and tissue dysfunction observed after exercise may be attributed to oxidative stress. Vitamin E is the most prevalent and potent lipid soluble antioxidant in the human body and may have an important role in decreasing oxidative stress associated with exercise by improving the oxidant/antioxidant balance. Concentrations of vitamin E can be significantly increased in muscle tissues within a relatively short period of time by dietary intake and/or supplementation. Concentrations of vitamin E may also vary among the muscle fibre types. Although rarely observed, vitamin E deficiency may induce muscle fibre degradation, affect skeletal muscle fibre type distribution, promote deposition of lipid granules, and result in other increased indications of oxidative stress. Oxygen consumption increases markedly during strenuous exercise and is believed to contribute to the increase in oxygen radical formation during exercise. Acute exercise and exercise training increase antioxidant enzyme levels in skeletal muscle of animals. Since exercise reduces vitamin E content of skeletal muscle, the requirement for vitamin E may be substantially above the current recommended dietary allowance (RDA), particularly for individuals who regularly exercise. Data suggesting a protective effect of vitamin E supplementation on human performance are limited; however, the preponderance of evidence indicates beneficial effects from long-term vitamin E supplementation in decreasing exercise-induced oxidative stress and associated risk factors.

Introduction

It is well accepted that physically active individuals are generally healthier than sedentary people. This can, in part, be attributed to their adoption of a lifestyle that includes regular exercise, proper and balanced nutrition, abstention from smoking and drug use, and moderation in alcohol intake. Among the athletes and individuals with regular physical regimens, general nutrition and the many individual components of daily foods have received great attention not only for maintaining energy balance but also for improving performance among the elite. In addition, compelling scientific evidence supports the importance of food and particular nutrients in the prevention of degenerative disease such as cancer and cardiovascular diseases. Scientists and health professionals have introduced the concept of an oxidant/antioxidant balance in order to explain some of the effects observed with changes in diet and exercise. This concept has also been translated into lay language to promote healthy lifestyle choices to the public.

Of the nutrients, vitamin E has received considerable attention for its antioxidant function in living organisms and its beneficial effects on health. It is the most effective natural antioxidant in the biological system, and has been found to play an important role in preventing cardiovascular disease and certain types of cancer, as well as improving the immune system (Meydani, 1995a, b). It has also been suggested that vitamin E may have an important role in reducing the oxidative stress associated with exercise, thereby improving the oxidant/antioxidant balance (Meydani et al.,

1993). Vitamin E deficiency in humans is rarely observed; however, marginal deficiency without apparent clinical symptoms may occur, especially among elderly people (Garry et al., 1982; Meydani and Blumberg, 1992), individuals engaged in strenuous physical activity or training, and in those consuming a high-carbohydrate, low-fat diet without adequate dietary intake of antioxidants, such as vitamins E and C and carotenoids. Vitamin C may spare vitamin E utilization in the body and recycle the oxidized form of vitamin E back to its reduced form (Niki, 1987).

High intake of vitamin E has been shown to reduce numerous indices of oxidative stress, prevent muscle damage, and possibly increase endurance during exercise in experimental animals through its ability to scavenge free radicals (Quintanilha and Packer, 1983; Ji and Leichtweis, 1997). In normal cell metabolism, about 2% of oxygen escapes the mitochondrial electron transport system resulting in the formation of reactive oxygen species (ROS) (Halliwell, 1987). Elevated levels of oxygen-derived, highly reactive free radicals that exceed the antioxidant capacity of tissues are known to damage several critical cellular components such as DNA, proteins and the polyunsaturated fatty acids (PUFAs) of membrane phospholipids. In lipid membranes, free radicals can trigger chain reactions leading to lipid peroxidation, release of toxic aldehydes, and loss of membrane integrity and organization. Exercise substantially increases total oxygen consumption in skeletal muscle, which already has one of the highest oxygen requirements of all tissues, thereby leading to enhanced production of free radicals which implicates ROS in the etiology of exercise-induced muscular dysfunction. During maximal exercise whole body oxygen consumption increases up to 20-fold whereas oxygen consumption within skeletal muscle may be elevated by as much as 100-fold with proportionate increases in ROS (Ji and Leichtweis, 1997). Vitamin E readily accumulates in skeletal muscle tissue where it functions as a potent, intramembrane antioxidant and protects muscle tissue from potential injury. Evidence suggests that vitamin E may be differentially distributed between muscle fibre types because type I fibres have been reported to contain higher α-tocopherol levels than muscle containing mostly type II fibres (Meydani et al., 1996). Brief, intense exercise has not been shown to alter vitamin E content in tissues significantly; however, vitamin E concentration has been demonstrated to decrease in human skeletal muscle after eccentric exercise (Meydani et al., 1993) and in rat skeletal muscle after endurance training (Tiidus and Houston, 1994; Tiidus et al., 1996). Substantial decreases of tissue vitamin E after training suggest increased free radical production during exercise, which ultimately consumes tissue stores of vitamin E. Thus, it is likely that vitamin E confers protection on skeletal muscle against exercise-induced free radical formation and, furthermore, increases of vitamin E levels in tissues through dietary supplementation may markedly augment this protection. This chapter briefly reviews the current knowledge about the characteristics of vitamin E, its dietary sources, and tissue distribution in humans, and discusses its relationship to other antioxidant defence systems in association with exercise-induced oxidative stress in skeletal muscle. Finally, a current overview of the role of vitamin E in exercise performance is provided.

Characteristics and sources of vitamin E

Vitamin E is a fat-soluble vitamin that is essential for normal metabolism. It includes a family of eight naturally occurring compounds, in two classes with different biological activities, designated as tocopherols and tocotrienols. Structurally, vitamin E contains a heterocyclic chromanol ring system with a phytyl side chain which may contain double bonds at the 4', 8' and 12' positions to form the tocotrienols (Fig. 1). Different structural isomers result from combinations of methyl and hydrogen groups on the R1, R2, and R3 positions of the chromanol ring and are important in conferring biological potency (Tab. 1). The presence of the hydroxyl group on the chromanol ring is crucial to the antioxidant activity of members of the vitamin E family. RRR-α-tocopherol (formerly d-α-tocopherol), which has the highest biological activity, is the most widely available

Table 1. The eight compounds found in nature that have vitamin E activity[*]

Compound	R1	R2	R3	Double bond on 4', 8', 12'	Biological activity (IU/mg) compared with d-α-tocopherol[+]	
d-α-Tocopherol	CH$_3$	CH$_3$	CH$_3$	None	1.49	100
d-β-Tocopherol	CH$_3$	H	CH$_3$	None	0.75	50
d-γ-Tocopherol	H	CH$_3$	CH$_3$	None	0.15	10
d-δ-Tocopherol	H	H	CH$_3$	None	0.05	3
d-α-Tocotrienol	CH$_3$	CH$_3$	CH$_3$	Yes	0.75	30
d-β-Tocotrienol	CH$_3$	H	CH$_3$	Yes	0.08	5
d-γ-Tocotrienol	H	CH$_3$	CH$_3$	Yes	Not known	
d-δ-Tocotrienol	H	H	CH$_3$	Yes	Not known	

[*] See Figure 1 for location of methyl groups and double bonds.
[+] Values in second column are percentages.

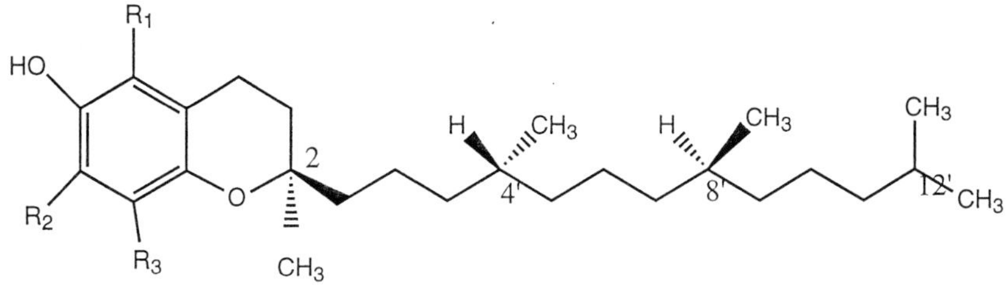

Figure 1. RRR-α-Tocopherol.

form of vitamin E in food. γ-Tocopherol, which has about one tenth of the biological activity of α-tocopherol, is also abundant in plant seed oils. Commercially available vitamin E supplements contain either the natural or the synthetic form of α-tocopherol. The synthetic forms of vitamin E consist of a roughly equal mix of eight stereoisomers, whereas the natural form contains only RRR-α-tocopherol alone or esterified with acetate or succinate. Investigation with deuterium-labeled ($[^2H]$) tocopherol has suggested that the natural form of vitamin E is more bioavailable than synthetic forms (Acuff et al., 1994). The higher binding capacity of the natural form of vitamin E to liver vitamin E-binding protein is believed to be responsible for the body's preference for the natural versus the synthetic form (Traber et al., 1993).

For practical purposes, 1 mg of the synthetic form, *all-rac*-α-tocopheryl acetate, has been set equal to 1 international unit (IU) of vitamin E. As a result, the natural form of RRR-α-tocopherol has a biopotency of vitamin E equal to 1.49 IU (Meydani et al., 1996). As foods may contain a mixture of tocopherol isomers, the term α-tocopherol equivalent (α-TE), which is equal to 1 mg of RRR-α-tocopherol, is used to estimate vitamin E activity in food.

Fats and seed oils are the major contributors of vitamin E to the average American diet, whereas fruits, vegetables, and breakfast cereals contribute a lesser but still substantial amount of vitamin E (Murphy et al., 1990). Nuts, whole grains, and wheat germ contain a relatively higher concentration of tocopherols, but these are not major components of the daily diet in Western society. Supplemental vitamin E is available in the form of soft gel capsules and as a component of multivitamins; however, multivitamins mostly contain a low concentration (about 30 IU). Most commonly available vitamin E supplements are in either a synthetic form (*dl*-α-tocopherol) or the esterified natural form (*d*-α-tocopherol), and are available in quantities of 100–400 IU. Cooking oils are also a good source of vitamin E, but most oils contain vitamin E in the form of γ-tocopherol. Therefore, to ingest a daily supplement of 100–400 IU of this vitamin from oil, consumption of one to two cups would be required and is not advised.

Dietary intake of vitamin E

The National Research Council has set the recommended daily allowance (RDA) for vitamin E at 10 mg for men and 8 mg for women (National Research Council, 1989). These levels are based primarily on representative intakes of this vitamin from US food sources. However, the daily requirement for this vitamin may increase with consumption of a diet containing a higher level of PUFAs. Vitamin E protects the susceptible double bonds of PUFAs from lipid peroxidation. Even though foods containing vegetable oils or margarine may be high in PUFAs, these often may not contain an adequate amount or the most effective form of tocopherol. Some of these products are relatively rich in α-tocopherol and may not provide enough α-TE to maintain a proper balance between PUFAs and vitamin E. To maintain the oxidant/antioxidant balance, over 0.4 mg RRR-α-tocopherol/g PUFA is desirable (Lehmann et al., 1986).

According to the second National Health and Nutrition Examination Survey (NHANES II), the vitamin E content of diets of most of the US population is slightly below the RDA (69% of men and 80% of women), and the diets of 20% of men and 32% of women contain less than 50% of the RDA (Murphy et al., 1990). Murphy et al. (1990) also reported that 23% of men and 15%

of women in that survey showed a ratio of vitamin E to PUFAs of less than 0.4 which is below the level believed to be critical for the maintenance of the oxidant/antioxidant balance.

Vitamin E concentration in skeletal muscle and other tissues

As vitamin E is fat soluble, its tissue concentration is dependent on the presence of lipids. Absorption of vitamin E from the gut is also dependent upon the digestion and absorption of fat. Vitamin E is transported to the liver by the lymphatic system. In the blood, it is principally carried in low-density lipoproteins (LDLs) and high-density lipoproteins (HDL) (Meydani et al., 1989). Newly absorbed vitamin E accumulates largely in adipose tissue, liver, and muscle (Tab. 2). Concentrations of vitamin E can be significantly increased in plasma and tissue within a relatively short period of time through supplemental intake of vitamin E (Meydani et al., 1993). Increase of

Table 2. Mean concentrations of vitamin E in plasma and tissues

Plasma	27 µM
Liver	23 nmol/g
Heart	23 nmol/g
Adipose tissue	230 nmol/g
Skeletal muscle	35 nmol/g

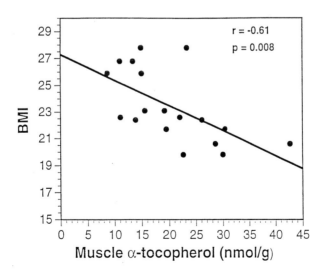

Figure 2. Relationship of BMI with muscle concentration of α-tocopherol.

vitamin E concentration in muscle tissue can be largely dependent on the oxidative metabolic state of muscle, the activity of lipoprotein lipase, and LDL receptors, as well as on the nonspecific exchange with plasma vitamin E (Traber et al., 1993).

The authors have found that intake of 800 IU of vitamin E by young healthy individuals for 4 weeks increased the α-tocopherol concentration in the vastus lateralis muscle from 37 to 57 nmol/g wet tissue (Meydani et al., 1993). Even though the increase in plasma level was much higher than in muscle (300% versus 55%), the increase in skeletal muscle vitamin E among these individuals correlated strongly with the increase of plasma vitamin E, indicating the presence of a close metabolic equilibrium between plasma and muscle vitamin E. The authors have also noted an inverse correlation of muscle vitamin E with body mass index (BMI) (Fig. 2), indicating that individuals with a higher BMI, and thus a higher level of adiposity and lower muscle lipoprotein lipase activity, may limit lipid oxidation and favor deposition of vitamin E in adipose tissue, resulting in lower levels of the vitamin in muscle tissue (Ferraro et al., 1993). Therefore, individuals with low levels of lean body mass, namely high adipose levels, which are supplemented with vitamin E, may accumulate this vitamin preferentially in fat deposits.

Vitamin E may have a differential role in the oxidative metabolism of different muscle fibres, the concentration of vitamin E in muscle possibly varying with the composition of fibre types. Human skeletal muscle consists of two main fibres: type I (slow-twitch) and type II (fast-twitch) fibres. Type I fibres are rich in myoglobin and mitochondrial enzymes and replenish their phosphocreatinine more efficiently via oxidative phosphorylation than type II fibres (Pette and Spamer, 1986). Relative to type II fibres, type I fibres also contain a higher catalase (CAT) activity (Riley et al., 1988) which is probably necessary to eliminate the harmful effects of ROS produced from a higher oxidative metabolism associated with a large number of mitochondria (Lammi-Keefe et al., 1980; Jenkins et al., 1982). Thus, type I fibres, which produce energy primarily via aerobic processes, may also utilize more vitamin E than type II fibres. Muscles composed mainly of type I fibre have been reported to contain a higher α-tocopherol concentration than muscle composed mainly of type II fibres (Fry et al., 1993). Similar observations have been made in rabbits injected with radiolabeled α-tocopherol (Salviati et al., 1980); however, a high dietary intake of or supplementation with vitamin E may have no effect on the increase of muscle fibre types. The authors found no effect of 800 IU of vitamin E supplementation for one month on the distribution of muscle fibre types in the vastus lateralis muscle of young volunteers.

The effect of vitamin E deficiency on skeletal muscles

Deficiency of vitamin E is rare in humans, and it is mainly observed in patients with chronic fat malabsorption syndrome (Neville et al., 1983; Federico et al., 1991). The manifestation of vitamin E deficiency in humans is similar to that found in animals (Van Fleet and Ferrans, 1976; Van Fleet and Ferrans, 1977; Neville et al., 1983; Lazaro et al., 1986). The affected muscle fibres contain ceroid and lipofuscin inclusions, possibly as a result of increased lipid peroxidation. A severe fibre loss with almost complete disappearance of large diameter fibres has been reported with vitamin E deficiency (Federico et al., 1991). Therapy with large doses of vitamin E has been reported to reverse some of the damage of vitamin E deficiency, including shift of fibres from type I to type II and reduction in the number of fibres containing lipid and ceroid granules (Lazaro et al., 1986).

As the type I fibres are more oxidatively active than type II fibres, they may be affected more by vitamin E deficiency than type II fibres. Pillari et al. (1994) reported that vitamin E deficiency appears to affect predominantly type I muscle fibres in rats. Neville et al. (1983) observed type I fibre predominance in one patient and fibre type grouping in another patient, with no excessive type II fibre production. Thus, it appears that vitamin E deficiency alters fibre type population; however, vitamin E supplementation beyond the deficiency does not increase type I fibre numbers above the normal level.

The exact mechanism by which vitamin E deficiency induces muscle degradation is not known. The muscle from vitamin E-deficient animals appears to be susceptible to free radical-mediated oxidation and may be responsible for the etiology of nutritional muscle distrophy. Shih et al. (1977) reported increased muscle proteolysis and a higher ratio of cystine:cysteine as an index of oxidative degradation of the muscle protein in dystrophic muscle. Vitamin E may be involved in regulation of the synthetic pathways of specific proteins required for normal muscle function (De Villers et al., 1973). Muscle fibres can also be degraded by apoptosis which is a form of programmed cell death that is instrumental in development, growth regulation, and the prevention of disease (Thompson, 1995). The level of ubiquitin, a protein indicator of apoptotic activity, increased in the muscles of mice after a night of spontaneous wheel-running followed by 2 days of rest. However, very little information is known about fully differentiated cells of the muscle fibre of adult animals (Sandri et al., 1995). Thiol compounds have been demonstrated to inhibit endonucleases that destroy DNA, which is a key characteristic of apoptosis (Cain et al., 1995). Trolox, a water-soluble analog of vitamin E and a powerful antioxidant in its own right, has been shown to prevent oxidative stress-induced apoptosis in murine models (Forrest et al., 1994). As immunological response in muscle after exercise is thought to involve free radical production by neutrophils, monocytes, and macrophages, depletion of vitamin E may contribute to muscle damage in humans.

Loss of calcium homoeostasis seems to be a key event in the damaging process. Depletion of cellular thiols and an increase of intracellular calcium may potentiate free radical generation. Vitamin E-deficient muscle appears to be more susceptible to intercellular calcium overload as a result of leakiness of membranes, permitting the influx of calcium which activates protease leading to muscle protein degradation (Dayton et al., 1979). It is also important to note that mitochondria are also very sensitive to increased intracellular calcium concentration which interferes with production of ATP. The increased protease activity caused by the increase of intracellular calcium in vitamin E deficiency may contribute to the thin filament degradation in the Z disk of muscle fibre.

In summary, severe deficiency of vitamin E is rarely observed in humans. However, vitamin E deficiency increases oxidative stress in skeletal muscle, alters muscle fibre types, and increases muscle degradation and inflammatory processes leading to dystrophic conditions.

Exercise and oxidative stress

Antioxidant properties and protection of PUFAs from lipid peroxidation are the most widely accepted biological functions of vitamin E. It is the most effective chain-breaking, lipid-soluble antioxidant in the biological membrane where it contributes to membrane stability, regulates fluidity, and protects critical cellular structures against damage from ROS and other free radicals (those containing an unpaired electron).

In normal cellular metabolism, most of the oxygen consumed is utilized in the mitochondrial electron transport system for substrate metabolism and ATP production. However, a small fraction of molecular oxygen ($<2\%$) may escape this process and produce a highly reactive molecules such as superoxide radical ($O_2^{-\cdot}$), hydroxyl radical (OH$\cdot$), or hydrogen peroxide (H_2O_2) (Ji and Leichtweis, 1997). These products can cause membrane lipid peroxidation and loss of membrane integrity. Thus, one of the important roles of vitamin E in the mitochondrial membrane and other phospholipid membranes is to counteract the deleterious effects of free radicals by preventing propagation of membrane lipid peroxidation.

The continuous and increased production of free radicals and their subsequent reaction with other critical cellular components such as DNA has been suggested as contributing to the ageing process and the pathogenesis of chronic diseases and injuries (Halliwell, 1987). These products as well as nitric oxide (NO$\cdot$) free radicals are also produced by activated immune and inflammatory cells. In addition, these products are involved in oxidant-sensitive signal transduction pathways which occur in gene activation.

In mitochondria, ROS are produced from ubiquinone oxidation in the respiratory electron transport system located on the inner mitochondrial membrane (Boveris et al., 1976). A substantial increase in whole-body consumption of oxygen, and particularly in skeletal muscle during exercise, increases the production of ROS. The increase in free radical production by exercise has been documented by direct measurements of free radical production. A marked reduction in respiratory control indices of mitochondrial enzymes was suggested by Davies et al. (1982) as evidence for mitochondrial involvement in the production of superoxide radicals during exercise. They showed that exhaustive exercise in rats resulted in increasing inner membrane leakiness to protons and decreasing energy coupling efficiency. The injection of spin-traps or vitamin E into rats before swimming to exhaustion significantly increased the time to exhaustion, providing indirect evidence for exercise-induced generation of free radicals and their elimination with antioxidants. Pentane, a product of fatty acid lipid peroxidation, has been shown in humans to increase during exercise (Dillard et al., 1978). A higher plasma creatine kinase (CK) concentration and a greater number of neutrophils in circulating blood were found in young subjects (<30 years) when compared with older subjects (>55 years) after an eccentric bout of exercise (Cannon et al., 1990). However, the level of plasma lipid peroxides was not affected by exercise. This was probably the result of rapid clearance of lipid peroxides from plasma. However, from the above study, the authors found that formation of conjugated dienes in muscle and excretion of lipid peroxides in urine increased in both age groups after exercise (Meydani et al., 1993).

Muscle injuries caused by mechanical force or excessive production of ROS from exhaustive exercise activate immune and inflammatory cells, leading to ROS production; this, in turn, contributes to further muscle damage after exercise. Immune cells infiltrate the site of injury to clear debris from injured fibres after strenuous exercise. Activation of chemotactic factors by superoxide radicals, similar to the early manifestation of the acute phase response in infection, has been suggested as occurring after exercise (Petrone et al., 1980). After exercise, effects similar to the initial events of an inflammatory response are induced, including activation of immune complement, subsequent mobilization and activation of neutrophils, production of acute phase proteins, and accumulation of monocytes and macrophages at the site of injury (Cannon and Kluger, 1983; Evans and Cannon, 1991). Neutrophils produce degradative enzymes such as elastase (Kokot et al., 1988) and lysozyme (Morozov et al., 1989) which further break down muscle fibres to be phagocytosed by macrophages and monocytes (Evans and Cannon, 1991); this then generates radicals such as $O_2^{-\cdot}$, $NO^{\cdot}$, $HO^{\cdot}$, H_2O_2, and hypochlorous radical ($HOCl^{\cdot}$) (Bast et al., 1991). The catabolic breakdown of protein and the anabolic utilization of amino acid products for remodeling and the generation of new fibres are continuous processes which may last for several weeks after initial muscle injury (Evans and Cannon, 1991). Therefore, the increased excretion of urinary thiobarbituric acid-reacting substances (TBARS), which are, in part, derived from lipid peroxidation, appear to parallel the proteolytic process which was also elevated 12 days post-exercise (Cannon et al., 1991). In a recent study, a significant increase of DNA strand breakage has been demonstrated in individuals 24 h after exhaustive exercise (Hartmann et al., 1995).

In addition to its antioxidant function, vitamin E has recently been shown to influence cellular response to oxidative stress through the modulation of signal transduction pathways in vascular and immune systems (Azzi et al., 1995). This mechanism of action of vitamin E has been suggested to contribute to its potential anti-cancer, anti-atherogenic effects and lends to its immunostimulatory effects. Some of the effects of vitamin E are believed to be independent of its antioxidant property. Current evidence indicates that vitamin E protects tissues from the harmful effects of ROS and, without doubt, it is an essential antioxidant for normal cell function during and after exercise.

Effect of physical training on antioxidant enzymes

In addition to dietary sources of antioxidants, a well-developed system of antioxidant enzymes exists within skeletal muscle which attenuates oxidative reactions in this metabolically active tissue and eliminates potential exercise-induced damage that arises from free radical formation. Superoxide dismutase (SOD), CAT, glutathione (GSH) peroxidase (GPX), and GSH reductase (GRS) all participate in eliminating ROS and preventing cellular injury.

The activity of antioxidant enzymes in skeletal muscle appears to be both tissue and fibre dependent, as well as closely related to the oxidative capacity. Antioxidant activities vary over a ten-fold range with elevated levels occurring in more metabolically active tissues (Meydani et al., 1996). Both exercise and exercise training in animal models generally increase antioxidant enzyme levels in skeletal muscle (Ji and Leichtweis, 1997).

In addition to enzymatic defences, the contribution of nonenzymatic antioxidants such as vitamin E is vital in prevention of exercise-induced oxidative stress and particularly in skeletal muscle. Tiidus and Houston (1994) determined the effect of endurance training on tissue antioxidant and oxidative enzymatic activities, in five different skeletal muscles of rats fed either normal vitamin E or vitamin E-deficient diets for 16 weeks and then exercised for the final 8 weeks by treadmill running. Vitamin E deprivation, training, or their combination did not generally affect the activities of the antioxidant enzymes, namely SOD, CAT, and GPX, in skeletal muscles. Tiidus et al. (1996) found in humans that although marked increases in maximal oxygen consumption occurred after 8 weeks of aerobic cycling (35 min; three times a week), activities of SOD, CAT and GPX in vastus lateralis skeletal muscle were not affected by the training protocol. In addition, training did not affect the concentrations of vitamin E (γ- and α-tocopherol) in skeletal muscles, suggesting that relatively inactive humans consuming a non-supplemented diet were not highly susceptible to vitamin E depletion. Powers et al. (1994) suggested that antioxidant enzyme adaptation may be dependent on the muscle involved, the intensity and duration of training employed, and the specific antioxidant enzyme measured in untrained humans.

Oxidative stress of exercise and vitamin E

Compelling evidence from animal and human studies has shown that oxidative stress is induced by a variety of exercise modalities (Dillard et al., 1978; Davies et al., 1982; Jackson et al., 1985; Alessio and Goldfarb, 1988; Meydani et al., 1993). Animal studies have demonstrated clearly that vitamin E has a profound protective effect on the oxidative damage of exhaustive exercise (Quintanilha, 1984; Reznick et al., 1992). In endurance trained experimental animals, the concentration of muscle vitamin E was found to be lower (Quintanilha, 1984; Starnes et al., 1989), indicating that vitamin E is needed more when an exercise regimen is adopted.

Running for 31 km has been shown to increase serum levels of conjugated dienes, an index of lipid peroxidation, in trained runners (Vasankari et al., 1997). The level of serum total antioxidant capacity and the concentration of α-tocopherol also rose during a 31-km run and after running a marathon. However, Viguie et al. (1993) reported that submaximal exercise for 3 consecutive days had no effects on plasma vitamin E or urinary 8-hydroxyguanosine (an index of DNA oxidative damage) in moderately trained men. There was a short-term reduction in blood glutathione (GSH) and an increase in oxidized glutathione (GSSG), as well as total and reduced ascorbate during exercise, which returned to baseline during recovery. They concluded that, in healthy young men, submaximal exercise has no long-term effect on blood antioxidants and does not result in DNA damage. Supplemental vitamin E has been shown to prevent exercise-induced free radical formation in cardiac muscle (Kumar et al., 1992). Young males receiving 300 mg α-tocopheryl acetate per day for 4 weeks had reduced leakage of enzymes and a lower blood malondialdehyde (MDA) level, another index of lipid peroxidation, at pre- and post-exercise compared with control (Sumida et al., 1989).

Cannon et al. (1990) reported that supplementation with 800 IU vitamin E/day in older subjects for 7 weeks tended to eliminate the differences from young subjects in CK release in plasma and increase the number of circulating neutrophils after eccentric exercise. Plasma lipid peroxides were not affected either by exercise or by vitamin E supplementation in the two age groups. From the same study, it was also found that vitamin E-supplemented subjects excreted a lower level of urinary TBARS 12 days post-exercise, indicating that vitamin E supplementation may have suppressed oxidative damage induced by eccentric exercise. Further, the protective effect of vitamin E was more prominent in older subjects. Recently, Hartmann et al. (1995) reported that an intake of 1200 mg vitamin E/day for 14 days suppressed the increase in DNA damage in four of five subjects after an exercise bout. It is worth noting that a few studies have found no protective effect of vitamin E supplementation on exercise-induced oxidative damage. Trained marathon runners were given α-tocopherol (400 IU/day) and ascorbic acid (200 mg/day) for 4.5 weeks before a marathon race (Rokitzki et al., 1994b). There were no differences in plasma lipid peroxides or lactate dehydrogenase (LDH) 24 h after the race compared with the subjects who had received a placebo. However, a significantly smaller increase of serum CK was observed in the supplemented groups compared with the placebo group.

In conclusion, oxidative stress is increased with exercise, which can be detected in body fluids and tissues. Animal and human studies have suggested that dietary vitamin E supplementation may prevent oxidative stress associated with exercise. Nevertheless, it appears that short-term vitamin E supplementation may not totally overcome the overwhelming oxidative stress produced from exhaustive exercise such as marathon running.

Vitamin E and exercise performance – does it help?

Coupled with proper training and sound coaching, "proper" nutrition is the cornerstone to success in athletic performance. Researchers and athletes alike have striven for decades for the perfect nutritional manipulations to achieve success in sports (for a complete review see Singh, 1992). With the well-described increases in ROS during exercise, coaches and athletes (as well as scientists) have speculated about the possible benefits of vitamin E supplementation during heavy training. Many nutritional supplements geared towards athletes purport that antioxidant vitamin requirements are significantly increased in athletic populations and have propagated unsubstantiated claims of improved performance. On the other hand, many high-caliber endurance athletes consume and require a diet high in carbohydrate (Costill, 1988), eliciting questions about the adequacy of their vitamin E intake.

Although nutritional survey methods for assessing micronutrient requirements can be problematic in the assessment of vitamin status, several studies have employed nutritional surveys to examine the nutrient intakes of athletes. A survey of 22 dietary intake studies of elite endurance and power athletes revealed generally adequate vitamin E intakes. However, 60–90% of the male athletes and 80–100% of the female athletes reported use of multivitamin and mineral supplementation (Economos et al., 1993). Other studies have also reported a high prevalence of vitamin supplement use among athletes (Nowak et al., 1988; Van Erp-Baart et al., 1989; Williams, 1989; Bazzarre et al., 1993). In a randomized, controlled, double-blind placebo trial of multivitamin supplementation on nutritional and running performance, Weight et al. (1988a, b) observed no improvements in nutritional status or running performance after 9 months of supplementation. Despite relatively high carbohydrate intake, particularly in endurance-trained athletes, vitamin E status appeared to be adequate.

Several clinical trials have assessed the effects of vitamin E supplementation on exercise or work performance. In male prisoners performing hard physical labor, a vitamin E-deficient diet induced a rapid drop in plasma tocopherol, suggesting an increased dietary requirement for vitamin E in individuals engaged in physically demanding occupations (Bunnell et al., 1975). However studies on supplementing individuals who already consume an adequate intake of vitamin E have not shown ergogenic effects. Two studies by Sharman and colleagues (Shephard et al., 1974; Sharman et al., 1976) reported no significant effects of vitamin E supplementation (400 mg α-tocopheryl acetate/day) on swimming (400 m) and running (1 mile run) performance in a group of boys who participated in 6 weeks of swim training, when compared with control subjects who participated in identical training with no supplementation. These results were also confirmed by Lawrence et al. (1975) who fed a group of well-trained swimmers 600 IU vitamin E (α-tocopheryl acetate)/day or placebo for 6 months and found no differences in 500-yard performance times.

Similar effects of vitamin E have been reported by other investigators (Shephard et al., 1974; Watt et al., 1974). Sumida et al. (1989) supplemented male college students with 300 mg vitamin E (α-tocopheryl acetate)/day for 4 weeks and observed no effects of vitamin E supplementation on maximal aerobic capacity or exercise time to exhaustion during an incremental cycle ergometer exercise test. Despite no effects on aerobic capacity or performance, after vitamin E supplementation the subjects had lower post-exercise increases in lipid peroxidation and serum β-glucuronidase levels, which suggests an enhanced radical quenching effect and less subsequent oxidative damage.

Low atmospheric pressures, such as those encountered on expeditions at high altitudes, may increase lipid peroxidation during exercise and increase antioxidant requirements. In one study, Simon-Schnass and Pabst (1988) observed significant reductions in expired pentane production in mountaineers supplemented with 200 mg vitamin E/day (α-tocopheryl acetate) during residence at high altitude (43 days at 5000 m). In addition, there was a small but statistically significant improvement in the percentage of maximum oxygen consumption ($VO_{2\,max}$) at which the anaerobic threshold occurred. The mechanisms by which vitamin E exerts a performance-enhancing effect at high altitude are not known. These results require further confirmation before widespread recommendations regarding vitamin E supplementation for athletes competing or training at elevated altitude are made.

Studies have also been conducted examining the effects of vitamin E supplementation on exercise performance in elite athletes. Rokitzki et al. (1994a) studied the effects of vitamin E supplementation (330 mg/day of α-tocopheryl acetate) on exercise tolerance in elite male cyclists. They reported no effects of vitamin E supplementation on the blood lactate response or the heart rate response to an incremental exercise test on a cycle ergometer, despite lower measures of lipid peroxidation during exercise in the vitamin E-supplemented group. More recently, Snider et al. (1992) have reported that ingestion of a multi-supplement (100 mg coenzyme Q_{10}, 500 mg cytochrome c, 100 mg inosine, and 200 IU vitamin E) for 4 weeks had no effect on exercise time to exhaustion at 70% $VO_{2\,max}$ (90 min of treadmill running, stationary cycling to exhaustion) in competitive triathletes. With the exception of one study conducted at high altitude (Simon-Schnass and Pabst, 1988), no studies have shown a performance-enhancing effect with vitamin E supplementation; their use by athletes to enhance performance cannot therefore be objectively confirmed and at present should not be recommended. However, in those studies that also measured indices of lipid peroxidation, vitamin E supplementation universally lowers peroxidation.

The long-term beneficial consequences of this reduced lipid peroxidation are not known and certainly warrant future study. In addition, individuals and athletes who, through their dietary habits, consume low levels of antioxidants may be at increased risk of the harmful effects of oxygen radicals.

Although there are no proven effects of vitamin E supplementation, the prevalence of vitamin use among athletes is high. Future studies are needed to assess whether antioxidant vitamins (specifically vitamin E) may speed recovery during heavy training or enhance performance in specific environmental extremes.

As with any nutritional supplement that has a potential interaction with metabolic functions during exercise, some studies have examined the role of vitamin E as a possible ergogenic agent. Although it is well established that during exercise measures of free radical generation and lipid peroxidation increase and vitamin E supplementation reduces free radical production and indices of lipid peroxidation during exercise, there is little evidence to suggest that the reduction in radical formation and/or lipid peroxide formation will enhance exercise performance or improve time for recovery from exhaustive exercise.

Conclusion

Vitamin E is essential for normal muscle function. Although deficiency of vitamin E is rare in humans, its presence may cause neuromuscular dysfunction. In animals, vitamin E deficiency results in muscle fibre loss and dystrophy. During and after strenuous exercise activity, this vitamin is utilized in concert with other antioxidant defences to neutralize the oxygen free radicals produced by skeletal muscles and/or immune cells in response to physical activity or injuries. ROS may also be responsible for muscle fibre loss through apoptosis. Vitamin E, as a strong antioxidant can modulate the oxidant/antioxidant balance in muscle fibres, influence muscle fibre metabolism, and possibly play an important role in the apoptotic process of fibres warranting further investigation.

Most of the population of the US consume foods that provide adequate amounts of vitamin E to meet the daily requirement. However, this requirement may be higher when oxidative stress is induced by higher physical activity or other conditions. There is an inverse relationship between BMI and muscle concentration of vitamin E. Individuals with sedentary lifestyles typically have a higher BMI, and thus more body fat, will accumulate most of the vitamin E in adipose tissues when they use supplements. However, this may not be the case in athletes and body builders. Even though they may have a high BMI, lean body mass is still a major component of body composition. On other hand, athletes who usually maintain a different level of total energy intake and have adapted to different dietary habits, may be at greater risk for antioxidant imbalance without supplement use, particularly those who maintain high-carbohydrate, low-fat regimens and engage in strenuous exercise. Further, even though training may increase endogenous antioxidant capacity, it may not be adequate to meet the challenge of increased oxidative stress. Therefore, vitamin E and other dietary antioxidants may have a complementary role in these situations.

Untrained individuals and elderly people may benefit even more from supplemental intake of vitamin E when engaged in physical activity such as weekend workouts to which they are unaccustomed. Compelling evidence from animal studies has indicated that vitamin E supplementation significantly reduces oxidative stress produced by physical activity. Very few controlled studies have shown this effect in humans and, at present, only limited evidence has demonstrated reductions in the incidence of oxidative stress with vitamin E supplementation. However, studies do not support the notion that vitamin E supplementation increases training capacity or improves performance in trained individuals. At present, the totality of evidence indicates beneficial effects from long-term vitamin E supplementation in lowering oxidative stress and associated risk factors which are believed to be involved in the pathogenesis of many disease processes. This aspect of the relationship between vitamin E and exercise warrants further investigation.

Acknowledgements
The authors would like to thank Timothy S. McElreavy, MA, for preparation of this manuscript. This project has been funded at least in part with Federal funds from the US Department of Agriculture, Agricultural Research Service under contract number 53-K06-01. The contents of this publication do not necessarily reflect the views or policies of the US Department of Agriculture, nor does mention of trade names, commercial products, or organizations imply endorsement by the US government.

References

Acuff RV, Thedford SS, Hidiroglou NN, Papas AM and Odom TAJ (1994) Relative bioavailability of RRR- and all-rac-α-tocopheryl acetate in humans: studies using deuteriated compounds. *Am J Clin Nutr* 60: 397–402.
Alessio HM and Goldfarb AH (1988) Lipid peroxidation and scavenger enzymes during exercise: adaptive response to training. *J Appl Physiol* 64: 1333–1336.
Azzi A, Boscoboinik D, Marilley D, Ozer NK, Stable B and Tasinato A (1995) Vitamin E: A sensor and an information transducer of the cell oxidation state. *Am J Clin Nutr* 62: 1337S–1346S.
Bast A, Haenin GRMM and Doelman CJA (1991) Oxidants and antioxidants: State of the art. *Am J Med* 91: 3C–2S.
Bazzarre TL, Scarpino A, Sigmon R, Marquart LF, Wu SM and Izurieta M (1993) Vitamin-mineral supplement use and nutritional status of athletes. *J Am Coll Nutr* 12: 162–169.
Boveris A, Cadenas E and Stoppani AOK (1976) Role of ubiquinone in mitochondrial generation of hydrogen peroxide. *Biochem J* 156: 435–444.
Bunnell RH, DeRitter E and Rubin SH (1975) Effects of feeding polyunsaturated fatty acids with a low vitamin E diet on blood levels of tocopherol in men performing hard physical labor. *Am J Clin Nutr* 28: 706–711.
Cain K, Inayat-Hussain SH, Kokileva L and Cohen GM (1995) Multi-step DNA cleavage in rat liver nuclei is inhibited by thiol reactive agents. *FEBS Lett* 358: 255–261.
Cannon JG and Kluger MJ (1983) Endogenous pyrogen activity in human plasma after exercise. *Science* 220: 617–619.
Cannon JG, Orencole SF, Fielding RA, Meydani M, Meydani SN, Fiatarone MA, Blumberg JB and Evans WJ (1990) The acute phase response in exercise I: The interaction of age and vitamin E on neutrophils and muscle enzyme release. *Am J Physiol* 259: R1214–R1219.
Cannon JG, Meydani SN, Fielding RA, Fiatarone MA, Meydani M, Farhangmehr M, Orencole SF, Blumberg JB and Evans WJ (1991) Acute phase response in exercise. II. Associations between vitamin E, cytokines, and muscle proteolysis. *Am J Physiol* 260: R1235–R1240.
Costill DL (1988) Carbohydrates for exercise: Dietary demands for optimal performance. *Int J Sport Med* 9: 1–18.
Davies KJA, Packer L and Brooks GA (1982) Free radicals and tissue damage produced by exercise. *Biochem Biophys Res Commun* 107: 1198–1205.
Dayton WR, Schollmeyer JV, Chan AC and Allen CE (1979) Elevated levels of a calcium-activated muscle protease in rapidly atrophying muscles from vitamin E-deficient rabbits. *Biochim Biophys Acta* 584: 216–230.
De Villers A, Simard P and Srivastava U (1973) Biochemical changes in progressive muscular dystrophy. X. Studies on the biosynthesis of protein and RNA in cellular fractions of the skeletal muscle of normal and vitamin E deficient rabbits. *Can J Biochem* 51: 450–459.
Dillard CJ, Litov RE, Savin WM, Dumelin EE and Tappel AL (1978) Effects of exercise, vitamin E, and ozone on pulmonary function and lipid peroxidation. *J Appl Physiol* 45: 927–932.

Economos CD, Bortz SS and Nelson ME (1993) Nutritional practices of elite athletes: Practical recommendations. *Sport Med* 16: 381–399.

Evans WJ and Cannon JG (1991) The metabolic effects of exercise-induced muscle damage. *In*: JO Holloszy (ed.): *Exercise and Sport Sciences Reviews*, Vol. 19. Williams and Wilkins, Baltimore, pp 99–126.

Federico A, Battisti C, Eusebi MP, de Stefano N, Malandrini A, Mondelli M and Volpi N (1991) Vitamin E deficiency secondary to chronic intestinal malabsorption and effect of vitamin supplement: A case report. *Eur Neurol* 31: 366–371.

Ferraro RT, Eckel RH, Larson DE, Fontvieilli A-M, Rising R, Jensen DR and Ravussin E (1993) Relationship between skeletal muscle lipoprotein lipase activity and 24-hour macronutrient oxidation. *J Clin Invest* 95: 1846–1853.

Forrest VJ, Kang Y-H, McClian DE, Robinson DH and Ramakrishnan N (1994) Oxidative stress-induced apoptosis is prevented by Trolox. *Free Radical Biol Med* 16: 675–684.

Fry JM, Smith GM and Speijers EJ (1993) Plasma and tissue concentration of alpha-tocopherol during vitamin E depletion in sheep. *Brit J Nutr* 69: 225–232.

Garry PJ, Goodwin JG, Hunt WC, Hooper EM and Leonard AG (1982) Nutritional status in a healthy elderly population: dietary and supplemental intakes. *Am J Clin Nutr* 36: 319–331.

Halliwell B (1987) A radical approach to human disease. *In*: B Halliwell (ed.): *Oxygen Radicals and Tissue Injury*. FASEB, Bethesda, MD, pp 139–143.

Hartmann A, Niess AM, Grunert-Fuchs M, Poch B and Speit G (1995) Vitamin E prevents exercise-induced DNA damage. *Mutat Res* 346: 195–202.

Jackson MJ, Edwards RHT and Symons MCR (1985) Electron spin resonance studies of intact mammalian skeletal muscle. *Biochim Biophys Acta* 847: 185–190.

Jenkins RR, Newsham D, Rushmore P and Tengie J (1982) Effect of disuse on the skeletal muscle catalase of rats. *Biochem Med* 27: 195–199.

Ji LL and Leichtweis S (1997) Exercise and oxidative stress: Sources of free radicals and their impact on antioxidant systems. *Age* 20: 91–106.

Kokot K, Schaefer RM, Teschner M, Gilge U, Plass R and Heidland A (1988) Activation of leukocytes during prolonged physical exercise. *Adv Exp Med Biol* 240: 57–63.

Kumar CT, Reddy VK, Prasd M, Thyagaraju K and Reddanna P (1992) Dietary supplementation of Vitamin E protects heart tissue from exercise-induced oxidant stress. *Mol Cell Biochem* 111: 109–115.

Lammi-Keefe CJ, Hegarty PVJ and Swan PB (1980) Effect of starvation and refeeding on catalase and superoxide dismutase activities in skeletal and cardiac muscles from 12-month-old rats. *Experientia* 37: 25–27.

Lawrence JD, Bower RC, Riehl WP and Smith JL (1975) Effects of α-tocopherol acetate on the swimming endurance of trained swimmers. *Am J Clin Nutr* 28: 205–208.

Lazaro RP, Dentinger MP, Rodichok LD, Barron KD and Satya-Murti S (1986) Muscle pathology in Bassen-Kornzweig Syndrome and vitamin E deficiency. *Am J Clin Pathol* 86: 378–387.

Lehmann J, Martin HL, Lashley EL, Marshall MW and Judd JT (1986) Vitamin E in foods from high and low linoleic acid diets. *J Am Diet Assoc* 86: 1208–1216.

Meydani M (1995) Antioxidant Vitamins. *Front Clin Nutr* 4: 7–14.

Meydani M (1995a). Vitamin E. *Lancet* 345: 170–175.

Meydani M and Blumberg JB (1992) Vitamin E. *In*: I Rosenberg and S Hartz (eds): *Boston Nutrition Status Survey*. Smith-Gordon, London, pp 103–109.

Meydani M, Cohn JS, Macauley JB, McNamara JR, Blumberg JB and Schaefer EJ (1989) Postprandial changes in the plasma concentration of α- and γ-tocopherol in human subjects fed fat-rich meal supplemented with fat-soluble vitamins. *J Nutr* 119: 1252–1258.

Meydani M, Evans WJ, Handelman G, Biddle L, Feilding RA, Meydani SN, Burrill J, J, F, Fiatarone MA, Blumberg JB and Cannon JG (1993) Protective effect of vitamin E on exercise-induced oxidative damage in young and older adults. *Am J Physiol* 33: R992–R998.

Meydani M, Fielding RA and Fotouhi N (1996) Vitamin E. *In*: I Wolinsky and JA Driskell (eds): *Sports Nutrition: Vitamins and Trace Elements*. CRC Press, New York, pp 119–135.

Morozov VI, Priatkin SA and Nazarov IB (1989) Secretion of lysozyme by blood neutrophils during physical exertion. *Fiziol Zh SSSR* 75: 334–337.

Murphy SP, Subar AF and Block G (1990) Vitamin E intake and sources in the United States. *Am J Clin Nutr* 52: 361–367.

National Research Council (1989) *Recommended Dietary Allowances*. National Academy Press, Washington DC, p. 284.

Neville HE, Ringel SP, Guggenheim MA, Wehling CA and Starcevich JM (1983) Ultrastructural and histochemical abnormalities of skeletal muscle in patients with chronic vitamin E deficiency. *Neurology* 33: 483–488.

Niki E (1987) Interaction of ascorbate and α-tocopherol. *Ann N Y Acad Sci* 498: 186–199.

Nowak RK, Knudsen KS and Schulz LO (1988) Body composition and nutrient intakes of college men and women basketball players. *J Am Diet Assoc* 88: 575–578.

Petrone WF, English DK, Wong K and McCord JM (1980) Free radicals and inflammation: superoxide dependent chemotactic factor in plasma. *Proc Natl Acad Sci USA* 77: 1159.

Pette D and Spamer C (1986) Metabolic properties of muscle fibers. *Fed Proc* 45: 2910–2914.

Pillari SR, Traber MG, Kayden HJ, Cox NR, Toivio-Kinnucan M, Wright JC, Braund KG, Whitley RD, Gilger BC and Steiss JE (1994) Concomitant brainstem anoxal dystrophy and necrotizing myopathy in vitamin E-deficient rats. *J Neurol Sci* 123: 64–73.

Powers SK, Criswell D, Lawler J, Ji LL, Martin D, Herb RA and Dudley G (1994) Influence of exercise and fiber type on antioxidant enzyme activity in rat skeletal muscle. *Am J Physiol* 266: R375–R380.

Quintanilha AT (1984) Effects of physical exercise and/or vitamin E on tissue oxidatve metabolism. *Biochem Soc* 12: 403–404.

Quintanilha AT and Packer L (1983) Vitamin E, physical exercise and tissue oxidative damage. *Biology of Vitamin E*. Pitman, London, pp 56–69.

Reznick AZ, Witt E, Matsumoto M and Packer L (1992) Vitamin E inhibits protein oxidation in skeletal muscle of resting and exercised rats. *Biochem Biophys Res Commun* 189: 801–806.

Riley DA, Ellis S and Bain JL (1988) Catalase-positive microperoxisomes in rat soleus and extensor digitorum longus muscle fiber types. *J Histochem Cytochem* 36: 633–637.

Rokitzki L, Logemann E, Huber G, Keck E and Keul J (1994a) α-Tocopherol supplementation in racing cyclists during extreme endurance training. *Int J Sport Nutr* 4: 253–264.

Rokitzki L, Logemann E, Sagredos AN, Murphy M, Wetzel-Roth W and Keul J (1994b) Lipid peroxidation and antioxidative vitamins under extreme endurance stress. *Acta Physiol Scand* 151: 149–158.

Salviati G, Betto R, Margreth A, Novello F and Bonetti E (1980) Differential binding of vitamin E to sarcoplasmic reticulum from fast and slow muscles of the rabbit. *Experientia* 36: 1140–1141.

Sandri M, Carraro U, Podhorska-Okolov M, Rizzi C, Arslan P, Monti D and Franceschi C (1995) Apoptosis, DNA damage and ubiquitin expression in normal and mdx muscle fibers after exercise. *FEBS Lett* 373: 291–295.

Sharman IM, Down MG and Norgan NG (1976) The effects of vitamin E on physiological function and athletic performance of trained swimmers. *J Sport Med* 16: 215–225.

Shephard RJ, Campbell R, Pimm P, Stuart D and Wright GR (1974) Vitamin E, exercise, and the recovery from physical activity. *Eur J Appl Physiol* 33: 119–126.

Shih JCH, Jonas RH and Scott ML (1977) Oxidative detrioration of the muscle proteins during nutritional muscular dystrophy in chicks. *J Nutr* 107: 1786–1791.

Simon-Schnass I and Pabst H (1988) Influence of vitamin E on physical performance. *Int J Vit Nutr Res* 58: 49–54.

Singh VN (1992) A Current Perspective on Nutrition and Exercise. *J Nutr* 122: 760–765.

Snider IP, Bazzarre TL, Murdoch SD and Goldfarb A (1992) Effects of coenzyme athletic performance system as an ergogenic aid on endurance performance to exhaustion. *Int J Sport Nutr* 2: 272–286.

Starnes JW, Cantu G, Farrar RP and Kehrer JP (1989) Skeletal muscle lipid peroxidation in exercised and food-restricted rats during aging. *J Appl Physiol* 67: 69–75.

Sumida S, Tanaka K, Kitao H and Nakadomo F (1989) Exercise-induced lipid peroxidation and leakage of enzymes before and after vitamin E supplementation. *Int J Biochem* 21: 835–838.

Thompson CB (1995) Apoptosis inthe pathogenesis and treatment of disease. *Science* 267: 1456–1462.

Tiidus PM and Houston ME (1994) Antioxidant and oxidative enzyme adaptation to vitamin E deprivation and training. *Med Sci Sports Exerc* 26: 354–359.

Tiidus PM, Pushkarenko J and Houston ME (1996) Lack of antioxidant adaptation to short-term aerobic training in human muscle. *Am J Physiol* 274: R832–R836.

Traber MG, Cohn W and Muller DPR (1993) Absorption, transport and delivery to tissues. *In*: L Packer and J Fuchs (eds): *Vitamin E in Health and Disease*. Marcel Dekker, New York, pp 35–51.

Van Erp-Baart AM, Saris WM, Binkhorst RA, Vos JA and Elvers JW (1989) Nationwide survey on nutritional habits in elite athletes. Part II. mineral and vitamin intake. *Int J Sport Med* 10: S11–16.

Van Fleet JF and Ferrans VJ (1976) Ultrastructural changes in skeletal muscle of selenium-vitamin E-deficient chicks. *Am J Vet Res* 37: 1081–1089.

Van Fleet JF and Ferrans VJ (1977) Ultrastructural alterations in skeletal muscle of ducklings fed selenium-vitamin E-deficient diet. *Am J Vet Res* 38: 1399–1405.

Vasankari TJ, Kujala UM, Vasankari TM, Vuorimaa T and Ahotupa M (1997) Effects of acute prolonged exercise on serum and LDL oxidation and antioxidant defenses. *Free Radical Biol Med* 22: 509–513.

Viguie CA, Frei B, Shigenaga MK, Ames BN, Packer L and Brooks GA (1993) Antioxidant status and indexes of oxidative stress during consecutive days of exercise. *J Appl Physiol* 75: 566–572.

Watt T, Romet TT, McFarlane I, McGuey D, Allen C and Goode RC (1974) Vitamin E and oxygen consumption. *Lancet* 2: 354–355.

Weight LM, Myburgh KH and Noakes TD (1988) Vitamin and mineral supplementation: effect on the running performance of trained athletes. *Am J Clin Nutr* 47: 192–195.

Weight LM, Noakes TD, Labadarios D, Graves J, Jacobs P and Berman PA (1988a). Vitamin and mineral status of trained athletes including the effects of supplementation. *Am J Clin Nutr* 47: 186–191.

Williams MH (1989) Vitamin supplementation and athletic performance. *Int J Vit Nutr Res* 30: 163–191.

Differential susceptibility of skeletal muscle proteins to free radical-induced oxidative damage *in vitro*

J.W. Haycock[1], G. Falkous[2] and D. Mantle[2]

[1]*University Department of Medicine, Clinical Sciences Centre, Northern General Hospital, Sheffield, S5 7AU, UK*
[2]*Neurochemistry Department, Regional Neurosciences Centre, Newcastle General Hospital, Newcastle upon Tyne, NE4 6BE, UK*

Summary. An alternative experimental approach to investigation of the role of free radicals in muscle tissue damage is described, based on studies of oxidative protein damage *in vitro* induced via ^{60}Co γ irradiation. Muscle proteins in purified form, muscle tissue extracts or muscle tissue sections were used as targets for oxidative damage, with analysis via electrophoresis, histochemistry/immunocytochemistry and electron microscopy. Particular attention was focused on the potential interplay between free radical damage and the activity of endogenous proteases.
Muscle proteins in purified form showed differing comparative susceptibility to oxidative damage by OH˙ or $O_2^{-˙}$, suggesting that oxidative susceptibility in proteins must in part be encoded by differences in structure (particularly secondary/tertiary). There was no evidence to suggest that pre-exposure of contractile/cytoskeletal muscle proteins to OH˙ or $O_2^{-˙}$ increases susceptibility to subsequent generalized proteolysis by endogenous cytoplasmic or lysosomal enzymes. These data support the concept that oxidatively damaged proteins are degraded by specific proteolytic enzyme types (e.g. macropain), rather than by more generalized proteolytic action. It was, however, possible to demonstrate that proteolytic enzymes as a group are broadly susceptible themselves to oxidative damage *in vitro*; thus the potential interaction of free radicals in intracellular protein turnover *in vivo* may be mediated via their effects on proteolytic enzyme activities, as well as on the protein substrates of these enzymes.
Histochemical/immunocytochemical analysis of irradiated muscle tissue sections (in which the three-dimensional architecture of the muscle cell is retained) showed differential susceptibility of proteins (differing in function/localization) to oxidative damage, with certain mitochondrial proteins showing particular susceptibility. In addition, the use of monoclonal antibodies to different regions of the protein dystrophin showed the latter to contain both resistant and susceptible regions to oxidative damage. At the ultrastructural level, mitochondria were identified as being particularly susceptible to oxidative damage.

Introduction

Oxygen free radicals, classified under the more general term of reactive oxygen species (ROS, which includes non-radical species such as hydrogen peroxide), are highly reactive transient chemical species formed in all tissues during normal aerobic metabolism. ROS have the potential to damage the various intracellular organelles and components (nucleic acids, lipids and proteins) on which normal cell functioning depends. During normal metabolism the major source of ROS results from leakage of electrons from the mitochondrial respiratory chain; in addition, ROS may be generated via the action of various oxidase enzymes, and by phagocyte activation during infection/inflammation. Cells are protected from ROS-induced damage by a variety of endogenous ROS-scavenging proteins, enzymes and chemical compounds (Del Maestro, 1980; Freeman and Crapo, 1982). Cellular damage arising from an imbalance between these ROS-generating and -scavenging systems has been implicated in the pathology of a diverse range of human disorders, including muscular dystrophy (Murphy and Kehrer, 1986; Halliwell and Gutteridge, 1990). Muscle tissue is unique in its requirement and ability to undertake very rapid and coordinated changes in energy supply and oxygen flux during contraction, and it has been suggested that this renders the tissue particularly prone to ROS-mediated damage, as a result of increased electron flux and corresponding leakage from the mitochondrial respiratory chain (Jackson and O'Farrell, 1993).

A potential role for ROS in the pathogenesis of human muscular dystrophy was first suggested after the recognition of a form of muscular dystrophy in animals induced by nutritional deficiency of vitamin E or selenium (Bradley and Fell, 1980). Duchenne muscular dystrophy (DMD) is a fatal degenerative disorder of muscle, resulting from an Xp-21-linked genetic defect, manifest by aberrant production of the sarcolemmal-associated protein dystrophin (Koenig et al., 1988; Evasti and Campbell, 1993). The mechanism by which absence of this protein results in degeneration of muscle, which is characteristic of DMD, is unknown. It has been suggested that absence of dystrophin and the results abnormal membrane structure causes altered homoeostasis of intracellular calcium, which, in turn, results in excessive generation of ROS via activation of the xanthine dehydrogenase/oxidase system, and subsequent oxidative damage to key intracellular components (Austin, 1990). Direct experimental confirmation of such hypotheses is difficult; because of their high reactivity (second-order rate constants $10^6 - 10^9$ $m^{-1}s^{-1}$) ROS have very short half-lives ($10^{-9} - 10^{-4}$ s), making direct analysis in tissue samples extremely difficult, particularly with human subjects (Jackson et al., 1984; Halliwell and Grootveld, 1987).

Much of the previous research into the potential role of ROS in the pathogenesis of muscle disease has therefore been based on indirect analytical methods for quantifying ROS-induced damage products in experimentally induced or inherited muscle degeneration in animal models of disease. However, a number of problems have become apparent in the interpretation of such data, on which the hypothesis of ROS involvement in muscular dystrophy has been based. These include: (1) the choice of unsuitable animal models of relevance to human muscular dystrophy, e.g. the dystrophic chicken (Muzuno, 1988) which is now known to result from a different genetic defect to that responsible for DMD; (2) the quantification of ROS-induced damage by estimation of lipid peroxidation products via coupled reaction with thiobarbituric acid (Mechler et al., 1984), the methodology of which is now known to be subject to artefact (Marshall et al., 1985); and (3) changes in antioxidant levels in muscle or blood of DMD patients are variously reported to show increases, decreases or unchanged levels (Austin et al., 1992). Such difficulties in data interpretation are further compounded by the fact that relatively little is known about the basic mechanisms by which muscle cells are damaged by ROS; in particular, although data are available on the interaction of ROS with nucleic acids and lipids, relatively little is known about the action of ROS on key structural and enzymic proteins on which normal cell functioning depends.

In view of the above problems, we have adopted an alternative experimental approach to investigation of the role of ROS in muscle cellular damage, based on ROS generation under controlled experimental conditions *in vitro* using a ^{60}Co γ radiology source. There are a number of potential methods for generating ROS *in vitro*, the relative advantages and disadvantages of which have been reviewed (Stadtman, 1990). For the work described in this investigation, generation of ROS via γ radiolysis was chosen because of the degree of radical specificity and quantification of dosage possible. A similar experimental approach has been successfully applied to the investigation of ROS-induced oxidative damage of model proteins (in purified form) *in vitro*, such as albumin, lysozyme and haemoglobin (Davies, 1987; Davies et al., 1987), but this has not been applied to the investigation of ROS-induced oxidative damage of muscle proteins (particularly cytoskeletal/contractile). Previous attempts to treat DMD patients with ROS-scavenging compounds or enzymes have proved unsuccessful (Stern et al., 1982), and it can be argued that these failures result in part from a lack of understanding of the basic mechanisms by which ROS damage muscle cells. The development of improved novel therapeutic strategies for DMD based on antioxidants would benefit in particular from a clearer understanding of the differential susceptibility of muscle proteins to ROS-induced oxidative damage, and whether susceptible proteins could be protected from oxidative damage using specific ROS-scavenging agents. Much of the research into treatment of DMD has centred on gene transfer and/or myoblast transfer therapy (Mulligan, 1993); however, in view of the technical difficulties that have become apparent with these approaches, it would be useful to have alternative, pharmacologically based, therapeutic strategies available which rely on administration of appropriate antioxidants. The objectives of the present investigation were therefore: (1) to determine the relative susceptibility of different categories of muscle proteins, in purified form or tissue homogenates, to oxidative damage by $OH^{.}$ or $O_2^{-.}$; (2) to determine whether exposure of muscle proteins (contractile, cytoskeletal) to ROS increases their susceptibility to subsequent degradation by cytoplasmic or lysosomal muscle proteases; (3) to determine whether muscle proteases themselves may have differential susceptibility to ROS-induced oxidative damage; and (4) to investigate ROS-induced oxidative damage to muscle proteins in tissue sections in which the three-dimensional architecture of the muscle cell is retained, via histological and ultrastructural analysis.

Materials and methods

All reagents, including muscle proteins in purified form, substrates for muscle protease assay, polyacrylamide gel electrophoresis, histological analysis and general reagents were obtained from Sigma Chemical Co., Poole, UK (unless otherwise specified) and were of analytical grade where available. Human muscle tissue (rectus abdominis) for experiments using tissue homogenates or tissue sections was obtained from patients undergoing elective surgery, with no history of neuromuscular disease, and stored at −196 °C. All solutions for irradiation were made up in water that had been distilled three times.

Preparation of samples for γ radiolysis

Samples were prepared in 50 mM phosphate buffer pH 7.4 (for subsequent generation of $OH^{.}$), or the same buffer containing 20 mM sodium formate (for subsequent generation of $O_2^{-.}$). Proteins in purified form (myosin, actin, α-actinin, troponin, tropomyosin, myoglobin, creatine kinase, catalase, choline acetyltransferase, and albumin) were dissolved in buffer at a concentration of 0.1–1 mg/ml. Samples of muscle tissue were homogenized 1:10 (w/v) tissue:buffer using an Ultra-Turrax homogenizer; after centrifugation (15 000 g, 20 min) pellets were washed three times and resuspended with an equivalent volume of extraction buffer. Pellets (comprising contractile and cytoskeletal proteins) and supernatants (soluble enzymes, including proteases) were retained for analysis. For histochemistry, 10-μm thick frozen transverse muscle sections were made using a cryostat (Leitz Wetzlar, model 1720) at $-20\ °C$, and mounted on chromic acid-cleaned slides. For immunocytochemistry, 6-μm thick frozen transverse muscle sections were made as above, mounted on gelatinized slides (0.05% w/v gelatin; 0.05% w/v chrome alum) and allowed to air dry at room temperature for 1 h. For electron microscopy, muscle samples were oriented and cut under tension as described previously (Cullen and Mastaglia, 1980).

ROS generation via ^{60}Co γ radiolysis

The method used for generation of OH$^{.}$ or $O_2^{-.}$ radicals *in vitro* by ^{60}Co γ radiolysis of aqueous solutions was based on that of Davies (1987), in which both theoretical and methodological aspects of the procedure are described in detail. Briefly, samples in buffer gassed to saturation with N_2O or O_2 (in the presence of 20 mM sodium formate) were used for exclusive generation of OH$^{.}$ or $O_2^{-.}$ species respectively. Samples were irradiated at a dose rate equivalent to 99 krads/hour (determined via standard dosimetric technique; Fricke and Hart, 1966) over a dosage range of about 10–2500 krads. For histological analysis, slide-mounted tissue sections were placed into plastic slide containers (Cellpath, Hemel Hempstead, UK) and completely immersed in buffer saturated with N_2O or O_2, and irradiated as above. Additional samples, processed as above but not irradiated, served as appropriate comparative controls for the effects of OH$^{.}$ or $O_2^{-.}$ species respectively.

Analysis of ROS-induced oxidative protein damage *in vitro*

Electrophoretic analysis

Oxidative damage to target proteins in purified form (myosin, actin, α-actinin, troponin, tropomyosin, myoglobin, and albumin) or muscle tissue homogenates by OH$^{.}$ or $O_2^{-.}$ after irradiation was assessed via sodium dodecyl sulfate (SDS) polyacrylamide gel electrophoresis (SDS-PAGE; Laemli, 1970), using 14×10 cm slab gels, with 5% acrylamide stacking gel and 5–20% linear acrylamide gradient in the separating gel. Samples were diluted 1:1 with treatment buffer and boiled for 5 min. Samples containing the equivalent of 10 μg purified protein or 150 μg tissue homogenate were run at 20 mA/gel slab for 1 h, followed by 40 mA/gel slab for 2 h (equivalent to $1\times$ tracking dye migration distance). After electrophoresis, gels were fixed by immersion in 20% (w/v) trichloroacetic acid, stained for 30 min in Coomassie brilliant blue (0.25% (w/v) in acetic acid/water/methanol, 1:5:5 by volume) and destained in 10% (v/v) acetic acid. Protein band staining intensities were quantified via scanning densitometry using a Camag gel scanner with version III programme software ($\lambda=492$ nm, reflectance mode).

Oxidative protein damage to purified forms of creatine kinase (Pearce et al., 1964), catalase (Cohen et al., 1970) and choline acetyltransferase (Tucek, 1982) induced by OH$^{.}$ or $O_2^{-.}$ after irradiation was assessed by changes in activity using specific assay methods above.

Proteolytic enzyme assays

To determine the effect of OH˙ or O_2^{-}˙ directly on protease activities, individual enzyme activities were determined in muscle tissue supernatants after irradiation for the following enzyme types, based on hydrolysis of specific fluorometric substrates as described previously (Blanchard et al., 1993): alanyl-, arginyl-, leucyl- and pyroglutamyl aminopeptidases; dipeptidyl aminopeptidase IV, tripeptidyl aminopeptidase; proline endopeptidase (cytoplasmic proteases); dipeptidyl aminopeptidases I and II; cathepsins B, D, H and L (lysosomal proteases). Extraction and irradiation of lysosomal proteases were carried out in 50 mM acetate buffer pH 5.5, because lysosomal proteases are unstable in tissue extracts at neutral pH. Briefly, enzyme sample (0.05 ml) was incubated with the appropriate assay medium (total volume 0.3 ml) at 37 °C (10 min to 2 h), and the reaction terminated by addition of 0.6 ml ethanol. The fluorescence of liberated aminoacyl-7-amino-4-methyl coumarin, was measured by reference to an appropriate fluorescent standard ($\lambda_{ex} = 380$ nm, $\lambda_{em} = 440$ nm). Assay blanks were run in which enzyme was added to the medium immediately before ethanol addition. Assay conditions were modified for samples with high activity such that the extent of substrate utilization never exceeded 15%.

To determine whether exposure of muscle proteins (contractile, cytoskeletal) to ROS increases their susceptibility to subsequent degradation by proteolytic enzymes *in vitro*, muscle pellet proteins obtained after centrifugation of muscle homogenate were irradiated (as described above), and then recombined with the original supernatants (containing non-irradiated proteases). Samples (total volume 0.5 ml) at pH 7.5 or 5.5, corresponding to the pH optima of the two principal endogenous protease groups of relevance to intracellular muscle protein catabolism (cytoplasmic and lysosomal), were then incubated at 37 °C over a time period of 5 h to 5 days; 3 mM NaN_3 was included in all incubated tubes to inhibit bacterial growth. Similar time course assays were run in which non-irradiated muscle supernatants were recombined with non-irradiated muscle pellets, to determine the relative rates of degradation of pellet proteins by endogenous proteases in the absence of ROS-induced oxidative damage. Time course assays were terminated by addition of 0.5 ml of 20% (w/v) SDS/5% 2-mercaptoethanol, the samples were boiled for 3 min, and subsequent analysis via SDS-PAGE was carried out as described above.

Histological analysis

For histochemical analysis of irradiated and non-irradiated control tissue sections, cytochrome oxidase was determined by the method of Seligman et al. (1968) using diaminobenzidine as electron donor and cytochrome c as substrate; NADH tetrazolium reductase activity was determined by the method of Pearse (1972) using MTT as an electron acceptor; succinate dehydrogenase activity was determined by the method of Johnson (1991). For immunocytochemical analysis of irradiated and control tissue sections, the following monoclonal antibodies (MABs) were obtained from Novocastra Laboratories (Newcastle upon Tyne, UK): DY 10/12B2 (dystrophin amino-terminal residues 308–351); DY 46D3 (dystrophin central rod domain residues 1181–1388); CDYS 3/21C1 (dystrophin central rod domain residues 2592–3026); DY 8/6C5 (dystrophin carboxyl-terminal residues); 43 DAG/8D5 (β-dystroglycan, carboxy-terminal residues 880–894); RBC 2/3D5 (β-spectrin from red cell membranes); and myosin heavy chain (fast and slow forms). Tissue sections were examined by indirect immunofluorescence. After incubation with undiluted MABs for 1 h at 25 °C, and subsequent washing with Tris Buffer Saline Tween

(TBST) (three times for 15 min), sections were incubated with rhodamine-conjugated, rabbit anti-mouse secondary antibody (Dakopats, Denmark) for 1 h (diluted 1:100 with TBST). Sections were washed with TBST (three times for 15 min) before mounting on glass coverslips using aqueous mounting media (Uvinert, BDH, UK). Fluorescent photomicrographs were obtained using a photomicroscope (Nikon HFX II) equipped with an epifluorescence illumination filter for rhodamine ($\lambda_{ex} = 555$ nm, $\lambda_{em} = 580$ nm) using a $\times 40$ oil immersion objective lens.

For ultrastructural analysis, muscle (held under tension) was cut into 2 mm$\times$1 mm blocks using a stereomicroscope, and fixed in 5% glutaraldehyde in 0.1 M phosphate buffer at pH 7.4 for 1 h. After rinsing in buffer for 15 min, the muscle was processed for secondary fixation in 1% osmium tetroxide in 0.1 M phosphate buffer for 2 h, rinsed in buffer (twice for 10 min), and stained with 4% aqueous uranyl acetate for 20 min. After serial dehydration in aqueous ethanol (50–90%), muscle blocks were embedded in araldite resin and allowed to set for 24 h. Sections for viewing under the electron microscope were cut at approximately 70 nm using a microtome (Reichert).

Quantification of protein carbonyl content in irradiated muscle tissue

Human skeletal muscle samples (0.2 g, rectus abdominis muscle) from patients undergoing elective surgery were solubilized by homogenization, and exposed to increasing dosages (up to 2500 krads) of OH⁻ or O_2^{--} via ^{60}Co γ-irradiation as described earlier. Quantification of protein carbonyl content was then determined by the method of Reznick and Packer (1994). Briefly, streptomycin sulphate (10%, w/v in 50 mM HEPES buffer pH 7.5) was added to muscle homogenates (equal to 1%, w/v final concentration) to precipitate contaminating nucleic acids, and insoluble material was removed by centrifugation (10000$\times$g, 15 min). Each sample was then divided into two equal volumes: 4 volumes of 2,4-dinitrophenylhydrazine (DNPH) (10 mM in 2 M HCl) were added to one of the sample pairs (incubated 1hr at 25 °C) and 4 volumes of 2 M HCl added to the other (as reagent blank). Chemical derivatization was terminated by addition of an equal volume of 20% (v/v) trichloroacetic acid. Precipitated protein pellets were washed several times with 1 ml ethanol/ethyl acetate (1:1, v/v) to remove unbound DNPH and lipid contaminants. Precipitated proteins were then dissolved in 1 ml 6 M guanidine-HCl/20 mM phosphate buffer pH 2.3, and the ultraviolet spectrum determined between 355 and 390 nm. The carbonyl content was calculated from the peak maximum absorbance (relative to reagent blank) using a value for ε_{max} of 22000 M^{-1}cm^{-1}, and expressed in terms of nanomoles of carbonyl per milligram of soluble protein (determined by the method of Lowry et al., 1951).

Results and discussion

Effect of ROS on purified muscle proteins

The results of a typical experiment used to determine the susceptibility of individual proteins, in purified form, to oxidative degradation by OH˙ or $O_2^{-\cdot}$ after irradiation (via SDS-PAGE analysis) are described for actin. Exposure of this protein to increasing dosage (up to 100 krads) of $O_2^{-\cdot}$ showed a progressive decrease in staining intensity of the 43-kDa actin band relative to the non-irradiated control sample; exposure to OH˙ showed complete destruction of the actin band with dosages greater than 20 krads. It is of note that there was no evidence for the formation of actin aggregates or fragments, as could be expected after exposure of proteins to OH˙ or $O_2^{-\cdot}$ in the absence or presence of oxygen respectively (Davies, 1987). There have been reports, in previous work on the effects of ROS generated by radiolysis on purified model proteins, of the formation of specific aggregates of lysozyme after exposure to OH˙ (Franzini et al., 1993) and of the formation of specific aggregates or fragments of haemoglobin after exposure to OH˙ or OH˙/$O_2^{-\cdot}$ (Puchela and Schaefler, 1993), respectively.

The ROS-induced oxidative degradation of albumin (arguably the most extensively characterized of the model proteins) has been compared with that for a wide range of individual proteins (Davies, 1987); in each case characteristic aggregation (OH˙) or fragmentation (OH˙/$O_2^{-\cdot}$) products were identified, with little oxidative damage occurring in the presence of $O_2^{-\cdot}$ alone. In the present study, exposure of albumin to OH˙ (up to 100 krads) showed increased loss of intensity

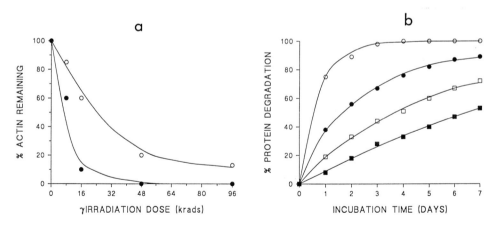

Figure 1. Free radical induced changes in muscle protein levels assessed via SDS-PAGE analysis. The preparation of samples, exposure to free radicals and SDS-PAGE protein band analysis via scanning densitometry were described under Material and methods. (a) Graphical representation of oxidative destruction of actin protein following exposure to increasing dosage of OH˙ (●) or $O_2^{-\cdot}$ (○), relative to non irradiated control sample. (b) Graphical representation of rate of degradation of myosin and actin via endogenous cytoplasmic and lysosomal muscle proteases: myosin, pH 5.5 (○); actin, pH 5.5 (●); myosin, pH 7.5 (□); actin, pH 7.5 (■).

Table 1. Relative susceptibility of purified muscle proteins to oxidative degradation by free radicals

Protein type	Dose required to degrade 1 nmol protein (krads)	
	OH·	$O_2^{-·}$
Myoglobin	0.51	2.72
Albumin	0.90	>100
Creatine Kinase	0.95	2.60
Actin	4.91	11.8
Troponin	5.05	50.3
Peroxidase	9.23	16.4
Tropomyosin	10.2	26.4
Myosin	40.6	88.1
Choline Acetyltransferase	49.0	99.0
Catalase	56.3	69.7

The preparation of target protein samples, exposure to free radicals, and SDS-PAGE protein band analysis via scanning densitometry were described in the Section on *Material and Methods*. Data showing the dose (krads) of OH· or $O_2^{-·}$ required to degrade 1 nmol of each protein type were calculated from the rate of protein degradation with increasing radical dosage (shown graphically for actin in Fig. 1a), and the concentration (mg/ml) of protein used.

of protein band staining and corresponding formation of high-molecular-weight protein aggregates, whereas exposure to similar dosages of $O_2^{-·}$ resulted in comparatively little oxidative protein damage, in general agreement with previously reported data (Davies, 1987). It is therefore presumed that the disappearance of the actin band, and absence of corresponding aggregates or fragments, must result from either the formation of very large protein aggregates or the break down of the protein into very small peptide fragments; neither of these is resolved via SDS-PAGE analysis. Quantitative analysis of the effect of OH· and $O_2^{-·}$ on actin band staining intensity via scanning densitometry is shown in Figure 1a. It is apparent that the extent of degradation (i.e. ROS dosage required for 50% destruction of initial protein at 0.2 mg/ml) of actin by OH· is about 2.5-fold greater than by $O_2^{-·}$ (although oxidative damage caused by the latter remains significant).

Comparative data for the effects of OH· and $O_2^{-·}$ on the extent of protein degradation (defined as above), determined by SDS-PAGE and densitometric analysis as for actin, are shown for other muscle protein types in Table 1. Also included are data showing comparative oxidative damage by OH· and $O_2^{-·}$ on enzymic proteins determined via specific assay procedures. It is apparent from the data shown in Table 1 that: (1) individual muscle protein types show different susceptibility to oxidative damage via OH· or $O_2^{-·}$ respectively; (2) oxidative damage to corresponding proteins induced by OH· is invariably greater than for $O_2^{-·}$, typically by a factor of two- to threefold; however, the extent of oxidative protein damage induced by $O_2^{-·}$ in absolute terms remains significant, in contrast to previously reported studies on model proteins using a similar format (Davies, 1987). It therefore appears that the differential susceptibility of proteins to ROS-induced oxidative damage may be determined at least in part by the structure (i.e. primary/secondary/tertiary) of the protein. In this regard, previous work has shown that the amino acids tryptophan, tyrosine, cysteine and histidine are particularly susceptible to ROS-induced oxidative modification after irradiation (Stadtman, 1990). The relative content of these amino acid residues (as a combined percentage of total amino acid composition) was therefore determined for muscle proteins investigated using a computerized database search programme (PC GENE) as follows: actin (43 kDa) = 9.0%; α-actinin (103 kDa) = 8.6%; myosin heavy chain (222 kDa) = 5.4%; tropomyosin (32.8 kDa) = 4.0%; troponin C (18.2 kDa) = 2.4%. Comparison of the above data with corresponding data for relative susceptibility of proteins to oxidative damage (Tab. 1) shows no correlation between amino acid content and damage susceptibility. It is therefore concluded that differences in susceptibility of the above proteins to oxidative damage must be determined by secondary/tertiary structural differences in protein folding and/or subunit orientation, rather than by the primary structural amino acid sequence.

Effect of ROS on muscle tissue homogenates

The results of a typical experiment to investigate ROS-induced oxidative damage to proteins in muscle tissue are described for the effects of $OH^{\cdot}$ and $O_2^{-\cdot}$ on contractile/cytoskeletal proteins in pelleted muscle homogenate (as described earlier). The overall SDS-PAGE protein fractionation pattern (Blanchard et al., 1993) showed a generalized progressive loss in band staining intensities for most proteins with increasing $OH^{\cdot}$ dose (0–3000 krads), with essentially complete degradation of all proteins at the highest $OH^{\cdot}$ dose used. (Similarly, SDS-PAGE analysis of muscle-soluble proteins (supernatant fraction) showed complete degradation of proteins for $OH^{\cdot}$ doses > 1500 krads). For corresponding dosages of $O_2^{-\cdot}$, the extent of contractile/cytoskeletal protein oxidative degradation was less than for $OH^{\cdot}$, e.g. the relative rates of degradation of myosin and actin by $OH^{\cdot}$ and $O_2^{-\cdot}$ were 1.8- and 3.0-fold respectively (via densitometric analysis). Thus, the relative susceptibility of muscle proteins to oxidative damage by $OH^{\cdot}$ or $O_2^{-\cdot}$ in tissue extracts parallels that determined for corresponding proteins in purified form. As noted for the latter, there was no evidence from the SDS-PAGE protein fractionation profile for the formation of aggregates or fragments of contractile/cytoskeletal proteins after exposure to $OH^{\cdot}$ or $O_2^{-\cdot}$. It is therefore concluded that the disappearance of individual bands from the SDS-PAGE profile again results from the formation of very large protein aggregates or small peptide fragments.

Effect of ROS exposure on subsequent degradation of proteins by endogenous muscle proteases

Previous work has shown that oxidative damage to proteins increases their susceptibility to subsequent proteolytic degradation (Davies et al., 1987). Much of this research has focused on one proteolytic enzyme in particular – the proteasome-associated multicatalytic protease macropain (Goldberg, 1992; Hershko and Ciechanover, 1992). Relatively little is known about increased susceptibility of oxidatively damaged proteins to proteolytic degradation by other protease types.

Table 2. Effect of free radical exposure on muscle protein degradation via endogenous proteases

Protein	pH	Degradation (50%) rate (days)		
		Control	OH˙	$O_2^{-˙}$
Myosin	5.5	0.63±0.11	0.51±0.10	0.70±0.15
Actin	5.5	1.62±0.17	1.50±0.32	1.83±0.29
Myosin	7.5	4.10±0.48	3.49±0.85	4.56±0.76
Actin	7.5	7.22±1.13	6.57±1.68	8.35±1.95

Data presented show the relative time (in days) for 50% degradation of the initial protein levels of myosin and actin by endogenous cytoplasmic (pH 7.5) or lysosomal (pH 5.5) proteases, after exposure (100 krads) to OH˙ or $O_2^{-˙}$, relative to non-irradiated control muscle pellet samples (shown graphically in Fig. 1b). Data (mean±SE) from three separate experiments are shown.

The objective of this experiment was therefore to determine *in vitro* whether pre-exposure of muscle proteins to OH˙ or $O_2^{-˙}$ increases their susceptibility to degradation via endogenous muscle cytoplasmic or lysosomal proteases (representing the two major pathways of intracellular protein degradation as described below).

Figure 1b shows the graphical analysis of the degradation of the two major contractile proteins, myosin and actin (muscle pellet), by endogenous cytoplasmic (pH 7.5) and lysosomal (pH 5.5) proteases (muscle supernatant) in non-irradiated muscle tissue using SDS-PAGE. It is apparent that the rate of protein degradation at pH 5.5 is greater than that at pH 7.5, and that at a given pH the degradation rate of myosin is greater than that of actin. Parallel experiments were carried out in which washed muscle pellets (contractile/cytoskeletal proteins) were irradiated using increasing dosage of OH˙ or $O_2^{-˙}$, and then recombined with non-irradiated muscle supernatants (containing endogenous proteases), and incubated at pH 7.5 or pH 5.5 (as described earlier). Data from these experiments are shown in Table 2; the rates of degradation of myosin and actin proteins (i.e. the time required for 50% degradation of the intact protein) by endogenous cytoplasmic or lysosomal proteases are shown to be essentially unaltered after pre-exposure of the proteins to OH˙ or $O_2^{-˙}$ (up to 100 krads). Thus there is no evidence, when using myosin or actin as endogenous protease substrates, that pre-exposure of proteins to ROS increases subsequent susceptibility to degradation by either cytoplasmic or lysosomal proteases (i.e. when grouped collectively). It is concluded that the above data support the concept that oxidatively damaged proteins are degraded by specific protease types (especially macropain), rather than by more generalized proteolytic action.

Effect of ROS on protease activities in vitro

Proteolytic enzymes are responsible for the degradation of proteins and peptides (ultimately to free amino acids) for reuse during generalized intracellular protein degradation, a process that is essential to the normal functioning of cells in all tissue types (specific proteases may also have additional more specialized functions, e.g. processing of neurotransmitter peptides, antigen processing in the immune response, etc.). Proteases are classified on the basis of their intracellular distribution, pH optimum for activity, size of substrate degraded, substrate bond specificity and active site type. The activity of proteases may be controlled via activation of proenzymic forms, changes in endogenous inhibitor levels or levels of enzyme effectors such as cations (Barrett and McDonald, 1980). In contrast to the complementary process of protein synthesis, comparatively little is known about the mechanism of protein degradation, i.e. which factors initiate protein degradation, the precise role of individual proteases in the various steps comprising the degradation cascade, or how the overall process is coordinated and controlled. However, it has been recognized that there are two major sites of intracellular protein degradation: lysosomal, comprising proteases with broad substrate specificity and acidic pH optima of activity (collectively known as cathepsins), and cytoplasmic, comprising enzymes with neutral pH optima of activity and more specific substrate specificities; – several degradation pathways have been identified in the cytoplasm, e.g. Ca^{2+}-dependent proteolysis (mediated by calpains) and ubiquitin dependent proteolysis (mediated by macropain) (Seglen and Bohley, 1992).

Abnormal intracellular protein processing resulting from altered protease activities has been implicated in the pathogenesis of a variety of disorders, including cancer, arthritis and Alzheimer's disease; in particular substantial increases in activity of a number of protease types has been reported in muscle tissue from patients with wasting disorders such as DMD (Pennington, 1988). Thus proteolytic enzymes, together with free radicals, represent two of the major classes of endogenous cellular damaging agents, the interplay between which (largely unexplored at present) may be of particular relevance in understanding the mechanism of tissue damage in degenerative disorders of tissues such as muscle. In this regard, it has been suggested that ROS may be one of the factors responsible for initiation of protein degradation, because there is evidence that oxidatively damaged proteins (with increased hydrophobicity after partial denaturation) are selectively and rapidly degraded via proteases, particularly macropain (Goldberg, 1992). However, one area that has been little investigated is the potentially damaging action of ROS on the activities of proteolytic enzymes themselves. Although the action of ROS on a wide range of different functional classes of enzyme has been described (Stadtman, 1990), there are few data available regarding the susceptibility of proteases to oxidative damage. The objective of this experiment was therefore to determine whether individual proteases comprising the cytoplasmic or lysosomal degradative pathways show differential susceptibility to oxidative damage by $OH^{.}$ or $O_2^{-.}$, and whether this susceptibility may be influenced by the cellular environment in different tissue types.

Human tissues (skeletal muscle, cerebral cortex, lumbar cervical cord and liver) were solubilized via homogenization in extraction buffer at pH 7.5 (cytoplasmic proteases) or pH 5.5 (lysosomal proteases) and the respective supernatants (containing >85% of total tissue protease activity) retained for analysis after centrifugation; tissue supernatants were then exposed to a fixed dosage (100 krads) of $OH^{.}$ or $O_2^{-.}$ via ^{60}Co γ radiolysis and the subsequent activity of various cytoplasmic and lysosomal protease types determined (as described earlier). The results of experiments to determine the relative susceptibility of proteases to oxidative damage by $OH^{.}$ and $O_2^{-.}$ in skeletal muscle and cerebral cortex tissue are shown in Table 3. It is apparent from these data that: (1) most protease types showed evidence of oxidative protein modification via reduced activity levels compared with non-irradiated control samples, with significantly greater oxidative damage by OH˙ compared with O_2^{-}; (2) several protease types (leucyl aminopeptidase, tripeptidyl aminopeptidase, and cathepsin D) were relatively resistant to oxidative damage by OH˙ or O_2^{-}, whereas other proteases (pyroglutamyl aminopeptidase and dipeptidyl aminopeptidase II) showed a relatively high degree of susceptibility to oxidative damage by both OH˙ and O_2^{-}; (3) there is no evidence for differential susceptibility of the enzymes making up cytoplasmic or lysosomal degradative pathways to oxidative damage; and (4) the pattern of susceptibility of individual proteases to oxidative damage found in skeletal muscle is similar to that found in other tissue types (data for spinal cord and liver tissues not shown).

In summary, evidence has been presented which shows that proteolytic enzymes as a group are broadly susceptible to ROS-induced oxidative damage *in vitro*. Thus the effect of oxidative modification by ROS on intracellular protein turnover may be twofold: first the increased resistance of protein substrates to proteolysis resulting, for example, from the formation of cross-linked aggregates, and, second a generalized reduction in proteolytic enzyme activities. It is therefore

Table 3. Effect of free radicals on endogenous protease activities

Protease type	Enzyme activity (% control)			
	Muscle		Brain	
	OH˙	O_2^{-}	OH˙	O_2^{-}
Cytoplasmic				
Alanyl aminopeptidase	64.5±5.7**	92.1±2.4	63.7±5.2**	91.4±4.4
Arginyl aminopeptidase	60.7±4.8**	97.9±4.9	66.0±3.2**	88.0±8.1
Leucyl aminopeptidase	99.0±10.1	95.0±10.9	97.3±4.7	92.2±5.0
Pyroglutamyl aminopeptidase	34.5±2.6**	60.0±4.9**	25.0±3.2**	40.0±2.2***
Dipeptidyl aminopeptidase IV	85.9±3.5	98.6±7.4	83.3±3.5	99.2±6.2
Tripeptidyl aminopeptidase	94.8±4.6	97.4±3.2	87.6±5.2	88.4±5.3
Proline endopeptidase	81.4±4.0	93.8±4.6	86.8±6.8	92.5±3.5
Lysosomal				
Dipeptidyl aminopeptidase I	81.1±5.7	80.1±6.0	83.2±8.7	90.1±8.8
Dipeptidyl aminopeptidase II	35.1±2.8**	25.8±1.6***	40.9±5.9**	50.9±5.4**
Cathepsin B	62.2±9.1	78.9±8.6	72.2±6.4	88.9±9.9
Cathepsin H	80.7±4.9	79.1±6.7	80.6±5.6	79.8±5.2
Cathepsin L	80.3±2.6*	80.1±6.2	74.8±3.1*	88.6±3.3
Cathepsin D	73.5±4.6*	89.6±7.7	84.0±5.5	93.1±6.3

The preparation of muscle extracts, exposure to free radicals and assay of proteolytic enzyme activities were described in the Section on *Material and Methods*. Data for individual protease types showing changes in activity following exposure (100 krads) to OH˙ or O_2^{-} are expressed in terms of percentage activity remaining, relative to that of the corresponding control in non-irradiated tissue extracts. Data (mean±SE) from three separate experiments are shown. Statistical significance of differences in activity levels of individual protease types in control and irradiated tissue samples was assessed via Student's t-test: *$p<0.05$; **$p<0.01$; ***$p<0.001$.

conceivable that aberrant intracellular protein degradation which results from the above factors may ultimately result in cell death *in vivo*. This hypothesis may be of particular relevance to the formation of characteristic protein deposits in brain tissue of Alzheimer's disease cases, the formation of which is thought to be of central importance in the pathogenesis of this disorder.

Effect of ROS on proteins in muscle tissue sections, assessed via histological/
immunocytochemical analysis

Muscle tissue sections were prepared and exposed to varying dosages of $OH^{\cdot}$ or $O_2^{-\cdot}$ as described earlier. Comparison of control and irradiated tissue sections stained with haematoxylin and eosin showed no evidence of gross changes in muscle morphology after exposure to $OH^{\cdot}$ or $O_2^{-\cdot}$. It was possible, however, to demonstrate ROS-induced damage in specific muscle proteins using appropriate histochemical or immunocytochemical visualization procedures. The results of a typical experiment to investigate the differential susceptibility of mitochondria-associated proteins to oxidative damage are shown in Figure 2. Thus the activity of succinate dehydrogenase was completely destroyed in tissue sections exposed to an equivalent dosage (2400 krads) of either $OH^{\cdot}$ (Fig. 2 c, f) or $O_2^{-\cdot}$ (Fig. 2 i, l); the activity of cytochrome oxidase was completely destroyed after exposure to $OH^{\cdot}$ (Fig. 2 b, e) and unaltered after exposure to $O_2^{-\cdot}$ (Fig. 2 h, k), whereas the activity of NADH tetrazolium reductase was relatively resistant to damage by both $OH^{\cdot}$ (Fig. 2 a, d) and $O_2^{-\cdot}$ (Fig. 2 g, j). Similarly, the results of an experiment to investigate ROS-induced oxidative damage to the sarcolemma-associated protein dystrophin, with immunocytochemical visualization of the target protein using MABs recognizing different protein domains, have been described (Haycock et al., 1996). Clear differences in reduced fluorescent staining intensity were evident between muscle sections exposed to $OH^{\cdot}$ compared with non-irradiated controls. This was particularly evident using the MAB DY10/12B2, which recognizes the N-terminal protein region, and to a lesser extent using MABs CDY S3/21C1 (recognizing part of the central rod region) and DY8/6C5 (recognizing the carboxy-terminal region) respectively. There was little difference in fluorescent staining intensity between irradiated and control tissue sections using the MAB DY4/6D3 to the central rod domain. After exposure to an equivalent dosage (2400 krad) of $O_2^{-\cdot}$, differences in fluorescent staining intensity between irradiated and non-irradiated tissue sections were markedly reduced compared with those noted above with $OH^{\cdot}$ radicals using corresponding MABs to label dystrophin. The greatest change in immunolabelling intensity was noted using the MAB DY8/6C5, compared with that using MABs recognizing other regions of the protein.

Data showing the susceptibility of different muscle proteins to oxidative damage by $OH^{\cdot}$ or $O_2^{-\cdot}$ assessed via histological or immunocytochemical analysis, are summarized in Table 4. From the data presented, it is concluded that: (1) all muscle proteins investigated were subject to oxidative damage by ROS in a dose dependent manner, with greater damage by $OH^{\cdot}$ than $O_2^{-\cdot}$ at a corresponding dosage; (2) different target proteins showed differing relative susceptibility to oxidative damage depending on function or intracellular localization of the protein; and (3) visualization of dystrophin using MABs to different molecular domains has suggested that the protein comprises molecular regions that are more or less susceptible to ROS-induced oxidative

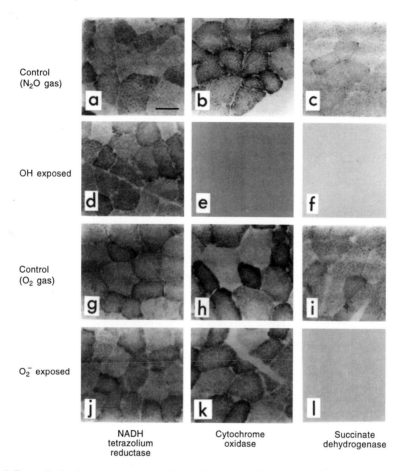

Control
(N₂O gas)

OH exposed

Control
(O₂ gas)

O₂⁻ exposed

| NADH tetrazolium reductase | Cytochrome oxidase | Succinate dehydrogenase |

Figure 2. Free radical-induced changes in histochemical enzyme staining in muscle tissue sections. The preparation of muscle tissue sections, exposure to free radicals and histochemical analysis were described under Material and methods. Changes in histochemical staining of mitochondrial associated enzymes NADH tetrazolium reductase (a, d, g, j), cytochrome oxidase (b, e, h, k) and succinate dehydrogenase (c, f, i, l) are shown after exposure of tissue sections to OH' or O₂⁻' (2500 krads); tissue sections exposed to OH' (d, e, f) and corresponding non-irradiated control sections (a, b, c), tissue sections exposed to O₂⁻' (j, k, l) and non-irradiated control sections (g, h, i).

damage, presumably arising from differences in amino acid composition and/or protein folding. The particular susceptibility of the dystrophin carboxy-terminal region to damage by OH' or O_2^{-} may result from association of the latter with the lipid environs of the plasma membrane.

Analysis of ROS-induced muscle ultrastructural damage via electron microscopy

The ultrastructural organization of non-irradiated control muscle in oxygen-saturated buffer is shown via electron microscopy in Figure 3a. This shows the distinct muscle tissue banding pat-

Table 4. Changes in immunocytochemical and histochemical labelling of proteins in muscle tissue sections after exposure to OH˙ or $O_2^{-\cdot}$ radicals

Protein	Relative change in protein labelling					
	2400 krads		1800 krads		1100 krads	
	OH˙	$O_2^{-\cdot}$	OH˙	$O_2^{-\cdot}$	OH˙	$O_2^{-\cdot}$
Dystrophin						
DY10/12B2	+++	−	+	−	−	−
DY4/6D3	−	+	−	−	−	−
CDYS3/21c1	+++	+	+	−	−	−
DY8/6C5	+++	++	++	+	−	−
Dystrophin associated glycoprotein						
43DAG/8D5	+	−	+	−	−	−
β–Spectrin						
RBC2/3D5	+++	−	++	−	−	−
Myosin						
heavy chain, fast	+++	++	++	+	−	−
heavy chain, slow	++	+	+	−	−	−
Cytochrome oxidase	+++	−	++	−	−	−
NADH tetrazolium reductase	−	−	−	−	−	−
Succinate dehydrogenase	+++	+++	++	++	−	−

Changes in protein labelling intensity, compared with non-irradiated control sections, were assessed semiquantitatively via visual inspection as follows: no change in labelling intensity (−), small change (+), moderate change (++), large change or complete loss of labelling intensity (+++).

tern corresponding to normal morphological criteria, i.e. with Z-lines, A-bands and I-bands in register and running at right angles to the myofibrillar long axis, with the myofibrils aligned in a regular manner and sarcomeres present in well-defined units. Mitochondria with intact inner and outer membranes lie between the myofibrils located either side of the Z-bands. The sarcoplasmic reticulum (SR) and T system ensheathe the myofibrils and form characteristic triads at the junction of the A- and I-bands. The plasma membrane is located on the outside of the muscle fibres immediately under the external basement lamina associated with the endomysial connective tissue. After exposure to $O_2^{-\cdot}$ (300 krads), the most striking changes in ultrastructure were those relating to the mitochondria. At a relatively low magnification (7500×) an apparent random distribution of alterations indicating swelling of this organelle is evident (Fig. 3b).

Considerable variations in mitochondrial size are present, ranging from those with the same cross-sectional area to those in control muscle, to those that are about 25 times the size in the latter. Closer examination at higher magnification reveals early disappearance of the characteristic shape of the cristae, with general disorganization of the inner membrane evident in the slightly swollen mitochondria. More severe damage is evident with greater swelling of the mitochondria

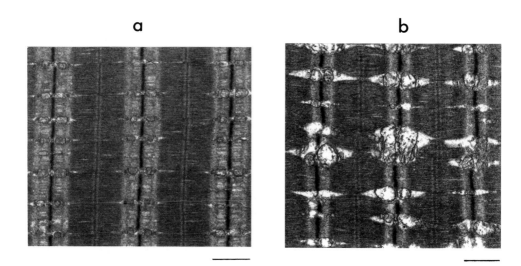

Figure 3. Free radical-induced changes in muscle ultrastructure assessed by electron microscopy. Muscle samples were prepared, exposed to free radicals and analysed via electron microscopy described in the Section on *Materials and Methods*. (a) Control non-irradiated tissue sample showing normal appearance of muscle ultrastructure (bar = 1 μm); (b) muscle sample exposed to $O_2^{-\cdot}$ (300 krads) showing widespread and selective damage to mitochondria (bar = 1 μm)

where myofibrillar distortion is evident. These larger, randomly isolated organelles still contain an outer membrane, but the inner membrane is completely destroyed, resulting in the formation of mitochondrial vacuoles. Other subcellular regions appear to have spaces in place of the mito-chondria once present. It is possible that a critical maximum volume is tolerated by mitochondria, with rupture of the outer membrane once this is exceeded, resulting in "bursting" and the formation of small vacuolar spaces. The action of $O_2^{-\cdot}$ on muscle ultrastructure appears selective for mitochondria, because alterations to other subcellular components of muscle are not evident. The position and size of the T tubules and SR are not altered, although swollen mitochondria are noted physically to displace the whole triadic structure.

Minor alterations in the regular banding pattern of Z-lines, M-bands and A-bands were noted, but the regular sarcomeric repeats and definitions were maintained, and no alteration in the struc-ture of the plasma membrane was evident. Investigation of ultrastructural modifications for muscle tissue by $O_2^{-\cdot}$ was restricted to a maximum dose of 300 krads, because incubation of muscle samples in oxygen-saturated buffer for longer than 4 h induced mitochondrial alterations (slight swelling, occasional vacuolar spaces) without irradiation. Similarly, ultrastructural investi-gation of ROS-induced damage was restricted to $O_2^{-\cdot}$, because incubation of samples in N_2O-saturated buffer (essential for generation of OH·) for more than 30 min induced widespread and extensive mitochondrial damage without irradiation. Alteration of mitochondrial ultrastructure after exposure to $O_2^{-\cdot}$ has been cited previously, using isolated rat liver mitochondria. Damage to these organelles by $O_2^{-\cdot}$ induces characteristic features which are apparently conserved in all of the studies described, irrespective of the method of $O_2^{-\cdot}$ generation or techniques used for damage assessment. Thus mitochondrial swelling and vacuolar formation in the cristae, after CCl_4 intoxication assessed via electron microscopy, have been reported (Janzen et al., 1990). Similarly, mitochondrial swelling identified by changes in absorbance has been described after exposure of a partially purified rat liver mitochondrial fraction to $O_2^{-\cdot}$ (Mehotra et al., 1991; Kakkar et al., 1992). These investigations are in agreement with the present findings that mitochondria are subject to a characteristic form of damage resulting from the action of $O_2^{-\cdot}$ radicals.

Effect of ROS on muscle protein carbonyl induction in vitro

Samples of rectus abdominis muscle obtained from patients undergoing elective surgery (age range 49–52 years) were solubilized via homogenization, exposed to various dosages of $OH^{\cdot}$ or $O_2^{-\cdot}$ via irradiation, and subsequent muscle protein carbonyl levels determined as described earlier. Data showing the correlation between increasing dosage of $OH^{\cdot}$ or $O_2^{-\cdot}$ and corresponding muscle protein carbonyl levels are shown in Table 5. Baseline protein carbonyl levels in non-irradiated muscle samples were comparable to those in muscles of chickens without muscle disease (Murphy and Kehrer, 1989), and in various tissues in other studies (Oliver et al., 1987). No increases in protein carbonyl levels were noted for control samples gassed with N_2O or O_2, indicating that ROS-induced protein oxidation does not occur *in vitro* for the duration of the irradiation experiment (i.e. up to 24 h) once tissue samples have been homogenized. Significant increases were found for introduction of carbonyl groups into muscle proteins exposed to increasing dosages of $OH^{\cdot}$ or $O_2^{-\cdot}$. It is of note that maximal introduction of carbonyl groups at the highest irradiation dosage was approximately fourfold higher after exposure to $OH^{\cdot}$, compared with $O_2^{-\cdot}$, implying that saturation of muscle proteins with carbonyl groups by respective oxidizing species is different. In muscle biopsy samples from patients with DMD, the protein carbonyl level (mean$\pm$SE) was 2.95 ± 0.62 nmol carbonyl/mg protein (range 1.74–5.13 nmol/ng), where-

Table 5. *In vitro* changes in skeletal muscle protein carbonyl levels following irradiation

Sample treatment	Protein carbonyl (nmol mg^{-1} protein)					
	Irradiation dose (krad)					
	0	196	392	784	1586	2352
O_2 gassed, not irradiated	2.23±0.09	2.20±0.10	2.07±0.12	2.13±0.14	2.17±0.03	2.20±0.12
$O_2^{-\cdot}$ exposed	2.10±0.10	2.07±0.09	2.50±0.26	4.27±0.46	4.77±0.15	4.83±0.19
N_2O gassed, not irradiated	2.13±0.18	2.13±0.18	2.07±0.12	2.07±0.03	2.30±0.12	2.37±0.09
$OH^{\cdot}$ exposed	2.23±0.18	2.47±0.18	3.27±0.17	9.77±0.43	18.8±0.52	20.9±0.44

The preparation of tissue extracts, exposure to free radicals and determination of protein carbonyl levels were described in the Section on *Material and Methods*. Data (mean$\pm$SE) are presented from three separate experiments.

as for normal controls, the corresponding value was 0.95 ± 0.22 nmol carbonyl/mg protein (range $0.56-1.77$ nmol/ng). Differences in protein carbonyl content between the latter controls (mean age 11.2 years) and controls used above for γ radiolysis experiments (mean age $= 51.3$ years) are probably attributable to the known increase in tissue protein carbonyl during the ageing process. The increase in protein carbonyl content, in muscle from DMD cases, of 2 nmol/mg protein (relative to controls) corresponds to equivalent dosages of about 500 krads $OH^{\cdot}$ and 800 krads $O_2^{-\cdot}$, based on values from *in vitro* irradiation experiments described above. Presumably these surprisingly high values reflect differences in muscle protein damage *in vivo* and *in vitro* based on: (1) differences in generalized and localized generation of ROS species *in vitro* and *in vivo*; and (2) time-dependent accumulative effects of free radical damage between radiolytic generation *in vitro* (short-term) and ROS generation during the disease process *in vivo* (long term).

Conclusion

Further investigations are now required to establish whether proteins identified as susceptible to oxidative damage *in vitro* show similar susceptibility to ROS-induced oxidative damage *in vivo*. In this regard, preliminary investigations have been carried out to identify oxidative damage to individual muscle proteins via SDS-PAGE and immunoblotting analysis (as described by Levine et al., 1994), using a primary MAB to the DNPH moiety (Sigma), with peroxidase-labelled anti-mouse IgE as second antibody (Southern Biotechnology, UK), and final visualization of protein bands via chemiluminescence/autoradiography (J. Haycock, S. MacNeil and D. Mantle, unpublished data). These studies have indicated that muscle biopsy samples from DMD cases contain proteins showing characteristic oxidative damage which are not present in tissue from other types of muscle disease or in normal control tissue, although the identity of these proteins remains to be established. The further identification of such proteins which show characteristic oxidative damage in DMD may, in the longer term, lead to the development of improved therapeutic strategies for DMD patients, based on identification of appropriate antioxidant compounds which specifically protect susceptible proteins from ROS-induced oxidative damage.

Acknowledgements
We thank Rapid Science Publishers Ltd. for permission to reproduce data shown in Table 5 (originally published in *Neuro Report*, 1996, 8: 357-361), and Springer-Verlag GmbH & Co. for permission to reproduce data shown in Table 4, and Figs 2 and 3 (originally published in *Acta Neuroopath*, 1996, 92: 331-340)

References

Austin L (1990) How the lack of dystrophin may upset calcium regulation and lead to oxidative damage. *In*: BA Kakulas and FL Mastaglia (eds): *Pathogenesis and Therapy of Duchenne and Becker Muscular Dystrophy*. Raven Press, New York, pp 69-82.
Austin L, de Niese M, McGregor A, Arthur H, Gurusinghe A and Gould MK (1992) Potential oxyradical damage and energy status in individual muscle fibres from degenerative muscle diseases. *Neuromusc Disord* 75: 27-33.
Barrett AJ and McDonald AJ (1980) *Mammalian Proteases*. Academic Press, London.

Blanchard P, Ellis M, Maltin C, Falkous G, Harris JB and Mantle D (1993) Effect of growth promoters on pig muscle structural protein and proteolytic enzyme levels *in vivo* and *in vitro*. *Biochimie* 75: 839–847.

Bradley R and Fell BF (1980) Myopathies in animals. In: JN Walton (ed.) *Disorders of Voluntary Muscle*, 3rd edn. Churchill Livingstone, London, pp 824–872.

Cohen G, Dembiec D and Marcus J (1970) Measurement of catalase activity in tissue extracts. *Anal Biochem* 34: 30–38.

Cullen MJ and Mastaglia FL (1980) Morphological changes in dystrophic muscle. *Brit Med Bull* 36: 145–152.

Davies KJA (1987) Protein damage and degradation by oxygen radicals. I. General aspects. *J Biol Chem* 262: 9895–9901.

Davies KJA, Lin SW and Pacifici RE (1987) Protein damage and degradation by oxygen radicals. IV. Degradation of denatured protein. *J Biol Chem* 262: 9914–9920.

Del Maestro RA (1980) An approach to free radicals in medicine and biology. *Acta Physiol Scand* 492: 153–168.

Ervasti JM and Campbell KP (1993) A role for the dystrophin glycoprotein complex as a transmembrane link between laminin and actin. *J Cell Biol* 122: 809–823.

Franzini E, Sellak H, Hakim J and Pasquier C (1993) Oxidative damage to lysozyme by the hydroxyl radical: comparative effects of scavengers. *Biochim Biophys Acta* 1203: 11–17.

Freeman BA and Crapo JD (1982) Biology of disease. Free radicals and tissue injury. *Lab Invest* 47: 412–426.

Fricke H and Hart EJ (1966) Radiation Dosimetry. In: FH Roesch and WC Roesch (eds): *Radiation Dosimetry*. Academic Press, New York, p. 167.

Goldberg AL (1992) The mechanism and functions of ATP dependent proteases in bacterial and animal cells. *Eur J Biochem* 203: 9–23.

Halliwell B and Grootveld M (1987) The measurement of free radical reactions in humans. Some thoughts for future experimentation. *FEBS Lett* 213: 9–14.

Halliwell B and Gutteridge JMC (1990) Role of free radicals and catalytic metal ions in human disease: an overview. *Methods Enzymol* 186: 1–85.

Haycock JW, Jones P, Harris JB and Mantle D (1996) Differential susceptibility of human skeletal muscle proteins to free radical induced oxidative damage: a histochemical, immunocytochemical and electron microscopical study *in vitro*. *Acta Neuropathol* 92: 331–340.

Hershko A and Ciechanover A (1992) The ubiquitin system for protein degradation. *Annu Rev Biochem* 61: 761–807.

Jackson MJ and O'Farrell S (1993) Free radicals and muscle damage. *Brit Med Bull* 49: 630–641.

Jackson MJ, Jones DA and Edwards RHT (1984) Techniques for studying free radical damage in muscular dystrophy. *Med Biol* 62: 135–138.

Janzen EG, Towner RA and Yamashiro S (1990) The effect of phenyl tert-butyl nitrone on CCl_4 induced rat liver injury detected by proton magnetic resonance imaging *in vivo* and electron microscopy. *Free Radical Res Commun* 9: 325–335.

Johnson MA (1991) Application of enzyme histochemistry in muscle pathology. In: PJ Stoward and AEG Pearse (eds): *Histochemistry*, vol. 3, 4th edn, Churchill Livingstone, London, pp 489–514.

Kakkar P, Mehotra S and PN Viswanath (1992) Interrelation of active oxygen species, membrane damage and altered calcium functions. *Mol Cell Biochem* 111: 11–15.

Koenig M, Monaco AP and Kunkel LM (1988) The complete sequence of dystrophin predicts a rod-shaped cytoskeletal protein. *Cell* 53: 219–228.

Laemli UK (1970) Cleavage of structural proteins during the assembly of the bacteriophage T4. *Nature* 227: 680–685.

Levine RL, Williams JA, Stadtman ER and Shacter E (1994) Carbonyl assays for determination of oxidatively modified proteins. *Methods Enzymol* 233: 346–357.

Lowry OH, Rosebrough NJ, Farr AL and Randall RJ (1951) Protein measurement with Folin phenol reagent. *J Biol Chem* 193: 265–275.

Marshall PJ, Warso MA and Lands WE (1985) Selective microdetermination of lipid hydroperoxides. *Anal Biochem* 145: 192–199.

Mechler F, Imre S and Dioszeghy P (1984) Lipid peroxidation and superoxide dismutase activity in muscle and erythrocytes in Duchenne muscular dystrophy. *J Neurol Sci* 63: 279–283.

Mehotra S, Kakkar P and Viswanath PN (1991) Mitochondrial damage by active oxygen species *in vitro*. *Free Radical Biol Med* 10: 277–285.

Mulligan RC (1993) The basic science of gene therapy. *Science* 260: 926–932.

Murphy ME and Kehrer JP (1986) Free radicals: A potential mechanism in inherited muscular dystrophy. *Life Sci* 39: 2271–2278.

Murphy ME and Kehrer JP (1989) Oxidation state of tissue thiol groups and content of protein carbonyl groups in chickens with inherited muscular dystrophy. *Biochem J* 260: 359–364.

Muzuno KOY (1988) Pathogenesis of progressive muscular dystrophy: studies on free radical metabolism in an animal model. *Acta Neurol Scand* 77: 108–114.

Oliver CN, Ahn B, Moerman EJ, Goldstein S and Stadtman ER (1987) Age related changes in oxidised proteins. *J Biol Chem* 262: 5488–5491.

Pearce JMS, Pennington RJT and Walton JN (1964) Serum enzyme studies in muscle disease: variations in serum creatine kinase activity in normal individuals. *J Neurol Neurosurg Psychiat* 27: 1–4.

Pearse AGE (1972) *Histochemistry, Theoretical and Applied*, 3rd edn. Churchill Livingstone, London.

Pennington RJT (1988) Biochemical aspects of muscle disease with particular reference to the muscular dystrophies. *In*: JN Walton (ed.): *Disorders of Voluntary Muscle*. Churchill Livingstone, Edinburgh, pp 455–486.

Puchala M and Schuessler H (1993) Oxygen effect in the radiolysis of proteins and haemoglobin. *Int J Radiat Biol* 64: 149–156.

Reznick AZ and Packer L (1994) Oxidative damage to proteins: spectrophotometric method for carbonyl assay. *Methods Enzymol* 233: 357–363.

Seglen PO and Bohley P (1992) Autophagy and other vacuolar protein degradation mechanisms. *Experientia* 48: 158–172.

Seligman AM, Karnovsky MJ, Wasserkrug HK and Hanker JS (1968) Non-droplet ultrastructural demonstration of cytochrome oxidase activity with a polymerising osmiophillic reagent, di-amino benzidine (DAB). *J Cell Biol* 38: 1–14.

Stadtman ER (1990) Metal ion-catalysed oxidation of proteins: biochemical mechanism and biological consequences. *Free Radical Biol Med* 9: 315–325.

Stadtman ER (1993) Oxidation of free amino acid residues in proteins by radiolysis and metal catalysed reactions. *Annu Rev Biochem* 62: 797–82.

Stern LZ, Ringel SP, Ziter FA, Menander-Hubeck K, Ionasescu V, Pellecirino RJ and Snyder RD (1982) Drug trial of superoxide dismutase in Duchenne muscular dystrophy. *Arch Neurol* 39: 342–346.

Tucek S (1982) The synthesis of acetylcholine in skeletal muscle of the rat. *J Physiol* 322: 53–69.

Oxidative stress and Ca²⁺ transport in skeletal and cardiac sarcoplasmic reticulum

V.E. Kagan[1], V.B. Ritov[1], N.V. Gorbunov[1], E. Menshikova[2] and G. Salama[2]

[1]Department of Environmental and Occupational Health and [2]Department of Cell Biology and Physiology, University of Pittsburgh, Pittsburgh, PA 15238, USA

Overview

Oxidative stress as a potential mechanism of Ca²⁺ disregulation in skeletal and cardiac muscles

Calcium is central to numerous cellular functions and impairment of membrane Ca²⁺ transport results in tissue injury (Rasmussen and Barrett, 1984; Thomas et al., 1996). In striated muscle, the signal that triggers force generation is a rise in cytoplasmic free Ca²⁺-concentration (Ca_i), whereas the removal of Ca_i imparts a state of relaxation. Based on sophisticated biochemical and biophysical studies, a generally accepted picture of force generation has emerged. In brief, force generation is produced from a cyclic interaction between actin filaments and cross-bridges projecting from the myosin filaments. This interaction is controlled by a regulatory protein complex (tropomyosin–troponin) bound to the actin filament. When Ca²⁺ is low ($< 10^{-7}$ M), the cross-bridge interactions are inhibited by the regulatory complex. As Ca_i rises, Ca²⁺ ions bind to the regulatory binding site of the troponin C subunit, thereby activating cross-bridge cyclic and tension development. In striated muscle, the regulation of Ca_i is controlled by a combination of influx and efflux processes. The firing of an action potential produces an influx of Ca_i which is predominantly the result of a release of Ca²⁺ from the sarcoplasmic reticulum via the opening of Ca²⁺ release channels of ryanodine receptors, with minor contributions from voltage-gated Ca²⁺ channels in the surface membrane and from the Na⁺/Ca²⁺ exchanger, in heart muscle. The removal of Ca_i is primarily driven by the active uptake of Ca²⁺/Mg²⁺ ATPases on the sarcoplasmic reticulum, with minor contributions from Ca²⁺ ATPases on the sarcolemma and in the heart, the Na⁺/Ca²⁺ exchanger on the sarcolemma (Bers, 1991; Sneyd et al., 1995; Sutko and Airey, 1996). In skeletal muscle, myocardium and aortic smooth muscle, these Ca²⁺-regulating mechanisms are known to be susceptible to oxidative stress produced by organic free radicals and reactive oxygen intermediates (Kagan, 1988; Comporti, 1989; Keith, 1993). Not surprisingly, oxidative injury of various cells has been shown to be associated with increased intracellular concentrations of Ca²⁺ (Forman et al., 1987; Livingston et al., 1992; Kumar et al., 1996).

Free radicals and reactive oxygen species in myocardial injury
The hypothesis that free radicals and other reactive oxygen species are responsible for myocardial injury (in particular, ischemic and post-ischemic injury, injury induced by anthracycline anti-biotics, chelated and non-chelated iron-dependent injury, etc.) has obtained much support over the

past few years (Meerson et al., 1982; Hess and Manson, 1984; Ferrari et al., 1996). Recent epide-miological studies indicate that the high levels of stored iron are associated with an increased risk of myocardial infarction, presumably as a result of the crucial involvement of iron in free radical production (Salonen et al., 1992). Conversely, higher dietary intake and plasma levels of the anti-oxidant vitamins and antioxidants are associated with reduced risk of mortality from ischemic heart disease (Salonen, 1991; Gey, 1993). These data, along with vitamin antioxidants' beneficial effects against experimentally induced oxidative damage to isolated hearts, suggest that a free radical components may be important in oxidation-induced myocardial damage.

Results of several studies performed in different laboratories suggest that the decrease of calcium uptake in cardiac sarcoplasmic reticulum caused by ischemia is not the result of a defect in calcium pumping capabilities but of an increased efflux through the ryanodine-sensitive Ca^{2+}-release channel (Davis et al., 1992). However, the molecular mechanisms responsible for the acti-vation of Ca^{2+}-release channels in the ischemic heart still need to be elucidated. Although the role of thiol oxidation as a potential mechanism of the activation of the Ca^{2+}-release channel in cardiac sarcoplasmic reticulum has been defined (Prabhu and Salama,1990; Salama et al., 1992), specific physiologically or pathologically relevant oxidants have not been identified so far.

In Langendorff-perfused hearts, direct electron spin resonance (ESR) measurements of semi-dehydroascorbyl radicals enhanced by iron, indicated that ascorbate and iron are released from cardiomyocytes into the perfusate during reperfusion of ischemic hearts (Liu et al., 1990; Nohl et al., 1991). Ascorbyl radical was likewise detected in the blood by its ESR signals during open-chest heart surgery (Tavazzi et al., 1992). These direct measurements of the generation of semidehydroascorbyl radicals provide robust evidence that the ascorbate/iron redox couple can be a major mechanism contributing to oxygen radical generation during ischemia–reperfusion, and thus might be partly responsible for ischemia–reperfusion injury (Liu et al., 1990; Nohl et al., 1991; Tavazzi et al., 1992). Of particular interest are the possible sites of action of the ascorbate/iron redox couple which can, in principle, induce lipid peroxidation and oxidize protein thiols. Studies on isolated sarcoplasmic reticulum (SR) vesicles and reconstituted ryanodine receptors in planar bilayers showed that the ascorbate radical directly activated ryanodine receptors of skeletal and cardiac muscles (Stoyanovsky et al., 1994). These direct effects of ascorbate/iron on SR Ca^{2+}-release channels may be important for understanding their potential contribution to ischemia–reperfusion injury and developing strategies to protect against such injuries.

Cardiotoxicity caused by thiol-specific oxidative stress-induced by phenolic anti-tumor drugs

Etoposide (VP-16), a semisynthetic derivative of podophyllotoxin, is increasingly used to treat cancer (Slevin, 1991). Since it was introduced in 1971, VP-16 has become the most widely used anticancer drug in the USA. It is frequently used as a first-line drug for treating small lung can-cer, germ-cell tumors, lymphomas, and more recently Kaposi's sarcoma associated with AIDS. The drug is also used to treat a variety of leukemias, including acute nonlymphocytic leukemia. Teniposide (VM-26), chemically a close relative of VP-16 and another derivative of podophyllo-toxin, is also a very potent antineoplastic drug. Both VP-16 and VM-26 are successfully used not only as single agents but also in combination chemotherapy with *cis*-platin and several other anti-tumor drugs, including cyclophosphamide, doxorubicin, and vincristine (Van Maanen et al., 1988), as well as in concurrent chemoradiotherapy (Goss et al., 1993). There is growing clinical experience documenting potential efficiency of both derivatives of podophyllotoxin in concurrent chemoradiotherapy of locally advanced non-small cell lung carcinoma, which is a major public health risk in the USA, accounting for almost 40 000 patient deaths yearly. The goal of many protocols of concurrent chemoradiotherapy (which includes VP-16 or VM-26) is to render operable those patients who were initially deemed unresectable (Langer et al., 1993).

Although myelosuppression is the major toxic effect of these two drugs, about 5% of patients experience cardiac arrhythmias and cardiac failures as side effects (Goss et al., 1993; Langer et al., 1993; Haak et al., 1993). In these patients, cardiotoxicity of VP-16 and VM-26 limits efficiency and intensification of cancer therapy. By using biochemical markers of cardiac injury (serum concentrations of cardiac lactate dehydrogenase and creatine kinase), Goren et al. (1988) demonstrated subclinical acute cardiotoxicity in 30–100% of children with leukemia who received VM-26 in combination with either methotrexate or amsacrine (m-AMSA). A recent study of human autopsy tissue distribution of the epipodophyllotoxins demonstrated that cardiac concentration of VP-16 may reach relatively high levels in some patients. For example, in a patient who received a cumulative dose of 1620 mg VP-16 (over 5 days), the concentration of VP-16 in the heart (200 ng/g tissue) was only 2.35 times lower than that in the tumor (473 ng/g) (Stewart et al., 1993).

Although the cardiotoxicity of some anti-tumor drugs has been associated with their ability to generate radicals, cardiotoxic mechanisms of podophyllotoxin derivatives – VP-16 and VM-26 – have not been investigated so far. Antracyclines are well-known examples of cardiotoxic anti-tumor drugs that generate oxygen radicals in cardiomyocytes via mitochondrial and microsomal electron transport-dependent mechanisms. Both VP-16 and VM-26 contain a hindered phenolic ring, the presence of which is a critical structural prerequisite for its anti-tumor activity. The cytotoxicity of VP-16 is considered to be dependent on a dual mechanism of DNA strand cleavage via inhibition of DNA topoisomerase II (Topo II) and/or direct DNA damage. It has been suggested that oxidative metabolic activation of etoposide by cytochrome P450-dependent monooxygenases, peroxidases, prostaglandin synthase, or tyrosinase is essential for its cytotoxicity. Phenoxyl radical is an inevitable intermediate in the oxidative activation of VP-16 by different oxidative enzymes. Significant activities of myeloperoxidase and prostaglandin synthase are characteristic of cardiac tissue (Golino et al., 1993; Raschke et al., 1993) and significant levels of cytochrome P450-dependent monooxygenases and tyrosinase activities have been measured in the heart. Bioavailability of VP-16 and occurrence of sufficiently high activities of VP-16-metabolizing enzymes in the heart may provide conditions for generation of VP-16 phenoxyl radicals in some patients. Previous studies demonstrated that redox cycling of VP-16 phenoxyl radicals which are in general highly reactive with protein thiols, are particularly potent at modifying ryanodine receptors and the Ca^{2+} pump in SR membranes from skeletal muscles.

Until now, the most robust evidence for the existence of hyperreactive sulfhydryls on the ryanodine receptor comes from studies using reactive disulfide compounds as thiol oxidants. Reactive disulfide compounds (e.g. 2,2'-dithiodipyridine and 4,4'-dithiodipyridine are absolutely specific thiol oxidants which are selectively attacked by low pK_a thiols, resulting in thiol–disulfide exchange reaction with the stoichiometric production of thiopyridone (Brocklehurst, 1979). These reactive disulfide compounds selectively triggered Ca^{2+} release from SR vesicles and activated ryanodine receptors reconstituted in planar bilayers; both effects were reversed by dithiothreitol (0.5–1 mM) or glutathione and inhibited ryanodine binding (Zaidi et al., 1989a, b). The stoichiometric production of thiopyridone indicated that only 2–4 nmol/mg protein of accessible protein thiols (out of 180 nmol/mg) were oxidized by reactive disulfide compounds whereas the remaining free thiols did not react with reactive disulfide compounds, but were detected through their interaction with dithiothreitol (Zaidi et al., 1989). These observations were confirmed with other thiol oxidants (Liu et al., 1994) and point out the unique feature of hyperreactive thiols and the strong likelihood of a functional role in the regulation of ryanodine receptors (Ca^{2+}-release channel) in cardiac and skeletal SR.

In contrast to anthracyclines, the cardiotoxic effects of VP-16 cannot easily be derived from their ability to generate oxygen radicals. Instead, high reactivity of VP-16 phenoxyl radicals towards protein thiol groups may be the major contributor to its cardiotoxicity. Thus, although free radicals may be involved in cardiotoxic effects for both anthracyclines and VP-16 (VM-26), the nature of radicals and mechanisms of radical-induced damage of cardiomyocytes are likely to be remarkably different. As a result, protection against cytotoxic effects of anthracyclines and VP-16 (VM-26) should use different strategies.

Oxidative stress in Duchenne muscular dystrophy and amyotropic lateral sclerosis (ALS)
Oxidative stress is suggested to be involved in the pathology of a variety of neural and muscular dystrophies, including amyotropic lateral sclerosis (ALS) and Duchenne muscular dystrophy. The antioxidant/ prooxidant imbalance induced by genetic and/or environmental factors may ultimately result in damage of critical biomolecules, triggering a chain of reactions that culminates in cell degeneration. Mutations in human copper zinc superoxide dismutase (CuZnSOD) are found in about 20% of patients with familial ALS (Rosen, 1993). Expression of high levels of human superoxide dismutase (SOD) containing a substitution of glycine for alanine at position 93 – a change that has little effect on enzyme activity – caused motor neuron disease in transgenic mice (Gurney et al., 1994). Surprisingly, transgenic mice with substitution of alanine at position 4 by valine, which results in a significantly decreased SOD activity, did not develop the stereotypical syndrome suggestive of motor neuron disease. Based on this, it was suggested that SOD-dependent generation of radicals, rather than the SOD deficiency, contributes to the pathogenesis of familial ALS (Gurney et al., 1994). Recent studies demonstrated that, in addition to its superoxide-scavenging function at low concentrations, SOD may act as a potent generator of hydroxyl radicals at higher concentrations (Sato et al., 1992; Yim et al., 1993). This emphasizes the significance of identifying potential endogenous enzymatic sources of reactive oxygen species (ROS) that may cause oxidative stress in neuromuscular dystrophies.

Potential involvement of oxidative stress in pathogenesis of the disease process of skeletal muscle has been implicated in numerous studies (Kanter, 1994; Louwerse et al., 1995; Tanaka et al., 1996). Unequivocal proof has, however, still to be obtained. One of the typical examples is Duchenne muscular dystrophy (DMD) which is associated with an abnormality of Ca^{2+} homoeostasis, which is considered to be a basic mechanism of cellular degeneration and necrosis (Bodensteiner et al., 1978; Reyes et al., 1994). Although a disturbance in Ca^{2+} homeostasis is suggested as critical for muscle cell degeneration and necrosis, its mechanism and association with oxidative stress are not clear. It has been suggested that oxidative stress can contribute to the muscle deterioration associated with DMD (Matcovics et al., 1982; Grinio et al., 1984; Hunter and Mohamed,1986; Murphy and Kehrer, 1989). Overproduction of end-products of free radical oxidation has been detected in DMD patients (Grinio et al., 1984; Reyes et al., 1994). Moreover, antioxidant intake was reported to be able to reduce the rate of damage in DMD patients (Clark, 1984; Edwards et al., 1984; Badalyan et al.,1985; Orndahl et al., 1986). The results of studies on accumulation of peroxidation products in DMD patients together with the beneficial effects of antioxidants, suggest that oxidative stress may be an important component of DMD pathology.

It seems that systematic *in vitro* studies should address two important questions in elucidating the role of oxidative stress in damage of Ca^{2+} regulation in skeletal and cardiac muscle: (1) What are the endogenous mechanism(s) involved in the enhanced production of ROS resulting in oxidative stress? (2) How does oxidative stress specifically translate into altered SR Ca^{2+} transport and/or deterioration of SR membranes in cardiomyocytes and skeletal fibres?

In vitro *studies of oxidative stress-induced damage to Ca^{2+} transport in SR membranes*

Oxidative modification of Ca^{2+} transport in SR membranes of skeletal muscle

Studies conducted in the early 1970s demonstrated that ROS, formed as a result of redox cycling of iron by ascorbate, inhibit net Ca^{2+} accumulation by SR vesicles isolated from rabbit skeletal muscles (Kozlov et al., 1973). Subsequent investigations showed that both the peroxidation of the membrane lipid bilayer (lipid peroxidation) and the oxidative modification of Ca^{2+} ATPase were involved in the inhibition of Ca^{2+} transport by SR membranes. However, the high concentrations of oxygen radicals and other ROS generated by experimental oxidation systems make them highly reactive and indiscriminate in their attack on a wide range of biomolecules (proteins, lipids, sugars and nucleic acids); they can oxidize low-molecular-weight thiols and protein thiols. The fact that ROS can interact with a particular biomolecule *in vitro* does not mean that these interactions actually occur in the physiological or pathological setting. Earlier studies did not address the issue of selectivity or elucidate the relative contribution of protein versus membrane lipid modification as sites of actions of ROS and their effects on intracellular Ca^{2+} regulation. Another key issue is which protein thiols are likely to be preferentially oxidized, given that cytosolic protein thiols exhibit a wide range of reactivity and accessibility (e.g. creatine kinase, ryanodine receptor, myosin ATPase, etc.). For example, studies on isolated SR vesicles demonstrated that not all protein thiols are alike; in fact, certain cysteine residues are highly accessible and their oxidation–reduction state can be involved in the regulation of channel activity. Sulfhydryl oxidizing agents were found to trigger Ca^{2+} release from SR vesicles by selectively acting at critical thiols on the ryanodine receptor, with little or no effects on the SR membrane or the Ca^{2+}-Mg^{2+} ATPase (Salama and Abramson, 1984; Trimm et al., 1986; Zaidi et al., 1989a).

The remarkable feature of the critical thiols on ryanodine receptors was that Ca^{2+} release was initiated by oxidizing 2–4 nmol of thiol (-SH) groups per milligram of SR protein, even though these thiols represented 1–2% of all the accessible thiols (Salama and Abramson, 1984). These measurements suggested that the first line of attack and cell injury caused by oxygen free radicals could be through the reversible oxidation of critical thiols on ion channel proteins; this could account for the free radical modification of Ca^{2+} transport. In support of this hypothesis, phenoxyl radicals (generated from phenolic compounds (-phenol or phenolic anti-tumor drug, such as etoposide on VP-16), were found to oxidize low-molecular-weight and protein thiols selectively without affecting membrane lipids (Kagan et al., 1994). An example of a thiol-specific effect of phenoxyl radicals on Ca^{2+} transport is shown on Figure 1. Ca^{2+} transport in SR membranes from rabbit skeletal muscles is inhibited by a specific oxidation of protein thiols by phenoxyl radicals which are generated through the tyrosinase-catalyzed one-electron oxidation of phenol. Low-molecular-weight antioxidants (ascorbate and glutathione) are able to prevent the phenoxyl radical-induced oxidative damage of Ca^{2+} transport in SR membranes via direct reduction of phenoxyl radicals.

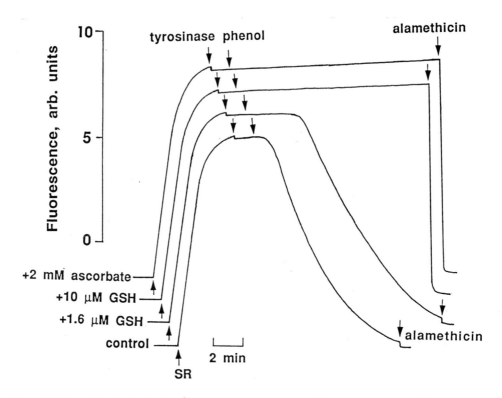

Figure 1. Phenoxyl radical-induced Ca^{2+} release from SR vesicles. Ca^{2+} accumulation and Ca^{2+} release were measured by chlortetracycline fluorescence in the medium containing 100 mM KCl, 1 mM $MgCl_2$, 15 μM $CaCl_2$, 1 mM ATP, 5 mM creatine phosphate, 5 IU/ml creatine kinase, 10 μM chlortetracycline and 10 mM HEPES (pH 7.0, 28 °C). The measurements were performed on a Perkin-Elmer LC-3 spectrofluorimeter in a thermostated cell, supplied with a vibrator. Addition of SR membranes (0.2 mg/ml) to the incubation medium containing ATP plus an ATP-regenerating system causes a time-dependent increase of the fluorescence. Tyrosinase (0.1 U/ml)/phenol (0.2 mM), when added to the incubation medium after the fluorescent response reached a plateau, completely eliminated the ATP-dependent increase of fluorescence. Ascorbate (2 mM) or glutathione (GSH, 1.6–10 μM) prevents the tyrosinase/phenol-induced release of Ca^{2+} from SR vesicles. Alamethicin (Alm, 5 μg/ml) was added to measure the total releasable pool of Ca^{2+}.

Oxidative modification of thiol-group(s) as a mechanism of activating SR Ca^{2+}-release channels
The reconstitution of Ca^{2+} channels by the fusion of SR vesicles with planar bilayer lipid membranes (BLMs) made it possible to characterize the single-channel properties of "native" Ca^{2+}-release channels (ryanodine receptor). The incorporation of purified 565-kDa proteins in BLMs indicated that isolated 565-kDa (junctional feet) proteins had almost identical properties to native Ca^{2+}-release channels in BLMs (Smith et al., 1988). Besides the large number of agents that impart pharmacological effects at SR Ca^{2+}-release channels, a variety of chemical oxidants and in particular, thiol-oxidizing reagents have been shown to activate Ca^{2+} release (Abramson et al., 1983; Zaidi et al., 1989). The actions of sulfhydryl oxidizing and reducing reagents were demonstrated in a variety of preparations, such as Ca^{2+}-release channels reconstituted in BLMs, SR vesicles, skinned skeletal muscle fibres, and papillary muscles. Substantial evidence indicated that sulfhydryl oxidizing reagents act at the physiological Ca^{2+}-release channel protein to increase single-channel open-time probability and sulfhydryl reducing agents promote the closure of the channel (Zaidi et al., 1989a; Prabhu and Salama, 1990). All the thiol-oxidizing reagents (Ag$^+$ and synthetic disulfides) that have been tested compete with ryanodine-binding to SR vesicles which indicates that these agents act at a specific site on the release channel, at or near the ryanodine-binding site (Hilkert et al., 1992; Pessah et al., 1987; Zaidi et al., 1989a). Agents known to stimulate (e.g. ATP, caffeine) or inhibit (e.g. ruthenium red, Mg^{2+}, tetracaine, procaine) SR Ca^{2+} release likewise stimulated or inhibited thiol-dependent activation of release.

In contrast to the actions of oxidizing reagents described above, the activation or opening of Ca^{2+}-release channels by sulfhydryl oxidizing reagents could be reversed by sulfhydryl reducing agents (e.g. cysteine, dithiothreitol) (Zaidi et al., 1989). Data from different laboratories indicate that calcium-induced calcium release from SR may play a dominant role in calcium overload and thus contribute to the onset of ischemic-reperfusion injury of the heart (Rassmussen and Barrett, 1984; Thomas et al., 1996). Although sulfhydryl oxidation–reduction has been shown to modify cardiac SR Ca^{2+} release, the role of this mechanism in physiological excitation–contraction coupling and in pathological states of skeletal and cardiac muscles remains unknown.

One approach to test whether or not thiol oxidation of the ryanodine receptor plays a role in cardiac ischemic injury is to compare Ca^{2+} transport of SR vesicles isolated form normal versus those from ischemic ventricular tissue, in the presence or absence of sulfhydryl reducing agents (Salama et al., 1996). As shown in Figure 2a, total SR Ca^{2+} uptake by vesicles isolated from normal (i.e. oxygenated) canine ventricular muscle was not modified by the addition of glutathione (GSH, 1 mM) indicating that the ryanodine receptor was normally in the fully reduced state and the further reduction did not improve Ca^{2+} sequestration by the vesicles. In contrast, Ca^{2+} transport by SR vesicles, isolated from canine ventricular muscle after a 30-min ischemic interval, accumulated considerably less Ca^{2+} than the control SR, and the rate and total Ca^{2+} uptake were markedly increased (50%) by the addition of glutathione (1 mM) (Fig. 2B). Transport measurements in dog and human SR isolated from ventricular tissue indicated that part of the ischemic injury is the activation of the ryanodine receptor through oxidation of critical sulfhydryl groups ($n = 6$).

Another approach is to show that physiologically relevant reductants, in the presence of transition metals, may act as endogenous regulators of ryanodine receptor through their ability to generate ROS and thus oxidize reactive thiol-groups. Indeed, ROS generated by ascorbate/iron redox

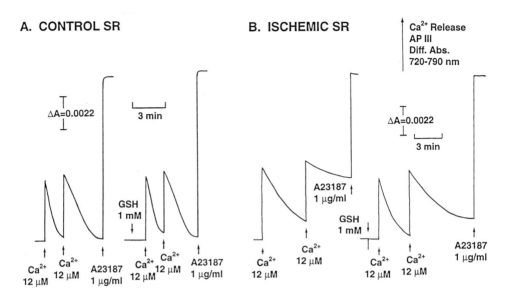

Figure 2. Effects of sulfhydryl reducing agents on Ca^{2+} transport by normal and "ischemic" SR vesicles. Sarcoplasmic reticulum vesicles were prepared from either normal or "ischemic" ventricular tissue, in the absence of sulfhydryl reagents throughout the procedure, as previously described (Prabhu and Salama, 1990). SR vesicles (0.2–0.4 mg protein/ml) were incubated in 100 mM KCl, 75 mM P_i, 1 mM $MgCl_2$, 0.1 mM antipyrylazo III (APIII), 0.1 mM ATP, 15 mM creatine phosphate, 10–15 U creatine kinase, 20 mM HEPES at pH 7.2 and 37 °C. The differential absorption changes of AP III were measured at 720–790 nm with a dual-wavelength spectrophotometer (University of Pennsylvania, Department of Biochemistry and Biophysics, model SBD3) to monitor extravesicular $[Ca^{2+}]$ with a 10 ms resolution. (A) Normal SR vesicles were Ca^{2+} loaded by sequential additions of Ca^{2+} until further additions of Ca^{2+} could no longer be sequestred by the SR vesicles. When glutathione (GSH, 1 mM) was added to the reaction medium, the rate and extent of Ca^{2+} uptake did not change, in six different preparations from six hearts ($n=6$). (B) When ischemic SR vesicles were Ca^{2+} loaded under identical conditions, the rate and extent of uptake was dramatically reduced compared with control SR vesicles. However, when GSH (1 mM) or L-cysteine (not shown) was added to the reaction medium, the rate and extent of Ca^{2+} uptake increased by more than 50%. These results were reproduced in six different ischemic hearts ($n=6$) and were consistently seen in sarcoplasmic reticulum isolated from human ischemic ventricular tissue procured at the time of transplantation but without carrying out the "ischemia" protocol.

cycling activated Ca^{2+}-release channels in SR vesicles reconstituted in artificial planar bilayers (Stoyanovsky et al., 1994). A specific oxidative attack on sulfhydryl groups of Ca^{2+} channels in SR membranes may be caused by phenoxyl radicals that are formed by enzymatic or nonenzymatic oxidation of different phenolic compounds (drugs, nutrients and environmental toxins). As shown in Figure 3, phenoxyl radicals generated from VP-16 by tyrosinase-catalyzed oxidation resulted in a pronounced activation of the Ca^{2+} channel (Fig. 3). In this case, the oxidative activation of the channel could be prevented by ascorbate which is an efficient scavenger of VP-16 phenoxyl radicals. Thus, in the absence of transition metals, ascorbate acts as an efficient protector of the Ca^{2+} transport against phenoxyl radical-induced activation of the Ca^{2+} channel.

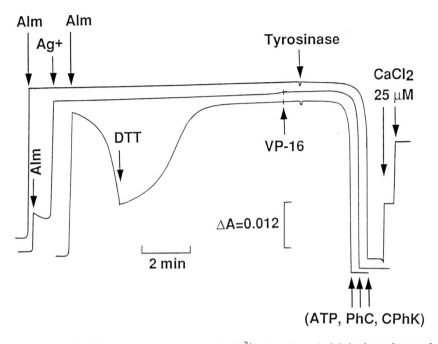

Figure 3. Effects of VP-16/tyrosinase and ascorbate on the Ca^{2+} release channel of skeletal muscle sarcoplasmic reticulum. Ca^{2+} transport across SR vesicles was determined by measuring extravesicular $[Ca^{2+}]_{free}$ through the differential absorption changes of antipyrylazo III (AP III). The recordings shown are prototypical of four separate experiments using SR membranes isolated at different days ($n=4$). Skeletal SR vesicles (0.2 mg protein) were suspended in the reaction medium, containing 100 mM KCl, 1 mM $MgCl_2$, 5 mM creatine phosphate, 2–3 IU/ml creatine kinase, 100 μM antipyrylazo III, and 20 mM HEPES (pH 7.0 at 28 °C). Two additions of Ca^{2+} were added to calibrate the response of AP III, then 100 μM ATP was added to initiate active Ca^{2+} uptake. An addition of the tyrosinase (1 U/ml) plus VP-16 (0.5 mM) elicited Ca^{2+} release which was effectively reversed by dithiothreitol (1 mM), indicating that Ca^{2+} was re-sequestered by Ca^{2+} pump on the SR. Alamethicin (Alm, 5 μg/ml) was added to measure the total releasable pool of Ca^{2+}.

Hydroperoxide-induced oxidative stress and damage of Ca^{2+} transport: Protective effects of nitric oxide

Interaction of hydrogen peroxide or organic hydroperoxides (including lipid hydroperoxides) with hemoproteins is known to produce oxoferryl hemoprotein species which act as very potent oxidants. As skeletal and cardiac muscle cells contain high concentrations of myoglobin this reaction may be an important mechanism of initiation or enhancement of oxidative stress which may impair their Ca^{2+} transport systems. As shown in Table 1, incubation of skeletal SR membranes in the presence of myoglobin and *tert*-butyl-hydroperoxide (*t*-BuOOH) caused inhibition of Ca^{2+} transport, which was dependent on the incubation time as well as concentrations of myoglobin and *t*-BuOOH. Electron proton resonance (EPR) measurements demonstrated the formation of free radical species derived from oxoferryl-myoglobin (Fig. 4). Incubation of

Table 1. NO protects the loss of Ca^{2+} uptake elicited by t-BuOOH/oxymyoglobin in SR

Time of incubation[*]	Ca^{2+} uptake rate (nmol Ca^{2+}/min/mg protein)	
(min)	Oxymyoglobin (100 μM) + t-BuOOH (150 μM)	Oxymyoglobin (100 μM) + t-BuOOH (150 μM) + PAPA-NONOate (500 μM)
0	85.0±2.3	92.0±8.0
5	68.5±5.6	80.6±7.1
15	43.4±6.4	79.2±3.2
20	23.8±3.2	76.3±5.0
30	13.4±4.2	73.0±6.0

[*]Skeletal SR vesicles were pre-incubated with 100 μM oxyMb and 150 μM t-BuOOH for 0–30 min at 37 °C. Ca^{2+} uptake was measured as in Figure 3. Initial Ca^{2+} uptake rates were calculated in the first 2 min.

Mb-FeII(O_2) (100 μM) with t-BuOOH (150 μM) gave rise to a broad two-line anisotropic EPR signal (Fig. 4b). The g values at the zero crossing point and the low-field maximum observed in the EPR spectra were 2.012 and 2.036, respectively. This EPR signal, obtained through the reaction of oxymyoglobin and t-BuOOH, had the profile and characteristic features of peroxyl radicals centered on a protein (Gibson and Ingram, 1956; King and Winfield, 1963; Sclick et al., 1985; Gunther et al., 1995; Kevan and Schlick, 1986; Patel et al., 1996). Apparently, formation of the protein-centered radicals proceeded in a two-step reaction. The first stage was the Fenton-type reaction which produced Mb-FeIII and butoxyl radicals (t-BuO˙) (EPR signal at $g=2.02$, Fig. 4). In the presence of excess t-BuOOH, Mb-FeIII can react with another t-BuOOH molecule, yielding oxoferryl species Mb-FeIV and t-BuO˙. At the second stage t-BuO˙ radicals react with apoprotein moieties of myoglobin to form – in the presence of O_2 – the protein-centered peroxyl radical (Mb-FeII(O_2)) (Rao et al., 1993; Kelman et al., 1994; Barr and Mason, 1995; Van der Zee et al., 1996). Addition of SR membranes (150 μg) to the mixture of Mb-Fe(O_2) and t-BuOOH caused a decrease in the intensity of peroxyl radical EPR signals (Fig. 4). Not surprisingly, this was accompanied by a significant accumulation of peroxidation products in the SR membranes.

Nitric oxide (NO) has been shown to reduce hydroperoxides at heme- and nonheme iron catalytic sites, thus preventing the formation of oxoferryl-derived radical species (Gorbunov et al., 1995, 1997). As shown in Figure 4, the features characteristic of oxoferryl-derived radicals were not present in the EPR spectrum obtained from SR membranes incubated with t-BuOOH and Mb-FeII(O_2) upon addition of NOC-15, an NO donor. This antiradical effect of NO correlated with the protection of SR membranes against oxidative damage (Tab. 1).

Activation of ryanodine receptors of skeletal and cardiac muscles by nitric oxide (NO)
The endothelial-derived relaxing factor, NO, has been shown to depress force in smooth and cardiac muscles through the activation of guanylate cyclase and an increase in cyclic GMP (cGMP). In fast skeletal muscle, NO (i.e. NO-related compounds) elicits a modest decrease in developed

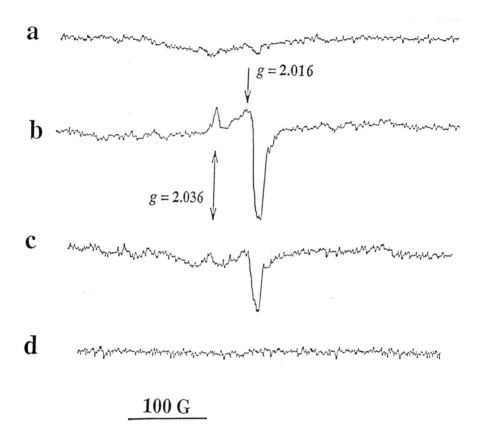

Figure 4. Low-temperature EPR spectra obtained from solutions of Mb-FeII(O$_2$) and/or t-BuOOH in the absence and presence of skeletal SR membranes and an NO donor, NOC-15. SR membranes from skeletal muscles and/or Mb-FeII(O$_2$), and/or an NO-donor, NOC-15 (PAPA-NONOate) were incubated with t-BuOOH for 15 min in 20 mM HEPES (pH 7.2) containing 100 mM KCl, 1 mM MgCl$_2$, 15 µM CaCl$_2$, 1 mM ATP, 5 mM creatine phosphate, 3 IU/ml creatine kinase, and 0.05 mM deferoxamine (DFO). Samples (250 µl) were placed into a Teflon tube (3.7 mm internal diameter) and frozen in liquid nitrogen, then removed from the tube to perform EPR measurements. For spectrum recording, each sample was placed in an EPR quartz tube (5 mm internal diameter) in such a way that the entire sample was within the effective microwave irradiation area. EPR measurements were performed on a JEOL-RE1X spectrometer with a variable temperature controller (Research Specialists, Chicago, IL). The spectra were recorded at –170 °C, 320 mT center field, 10 mW power, 0.1 mT field modulation, 25 mT sweep width, 500 receiver gain, 0.1 s time constant. The g-factor values were determined relative to external standards, containing Mn^{2+} (in MgO). Analog signals were converted into digital form and imported to an IBM computer. Intensity of the signals was calculated using a program developed by David Duling. All spectra (except "c") were recorded at a scan rate of 10.25 mT/min. Spectrum (c) was recorded at a scan rate of 6.25 mT/min. (a) Control solution (150 µg/ml SR membranes and 500 µM t-BuOOH); (b) same as (a), but in the presence of 100 µM oxyMb; (c) same as (b), except in the presence of 150 µM t-BuOOH; (d) same as (b), but in the presence of 500 µM NOC-15.

force, but, in contracting muscle, NO increases force by a mechanism independent of cGMP (Kobzik et al., 1994). In cardiac muscle, exogenously added NO produced negative ionotropic effects (Balligand et al.,1992; Finkel et al., 1992), but recent re-examination of this issue produced the opposite results (Mohan et al., 1996). Thus, the effects of NO appear to be multifaceted; there is increasing evidence that the biologically related actions of NO cannot be entirely attributed to cGMP mechanisms. In principle, authentic NO (i.e. NO gas) is highly reactive and, in the presence of oxygen, will readily react to form N_2O_3 which can either neutralize NO to NO_2^- or interact with low-molecular weight thiols to form S-nitrosothiols (R-SNO). Nitrosothiol compounds can, in turn, impart biological activity through transnitrosylation of protein thiols. Based on findings that ryanodine receptors contain hyperreactive thiols, NO and NO donors were tested as possible activators of SR Ca^{2+} release. In skeletal and cardiac SR, NO gas and NO donors (S-nitrosocysteine, S-nitrosopenicillamine [SNAP] and nonthiol NO donors, PAPA-NONOate or NOC-15) were found to trigger Ca^{2+} release from Ca^{2+}-loaded SR vesicles (Stoyanovsky et al., 1997). Nitrosothiols were found to be potent activators of SR Ca^{2+} release and increased the open probability of ryanodine receptors being reconstituted in planar bilayers (Stoyanovsky et al., 1997). Based on the relative potency of various NO donors and the reversibility of their effects with dithiothreitol, the most likely mechanism is a transnitrosylation of NO from R-SNO donors to the ryanodine receptors. In principle, NO can form peroxynitrite in the presence of superoxide ions; peroxynitrite has been shown to oxidize low-molecular-weight and protein thiols readily and could therefore act on regulatory thiols on SR ryanodine receptors. Direct tests of peroxynitrite and a peroxynitrite donor SIN-1 indicated that, indeed, both agents elicited Ca^{2+} release from SR vesicles, but their effects could not be reversed by sulfhydryl-reducing agents. Moreover, peroxynitrite added to planar bilayers promoted lipid per-oxidation which was tracked as a decrease in resistance and increase in nonselective transmembrane current (Stoyanovsky et al., 1997).

In striated muscle, NO is known to depress force and it might therefore seem paradoxical that it does so by promoting SR Ca^{2+} release via ryanodine receptors. To avoid this apparent inconsistency, Meszaros et al. (1996) proposed that NO inhibited SR Ca^{2+} release (by partially reversing the effect of caffeine) and decreased the open probability of ryanodine receptors being reconstituted in planar bilayers. Both effects were reversed by the sulfhydryl reducing agent mercaptoethanol. However, this scheme is inconsistent with extensive observations that oxidation opens and reduction closes the Ca^{2+}-release channel.

A closer examination of NO-induced Ca^{2+} release from the sarcoplasmic reticulum reveals that this mechanism is entirely consistent with force depression when combined with cGMP stimulation of Ca^{2+} pumps on the plasma membrane. R-SNO-induced activation of ryanodine receptors will have different effects on force depending on the muscle type, as a result of the differences in Ca^{2+} handling and excitation-contraction mechanisms. In fast skeletal muscle, twitch and tetanic forces are entirely dependent on intracellular stores of Ca^{2+}, and Ca^{2+} is recycled in and out of the SR with little or no contribution from extracellular influx. In this case, activation of ryanodine receptors by R-SNO elicits greater Ca^{2+} transients, increases developed force and may elevate the background levels of Ca^{2+} and baseline force. The simultaneous NO-dependent stimulation of guanylate cyclase and elevation of cGMP will promote the efflux of the excess Ca^{2+} released from the sarcoplasmic reticulum, thus causing a net decrease in SR Ca^{2+}

load and depression of force (Murphy et al., 1997). This mechanism is demonstrated in Figure 5 in mouse diaphragm muscles, a fast skeletal muscle stimulated with electrical pulses (1 ms duration and 2× threshold voltage at 0.2 Hz). In Figure 5a, force is monitored before and during the addition of the NO donor cys-SNO (S-nitrosocysteine) which causes the expected gradual decrease in force. On the other hand, when cys-SNO is added more abruptly as a bolus, there is a highly reproducible transient increase in baseline force indicative of SR Ca²⁺ release elicited by

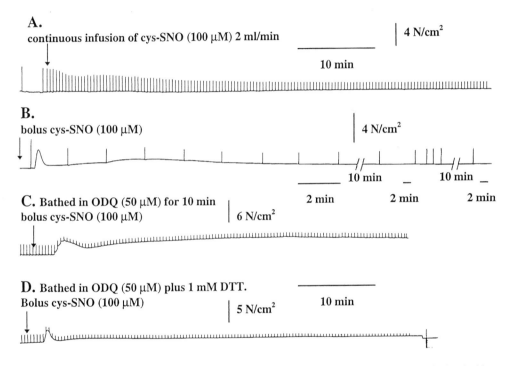

Figure 5. Dual actions of NO on mouse diaphragm muscle. Mouse diaphragmatic muscles were isolated with a tendon at one end and the intercostal rib bone at the other and were mounted between a fixed clip and a tension transducer. The muscles were paced at 0.2 Hz with stimuli of 2 ms duration and 2× threshold voltage with platinum wire electrodes. L_0: the length at which maximum peak twitch force is generated was determined by gradually stretching the muscle; then the muscle was allowed to stabilize for 30 min before starting the measurements. Baseline force and twitch force were measured continuously on a fast response chart recorder (Gould). (A) The infusion of cys-SNO (S-nitrosocysteine) in the bath produced the expected force depression. (B) The abrupt addition of a bolus of cys-SNO to the bath produced a large rise in baseline force (40–60% of peak twitch) which recovered in minutes ($n=12$ preparations). Then twitch force decreased, as observed in (A). These concentrations of cys-SNO increased the cGMP levels in diaphragm muscles by twofold, according to radioimmunoassay detection of cGMP ($n=4$). (C) To elucidate the role of cGMP-dependent processes, the muscle was first bathed in ODQ (50 μM), a selective inhibitor of NO-sensitive guanylate cyclase. ODQ alone did not alter force generation or cGMP levels ($n=5$). The subsequent addition of cys-SNO produced a rapid rise in baseline force which did not recover for more than 1 h and was associated with a decrease in twitch force ($n=12$). As expected, ODQ blocked the rise of cGMP induced by cys-SNO ($n=5$). (D) In the presence of DTT and ODQ, a bolus of cys-SNO produced a transient increase in baseline force and decreased twitch force. Thus the main effect of the sulfhydryl reducing agent was to prevent the long-lasting maintenance of baseline force ($n=12$).

cys-SNO (Fig. 5b). To separate the cGMP and ryanodine receptor effects of NO, the mouse diaphragm was bathed in ODQ (50 µM), a highly selective blocker of NO-sensitive guanylate cyclase
(i.e. soluble guanylate cyclase). ODQ, in the absence of NO donors, had no effect on force
generation. The subsequent addition of cys-SNO produced an increase in baseline force which
reached a steady state of 50–60% of twitch force. Such long-lasting, high levels of intracellular
Ca^{2+} compromise the muscle, but can be reversed by either dithiothreitol or ryanodine (100 µM).
The effects of cys-SNO on the ryanodine receptors were blocked by sulfhydryl reducing agents
(e.g. 1–10 mM dithiothreitol) (Fig. 5d) which confirmed the mechanic link of transnitrosylation
of ryanodine receptors to the rise in baseline force. Thus, NO-induced Ca^{2+} release from the SR
is entirely compatible with excitation-contraction coupling in skeletal muscle, because NO has
dual actions: one is to release Ca^{2+} from the sarcoplasmic reticulum to regulate SR Ca^{2+} load
(and subsequent release) and elevate Ca^{2+} in the cytosol; the second is to stimulate guanylate
cyclase and remove excess cytosolic Ca^{2+} by cGMP-dependent pathways (Murphy et al., 1997).

A. Control

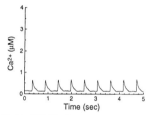

B. CysNO 50 µl

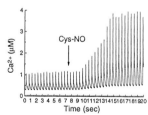

C. 25 min after

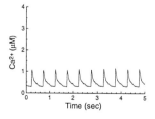

Figure 6. The NO donor cys-SNO elicits SR Ca^{2+}
release in rabbit hearts. Langendorff-perfused rabbit
hearts were loaded with the Ca^{2+} indicator dye
Rhod-2/AM by adding 0.3 mg dye in dimethyl-
sulfoxide to the coronary perfusate. Rhod-2 exhibits
over 100-fold increase in fluorescence upon
binding to Ca^{2+} and can be used to measure rapid
Ca^{2+} transients during a cardiac beat. (A) Control
Ca^{2+} transients. (B) Free intracellular Ca^{2+} during
an addition of cys-SNO. The NO donor caused an
increase in baseline and peak Ca^{2+}. (C) The effects
of a bolus addition of cys-SNO was transient as
both baseline Ca^{2+} and Ca^{2+} transients recovered
within 25 min.

In cardiac muscle, force development requires intra- and extracellular sources of Ca^{2+} coming from the SR and through voltage-gated L-type Ca^{2+} channels, respectively. The activation of ryanodine receptors by R-SNO would predict an increase in background Ca^{2+} and developed force. Figure 6 shows that these predictions are confirmed by direct measurements of intracellular Ca^{2+} in Langendorff-perfused rabbit hearts. The perfused rabbit heart was loaded with the Ca^{2+} indicator, Rhod-2AM, which diffuses across the cell membrane and is trapped in the cytosol after hydrolysis with endogenous esterases. The fluorescence of Rhod-2 is negligible but its quantum efficiency increases by more than 100-fold upon binding to Ca^{2+}. Ca^{2+} transients measured with Rhod-2 show a rapid rise and a biphasic recovery of Ca_i, with baseline values of 250 nM and peak transients of 680–800 nM. The addition of cys-SNO to the coronary perfusate produced a rise in baseline Ca^{2+} and peak Ca^{2+} within 20–25 beats, which gradually decreased in 10–15 min, after which, force remained depressed for 30–45 min. The initial rise in Ca^{2+} was not blocked by L-type channel blockers but was blocked by sulfhydryl reducing agents. These recent observations are again consistent with a dual role of NO in muscle.

Enzymatic lipid peroxidation in SR membranes
Enzymatic NAD(P)H-dependent systems of lipid peroxidation have previously been reported in SR membranes isolated from skeletal muscles of chicken and flounder (Lin and Hultin, 1976; Shewfelt and Hultin, 1983). It has now been demonstrated that enzymatic NADH- and NADPH-dependent lipid peroxidation occurs in an SR fraction isolated from rabbit skeletal muscle, as measured by accumulation of lipid peroxidation products (thiobarbituric and reactive substance, TBARS) (Tab. 2). Although the rates of the enzymatic lipid peroxidation are lower than the rate

Table 2. Enzymatic NADH- and NADPH-dependent lipid peroxidation in rabbit skeletal muscle SR membranes (nmol MDA/mg protein)

Additions[*]	Ascorbate (0.5 mM)	NADH (0.5 mM)	NADPH (0.5 mM)
Control	5.3±1.6	0.60±0.11	0.41±0.08
Thermal treatment			
(5 min, 95 °C)	5.5±1.4	0.02±0.01	0.02±0.01
NaN₃ (3×10^{-4} M)	5.3±1.6	0.56±0.08	0.39±0.07
4-Chloromercuribenzoate			
(10^{-3} M)	5.0±1.5	0.08±0.03	0.09±0.03
KCN (10^{-4} M)	–	0.57±0.08	0.42±0.06
CO	–	0.58±0.07	0.39±0.05
BHT (10^{-4} M)	0.03±0.02	0.04±0.02	0.03±0.02
α-tocopherol (10^{-4} M)	0.05±0.02	0.06±0.03	0.05±0.02
without Fe^{2+}	–	0.05±0.04	0.03±0.02
Storage at 0 °C for 4 days	5.5±1.5	0.20±0.05	0.07±0.01

[*]SR suspensions (2.0 mg protein/ml) were incubated with the inducers of lipid peroxidation, inhibitors and/or antioxidants (as indicated) in 50 mM phosphate buffer (pH 7.4 at 37 °C, 15 min).

of non-enzymatic process (initiated by ascorbate + iron), both NADH and NADPH produced pronounced accumulation of lipid peroxidation products. Both enzymatic activities were sensitive to thermal denaturation, thiol-reagents, storage, and antioxidants. Inhibitors of mitochondrial respiration and cytochrome P450-related activities did not affect the enzymatic peroxidation in SR membranes.

Conclusion

Ca^{2+} transport systems of cardiac and skeletal muscle SR membranes show high sensitivity to oxidative stress, suggesting that ROS and/or peroxidation intermediates or products may act as either signaling molecules or damaging agents which disturb Ca^{2+} homoeostasis. Both intracellular reductants (e.g. ascorbate, glutathione) and prooxidants (e.g. nitric oxide) may act as a double edged sword in either inducing or protecting against oxidative modification of Ca^{2+}-transport systems. Hence, mechanism-based antioxidant strategies need to be developed for effective control of the redox-driven modification of Ca^{2+}-transport systems. This cannot be achieved without identification of endogenous generators of ROS in muscle cells. In this respect, NO may act as an important regulator which controls SR Ca^{2+} load and force generation in striated muscle.

References

Abramson J, Trimm JL, Weden L and Salama G (1983) Heavy metals induce rapid calcium release from sarcoplasmic reticulum vesicles isolated from skeletal muscle. *Proc Natl Acad Sci USA* 80: 1526–1530.

Badalyan LO, Grinio LP, Islamova IB, Belousova LV, Rafanov VS, Prilipko LL and Kagan VE (1985) On the pathology of membrane structures in Duchenne myodystrophy in humans. *Sov J Neuropathol Psychiat* 85: 1631–1633.

Balligand JL, Kelly RA, Mardsen PA, Smith TW and Mitchel T (1992) Control of cardiac muscle cell function by an endogenous nitric oxide signaling system. *Proc Natl Acad Sci USA* 90: 347–351.

Barr DP and Mason RP (1995) Mechanism of radical production from the reaction of cytochrome C with organic hydroperoxides. *J Biol Chem* 270: 12709–12716.

Bers DM (1991) *Excitation-Contraction Coupling and Cardiac Contractile Force.* Kluwer Dordrecht.

Bodensteiner JB, Andrew G and Engel AG (1978) Intracellular calcium accumulation in Duchenne dystrophy and other myopathies: a study of 567000 muscle fibers in 114 biopsies. *Neurology* 28: 439–445.

Brocklehurst K (1979) Specific covalent modification of thiols: applications in the study of enzymes and other biomolecules. *Int J Biochem* 10: 259–274.

Clark IA (1984) Proposed treatment of Duchenne muscular dystrophy with desferrioxamine. *Med Hypoth* 13: 153–156.

Comporti M (1989) Three models of free radical-induced cell injury. *Chem-Biol Interact* 72(1–2): 1–56.

Davis MD, Lebolt W and Feher JJ (1992) Reversibility of the effects of normothermic global ischemia on the ryanodine-sensitive and ryanodine-insensitive calcium uptake of cardiac sarcoplasmic reticulum. *Circ Res* 70: 163–171.

Edwards RHT, Jones DA and Jackson MJ (1984) An approach to treatment trials in Muscular Dystrophy with particular reference to agents influencing free radical damage. *Med Biol* 62: 143–147.

Ferrari R, Ceconi C, Curello S, Benigno M, Lacanna G and Visioli O (1996) Left ventricular dysfunction due to the new ischemic outcomes – stunning and hybernation. *J Cardiovasc Pharmacol* 28(suppl 1): S18–S26.

Finkel MS, Oddis CV, Jacob TD, Watkins SC, Hattler BG and Simmons RL (1992) Negative ionotropic effects of cytokines on the heart mediated by nitric oxide. *Science* 257: 387–389.

Finkel MS, Shen L, Oddis CV, Romeo RC and Salama G (1993) Positive ionotropic effect of acetylcysteine in cardiomyopathic Syrian hamsters. *J Cardiovasc Pharmacol* 21: 29–34.

Forman HJ, Dorio RJ and Skelton DC (1987) Hydroperoxide-induced damage to alveolar macrophage function and membrane integrity: alterations in intracellular-free Ca^{2+} and membrane potential. *Arch Biochem Biophys* 259: 457–65.

Gey KF (1993) Prospects for the prevention of free radical disease, regarding cancer and cardiovascular disease. *Brit Med Bull* 49: 679–99.

Gibson JF and Ingram DJE (1956) Location of free electrons in porfirin ring complexes. *Nature* 178: 871–872.

Golino P, Ambrosio G, Villari B, Ragni M, Focaccio A, Pace L, Declerk F, Condorelli M and Chiariello M (1993) Endogenous prostaglandin endoperoxides may alter infarct size in the presence of thromboxane synthase. inhibition – studies in a rabbit model of coronary-artery occlusion reperfusion. *J Am Coll Cardiol* 21: 493–501.

Gorbunov, N.V., Osipov, A.N., Day, B.W., Zayas-Rivera, B., Kagan, V.E. and Elsayed, N.M. Reduction of ferrylmyoglobin and ferrylhemoglobin by nitric oxide: a protective mechanism against ferryl hemoprotein-induced oxidations. *Biochemistry* 34: 6689–6699.

Gorbunov NV, Yalowich JC, Gaddam AS, Thampatty P, Kisin ER, Ritov VB, Elsayed NM and Kagan VE (1997) Nitric oxide prevents oxidative damage produced by tert-butyl hydroperoxide in erythroleukemia cells via nitrosylation of heme and non-heme iron: electron paramagnetic resonance evidence. *J Biol Chem* 272: 12328–12341.

Goren MP, Li JL, Mirro Jr J and Ochs J (1988) Detection of subclinical acute cardiotoxicity by changes in lactate dehydrogenase and creatine kinase serum isoenzymes in patients receiving concomitant teniposide (VM-16) and methotrexate or AMSA therapy. *Proc Am Assn Cancer Res* 29: 229–236.

Goss GD, Vincent M, Germond C, Rowen J, Dhaliwal H and Corringham R (1993) Combination chemotherapy with teniposide (VM-26) and Carboplatin in small cell lung cancer. *Amer. J. Clin.Oncol.* 16: 295–300.

Grinio LP, Orlov ON, Prilipko LL and Kagan VE (1984) Lipid peroxidation in children with inherited Duchenne myopathy. *Bull Exp Biol Med USSR* 10: 423–425.

Gunther MR, Kelman DJ, Corbett JT and Mason RP (1995) Self-peroxidation of metmyoglobin results in formation of an oxygen-reactive tryptophan-centered radical. *J Biol Chem* 270: 16075–16081.

Gurney ME, Pu H, Chiu AY, Dal Canto MC, Polchow CY, Alexander DD, Caliendo J, Hentati A, Kwon YW, Deng H-X et al. (1994) Motor neuron degeneration in mice that express a human Cu,Zn superoxide dismutase mutation. *Science* 264: 1772–1775.

Haak HL, Gerrits WBJ, Wijermans PW and Kerkhofs H (1993) Mitoxantrone, teniposide, chlorambucil and prednisone (MVLP) for relapsed non-Hodgkin's-lymphoma: the impact of advanced age and performance status. *Neth J Med* 43: 122–127.

Hess ML and Manson NH (1984) Molecular oxygen: friend and foe. The role of the oxygen free radical system in the calcium paradox, the oxygen paradox and ischemia/reperfusion injury. *J Mol Cell Cardiol* 16: 969–985.

Hilkert R, Zaidi N, Shome K, Nigam M, Lagenaur C and Salama G (1992) Properties of immunoaffinity purified 106-kDa Ca^{2+} release channels from the skeletal sarcoplasmic reticulum. *Arch Biochem Biophys* 292: 1–15.

Hunter MIS and Mohamed JB (1986) Oxidative stress and muscular dystrophy. *Clin Chim Acta* 155: 123–131.

Kagan VE (1988) *Lipid Peroxidation in Biomembranes*. CRC Press, Boca Raton, FL, pp 55–146.

Kagan, V.E., Yalowich, J.C., Day, B.W., Goldman, R., Gantchev, T.G. and Stoyanovsky, D.A. Ascorbate is the primary reductant of the phenoxyl radical of etoposide (VP-16) in the presence of thiols both in cell homogenates and in model systems. *Biochemistry* 33: 9651–9660.

Kanter MM (1994) Free radicals, exercise, and antioxidant supplementation. *Int J Sport Nutr* 4: 205–220.

Keith F (1993) Oxygen free radicals in cardiac transplantation. *J Cardiac Surg* 8(2 suppl): 245–248.

Kelman DJ, DeGray JA and Mason RP (1994) Reaction of myoglobin with hydrogen peroxide forms a peroxyl radical which oxidizes substrates. *J Biol Chem* 269: 7458–7463.

Kevan L and Schlick S (1986) Peroxyl spin probes as motional probes in polymers and on oxide surfaces. *J Phys Chem* 90: 198–207.

King NK and Winfield ME (1963) The mechanism of metmyoglobin oxidation. *J Biol Chem* 238: 1520–1528.

Kobzik L, Reid MB, Bre DS and Stammler JS (1994) Nitric Oxide in skeletal muscle. *Nature* 372: 546–548.

Kozlov YP, Ritov VB and Kagan V (1973) Ca^{2+}-transport and free radical lipid oxidation in sarcoplasmic reticulum membranes. *Proc Natl Acad Sci USA* 212: 216–219.

Kumar R, Agarwal AK and Seth PK (1996) Oxidative stress-mediated neurotoxicity of cadmium. *Toxicol Lett* 89: 65–69.

Langer CJ, Curran WJ, Keller SM, Catalano R, Fowler W, Blankstein K, Litwin S, Bagchi P, Nash S and Comis R (1993) Report of phase II trial of concurrent chemoradiotherapy with radical thoractic irradiation (60 Gy), infusional fluorouracil, bolus cisplatin and etoposide for clinical stage IIIB and bulky IIIA non-small cell lung cancer. *Int J Radiat Oncol Biol Phys* 26: 469–478.

Lin TS and Hultin HO (1976) Ezymatic lipid peroxidation on microsomes of chicken skeletal muscle. *J Food Sci* 41: 1488–1493.

Liu XK, Prasad MR, Engelman RM, Jones RM and Das DK (1990) Role of iron on membrane phospholipid breakdown in ischemic-reperfused rat heart. *Am J Physiol* 259: H1101–H1107.

Liu GH, Abramson JJ, Zable AC and Pessah IN (1994) Direct evidence for the existence and functional role of hyperreactive sulfhydryls on the ryanodine receptor-triadin complex selectively labeled by the coumarin maleimide 7-diethylamino-3(4'-maleimidylohenyl)-4-methylcoumarin. *Mol Pharmacol* 45: 189–200.

Livingston FR, Lui EM, Loeb GA and Forman HJ (1992) Sublethal oxidant stress induces a reversible increase in intracellular calcium dependent on NAD(P)H oxidation in rat alveolar macrophages. *Arch Biochem Biophys* 299: 83–91.

Louwerse ES, Weverling GJ, Bossuyt PM, Meyjes FE and de Jong JM (1995) Randomized, double-blind, controlled trial of acetylcysteine in amyotrophic lateral sclerosis. *Arch Neurol* 52: 559–564.

Matcovics B, Laszlo A and Szabo L (1982) A comparative study of superoxide dismutase, catalase and lipid peroxodation in red blood cells from muscular dystrophy patients and normal controls. *Clin Chim Acta* 118: 289–292.

Meerson FZ, Kagan VE, Kozlov YuP, Belkina LM and Arkhipenko YuV (1982) The role of lipid peroxidation in pathogenesis of ischemic damage and the antioxidant protection of the heart. *Basic Res Cardiol* 77: 465–485.

Meszaros LG, Minarovic I and Zahradnikova A (1996) Inhibition of the skeletal muscle ryanodine receptor calcium release channel by nitric oxide. *FEBS Lett* 380: 49–52.

Mohan P, Bruseart DL, Paulus WJ and Sys SU (1996) Myocardial contractile response to nitric oxide and cGMP. *Circulation* 93: 1223–1229.

Murphy ME and Kehrer JP (1989) Oxidative stress and muscular dystrophy. *Chem-Biol Interact* 69: 161–173.

Murphy TD, Anno PR and Salama G (1997) Nitric oxide depresses force in mouse diaphragm by eliciting Ca^{2+} release from the sarcoplasmic reticulum. *Am J Respir Crit Care Med* 155: A448.

Nohl H, Stolze K, Napetschnig S and Ishikawa T (1991) Is oxidative stress primarily involved in reperfusion injury of the ischemic heart? *Free Radical Biol Med* 11: 581–588.

Orndahl G, Sellden U, Hallin S, Wetterquist H Rindby A and Selin E (1986) Myotonic dystrophy treated with selenium and vitamin E. *Acta Med Scand* 219: 407–414.

Patel RP, Svistunenko DA, Darley-Usmar VM, Symons MCR and Wilson MT (1996) Redox cycling of human methaemoglobin by H_2O_2 yields persistent ferryl iron and protein based radicals. *Free Radical Res* 25: 117–123.

Pessah IN, Stambuk RA and Casida JE (1987) Ca^{2+}-activated ryanodine binding: mechanisms of sensitivity and intensity modulation by Mg^{2+}, caffeine, and adenine nucleotides. *Mol Pharmacol* 31: 232–238.

Prabhu SD and Salama G (1990) Reactive disulfide compounds induce Ca^{2+} release from cardiac sarcoplasmic reticulum. *Arch Biochem Biophys* 282: 275–283.

Rao SI, Wilks A and Ortiz de Montellano PR (1993) The role of His-64, Tyr-103, Tyr-146, and Tyr-151 in the epoxidation of styrene and beta-methylstyrene by recombinant sperm whale myoglobin. *J Biol Chem* 268: 803–809.

Raschke P, Becker BF, Leipert B, Schwartz LM, Zahler S and Gerlach E (1993) Postischemic dysfunction of the heart induced by small numbers of neutrophils via formation of hypoclorous acid. *Basic Res Cardiol* 88: 321–339.

Rasmussen H and Barrett PQ (1984) Calcium messenger system: an integrated view. *Physiol Rev* 64: 938–984.

Reyes J, Salim-Hanna M, Lissi EA, Videla LA and Holmgren J (1994) Enhanced urinary spontaneous luminescence in Duchenne muscular dystrophy. *Free Radical Biol Med* 16: 851–853.

Rosen DR (1993) Mutations in Cu/Zn superoxide dismutase gene are associated with familial amyotrophic lateral sclerosis. *Nature* 362: 59.

Salama G and Abramson J (1984) Silver ions trigger Ca^{2+} release by acting at the apparent physiological Ca^{2+}-release site in sarcoplasmic reticulum vesicles. *J Biol Chem* 259: 13363–13369.

Salama G, Abramson JJ and Pike GK (1992) Sulphydryl reagents trigger Ca^{2+} release from the sarcoplasmic reticulum of skinned rabbit psoas fibres. *J Physiol* 454: 389–420.

Salama G, Choi B-R, Hein MC, Menshikova E and Abramson JJ (1996) The ACE inhibitor Captopril inhibits Ca^{2+} release from cardiac and skeletal sarcoplasmic reticulum (SR) by reducing critical thiols on ryanodine receptors (RyR). *Biophys J* 70: A257.

Salonen JT (1991) Dietary fats, antioxidants and blood pressure. *Ann Med* 23: 295–298.

Salonen, J.T., Nyyssonen, K., Korpela, H., Tuomilehto, J., Seppanen, R. and Salonen, R. (1992) High stored iron levels are associated with excess risk of myocardial infarction in eastern Finnish men. *Circulation* 86: 803–811.

Sato K, Akaike T, Kohno M, Ando M and Maeda H (1992) Hydroxyl radical production by H_2O_2 plus Cu,Zn-superoxide dismutase reflects the activity of free copper released from the oxidatively damaged enzyme. *J Biol Chem* 267: 25371–25377.

Sclick S, Chamulitrat W and Kevan L (1985) Electron spin resonance study of peroxy labels in a copolymer of tetrafluoroethylene-hexafluoropylene. *J Phys Chem* 89: 4278–4282.

Shewfelt RL and Hultin HO (1983) Inhibition of enzymic and non-enzymic lipid peroxidation of flounder muscle sarcoplasmic reticulum by pretreatment with phospholipase A_2. *Biochim Biophys Acta* 751: 432–437.

Slevin ML (1991) The clinical pharmacology of etoposide. *Cancer* 67: 319–329.

Smith JS, Imagawa J Ma Fill M, Campbell KP and Coronado R (1988) Purified ryanodine receptor from rabbit skeletal muscle is the calcium-release channel of sarcoplasmic reticulum. *J Gen Physiol* 92: 1–26.

Sneyd J, Keizer J and Sanderson MJ (1995) Mechanisms of calcium oscillations and waves: a quantitative analysis. *FASEB J* 9: 1463–1472.

Stewart DJ, Grewaal D, Redmond MD, Mikhael NZ, Montpetit VA and Goel R (1993) Human autopsy tissue distribution of the epipodophyllotoxins etoposide and teniposide. *Cancer Chemother Pharmacol* 32: 368–372.

Stoyanovsky DA, Salama G and Kagan VA (1994) Ascorbate/iron activates Ca^{2+}-release channels of skeletal sarcoplasmic reticulum vesicles reconstituteed in lipid bilyars. *Arch Biochem Biophys* 208: 214–221.

Stoyanovsky D, Murphy T, Anno PR, Kim Y-M and Salama G (1997) Nitric Oxide Activates skeletal and cardiac ryanodine receptors. *Cell Calcium* 21: 19–29.

Sutko JL and Airey JA (1996) Ryanodine receptor Ca^{2+} release channels: does diversity in form equal diversity in function? *Physiol Rev* 76: 1027–1071.

Tanaka M, Kovalenko SA, Gong JS, Borgeld HJ, Katsumata K, Hayakawa M, Yoneda and Ozawa T (1996) Accumulation of deletions and point mutations in mitochondrial genome in degenerative diseases. *Ann N Y Acad Sci* 786: 102–111.

Tavazzi B, Lazzarino G, Di Pierro D and Giardina B (1992) Malondialdehyde production and ascorbate decrease are associated to the reperfusion of the isolated postischemic rat heart. *Free Radical Biol Med* 13: 75–78.

Thomas AP, Bird GS, Hajnoczky G, Robb-Gaspers LD and Putney JW Jr (1996) Spatial and temporal aspects of cellular calcium signaling. *FASEB J* 10: 1505–1517.

Trimm J, Salama G and Abramson JJ (1986) Sulfhydryl oxidation triggers Ca^{2+} release from sarcoplasmic reticulum vesicles. *J Biol Chem* 261: 16092–16098.

Van der Zee J, Barr DP and Mason RP (1996) ESR spin trapping investigation of radical formation from the reaction between hematin and tert-butyl hydroperoxide. *Free Radical Biol Med* 20: 199–206.

Van Maanen JMS, Retel J, de Vries J and Pinedo HM (1988) Mechanisms of action of antitumor drug etoposide: a review. *J Natl Cancer Inst* 80: 1526–1533.

Yim MB, Chock PB and Stadtman ER (1993) Enzyme function of copper, zinc superoxide dismutase as a free radical generator. *J Biol Chem* 268: 4099–4105.

Zaidi NF, Lagenaur C, Pessah IN, Abramson JJ and Salama G (1989a) Reactive disulfide reagents trigger Ca^{2+} release from skeletal sarcoplasmic reticulum. *J Biol Chem* 264: 21725–21736.

Zaidi NF, Lagenaur C, Xiong H, Abramson JJ and Salama G (1989b) Disulfide linkage of biotin identifies a 106 kDa Ca^{2+} channel in skeletal sarcoplasmic reticulum. *J Biol Chem* 264: 21737–21747.

Oxidative stress in skeletal muscle atrophy induced by immobilization

H. Kondo

Faculty of Medicine, Kyoto University, Kyoto 606, Japan

Introduction

The production of free radicals in the mitochondria has been investigated more than that in other organelles, and is believed to play a major role in oxidative stress in the cell. In the muscle, the increase in muscular activity enhances the oxygen consumption and increased oxygen consumption accelerates the production of free radicals in the mitochondria (Asayama and Kato, 1990). In view of this, it was believed, with no consideration for the production of free radicals in the cytoplasm, that the oxidative stress in the muscle might parallel muscular activity, i.e. the oxidative stress might decrease in muscular atrophy. This may be the reason for the small number of reports about oxidative stress during muscular atrophy of disuse, although numerous studies have been carried out in this decade on oxidative stress during exercise.

First enhanced oxidative stress is described during muscular atrophy of disuse and its mechanism explained from a cell biological view. Its role in muscular atrophy is then discussed. Recovery from muscular atrophy is also accompanied by oxidative stress, which is explained at the end.

Oxidative stress in skeletal muscle atrophied by immobilization

Skeletal muscle atrophy resulting from disuse has been an important problem in the field of rehabilitation and space biology. This atrophy has been known for hundreds of years and investigations have been carried out for over 100 years. These investigations have been reviewed by others (Appell, 1990; Thomason and Booth, 1990). The present author used the immobilization model, because it causes muscular atrophy rapidly and reversibly, and produces exactly the same conditions for the different muscles involved. Many biochemical changes have been investigated using this model. Nick et al. (1989) reported that atrophy produced by immobilization is the result of atrophy of the muscle cell without a decrease in muscle cell numbers. The infiltration of other cells, such as phagocytes, are rarely found in the early phase of atrophy (Cooper, 1972). In addition, as the number of mitochondria are maintained in the soleus muscle during disuse (Nemeth et al., 1980), the activities of enzymes in mitochondria, such as Mn-containing superoxide dismutase, cannot be considered as reflecting their numbers, but rather indicate the amount of enzyme per mitochondrion. Taking these into consideration, the muscular atrophy produced by immobilization is regarded as a good model for studying cellular atrophy.

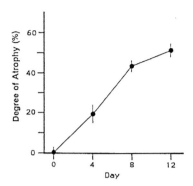

Figure 1. Changes in degree of atrophy as function of number of days after start of immobilization. Each point is mean±SE ($n=5$).

Male Wistar rats (14–16 weeks old) were used as animal models; one of their ankle joints was immobilized in the fully extended position (i.e. with the soleus muscle in a shortened position) (Kondo et al., 1991b). The periods of immobilization were 4, 8 and 12 days, in order to investigate the metabolic changes occurring as the atrophy proceeds rapidly. After immobilization, the typical slow red muscles of soleus muscle from immobilized and intact hindlimbs were collected

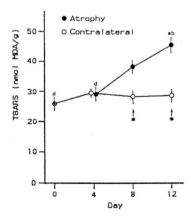

Figure 2. Changes in thiobarbituric acid-reactive substance (TBARS) level in atrophied and contralateral muscles as function of number of days after immobilization began. Day 0 is control. Each point is mean±SE ($n=5$). (a–d) indicate significant differences at $p<0.05$ compared with 0, 4, 8 and 12 days, respectively. * Significant difference at $p<0.05$ between atrophied and contralateral muscles.

(atrophied and contralateral control muscles, respectively). The degree of atrophy is shown in Figure 1. In the model, atrophy proceeded rapidly until day 8 and slowly from days 8 to 12.

Thiobarbituric acid-reactive substance (TBARS), total glutathione (total GSH) and oxidized glutathione (GSSG) were measured to estimate the level of oxidative stress.

As shown in Figure 2, the level of TBARS in atrophy did not change until day 4 and increased significantly from day 8; the TBARS concentration on day 12 was higher by about 60% compared with that of the control. The increase in TBARS level strongly suggests an increase in lipid peroxidation.

The levels of total GSH and GSSG and the ratio GSSG:GSH in 12-day atrophy are shown in Figure 3. In atrophy, total GSH concentration significantly decreased, and GSSG concentration significantly increased by 43% compared with those in the control. The activity of glutathione reductase (GSSGRx), which plays a main role in the conversion of oxidized to reduced glutathione, increased in atrophy (Fig. 4). It is thought that the increase in oxidative reactions exceeded that of GSSGRx activity, which resulted in increased levels of GSSG in atrophy. Thus, increases of GSSG and TBARS are proof of the enhanced oxidative stress in skeletal muscle that has been atrophied by immobilization. Furthermore, the increased activity of GSSGRx on day 4 suggests that oxidative stress had already increased on that day, although the TBARS level did not increase simultaneously.

On the other hand, Gilbert (1982) has reported the possibilities of enzyme regulation by thiol–disulphide exchange and modulation of the thiol/disulphide ratio *in vivo* to serve as a "third messenger" in response to 3',5'-cyclic adenosine monophosphate (cAMP) levels. The increased

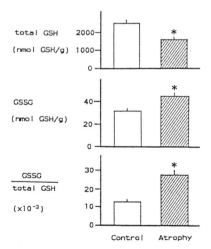

Figure 3. Level of total glutathione (GSH), oxidized GSH (GSSG), and ratio of GSSG to total GSH in 12-day atrophied and contralateral (control) muscles. Data are means±SE ($n=5$). *Significant difference at $p<0.05$ compared with control.

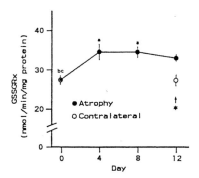

Figure 4. Changes in glutathione reductase (GSSGRx) activities (ordinate) as function of number of days after start of immobilization. Symbols explained in Figure 2. Data are means±SE ($n=6$).

ratio GSSG:total GSH in atrophy (see Fig. 3) indicates one of the causes of metabolic change in atrophied muscle.

Mechanism of oxidative stress in atrophied muscle

In the previous section, oxidative stress is shown to increase in skeletal muscle atrophied by immobilization, it has also been reported to increase during exhaustive exercise. Both increased and decreased muscle activities cause elevated oxidative stress in skeletal muscle. The mechanism of oxidative stress during exercise is explained in detail in the chapters by Komulainen and Vihko (this volume) and Jackson (this volume). In this section, the metabolism of active oxygen species and transition metals is described to clarify the mechanism of oxidative stress in muscular atrophy (Kondo et al., 1992a, 1993a, b).

Antioxidant enzyme systems

Superoxide dismutase (SOD), which acts in converting superoxide anion to hydrogen peroxide, has two forms, namely, CuZnSOD in the cytoplasm and MnSOD in the mitochondria. As shown in Figures 5 and 6, the two forms of SOD showed entirely different responses during atrophy. The activity of CuZnSOD increased throughout the 12 days; its activity on days 8 and 12 was significantly higher than that of the control. The activity of MnSOD tended to increase from day 8, and its activity on the day 12 decreased significantly to 60% of the control value. As it is generally known that SOD is induced by superoxide anions (Hassan, 1988), the level of SOD may be believed to reflect the generation of superoxide anions. It is also known that superoxide anions usually cannot cross biological membranes (Halliwell and Gutteridge, 1986). Taking these aspects into consideration, the increased levels of the cytoplasmic form of SOD (CuZnSOD) and

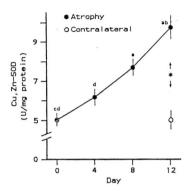

Figure 5. Changes in CuZnSOD activities as function of number of days after start of immobilization. Data are means±SE (*n*=6). Symbols are explained in Figure 2.

the decreased level of the mitochondrial form of SOD (MnSOD) may reflect an increased production of superoxide anions in the cytoplasm and decreased production in the mitochondria.

The elevated CuZnSOD level also indicates that the generation of hydrogen peroxide, the product of the reaction catalysed by SOD, might be enhanced in the cytoplasm. It is generally known that Se-dependent glutathione peroxidase (Se-GSHPx) and catalase have the ability to degrade hydrogen peroxide (Halliwell and Gutteridge, 1986). The activities of Se-GSHPx did not change significantly throughout the 12 days (Fig. 7). The glutathione level is known to limit the function of Se-GSHPx *in vivo* (Flohe and Günzuler, 1984) and, as shown in Figure 3, its level decreased in atrophied muscle. Accordingly, although the activity of Se-GSHPx did not change,

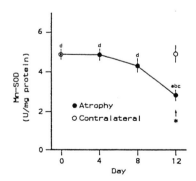

Figure 6. Changes in MnSOD activities as function of number of days after start of immobilization. Data are means±SE (*n*=6). Symbols are explained in Figure 2.

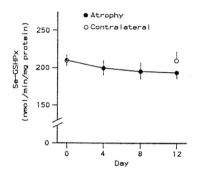

Figure 7. Changes in Se-dependent glutathione peroxidase (Se-GSHPx) activities as function of number of days after start of immobilization. Data are means±SE ($n=6$).

its function *in vivo* could be anticipated to decrease in atrophied muscle. The catalase activity also did not increase until the day 8 and increased slightly on the day 12 (Fig. 8). Although the activity of glutathione S-transferase (GST), which also has some GSP activity, rose in atrophy (Kondo et al., 1993b), this enzyme is unable to utilize H_2O_2 as a substrate (Habig et al., 1974). Hence, the level of H_2O_2 might well be increased, especially in the cytoplasm.

Cytochemical study of hydrogen peroxide

Is the prediction in the above section actually brought about? To answer this question, the author's group assessed the level of hydrogen peroxides in the atrophied muscle cell by

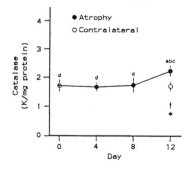

Figure 8. Changes in catalase activities as function of number days after start of immobilization. Data are means±SE ($n=6$). Symbols are explained in Figure 2.

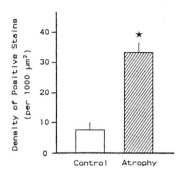

Figure 9. Densities of positive stains revealed by electron-microscopic cytochemistry per $1\,000\ \mu m^2$ in 8-day atrophy and control. Data are mean$\pm$SE ($n=6$). *Significant difference at $p<0.05$ compared with control.

cytochemical study using an electron microscope (Kondo et al., 1993b).

The cytochemical study was carried out according to the method of Babbs et al. (1991). Briefly, samples are incubated in short-term organ culture in the presence of 3,3'-diaminobenzidine (DAB), and H_2O_2 generated is allowed to react with DAB via the catalysis of the endogenous peroxidase to produce oxidized DAB; this deposits as an osmiophilic polymer in the cell. The number of the deposits of oxidized DAB reflects the level of H_2O_2 in the cell and, thereby, the deposits are scored using the electron microscope to estimate the level of H_2O_2 generated in organ culture.

As shown in Figure 9, the number of positive stains increased in atrophy. As the activities of peroxidase in 8-day atrophy were the same as those in the control (see earlier), the density of stains could be thought to reflect the level of H_2O_2. Taking these into consideration, the present cytochemical study indicated the increased generation of H_2O_2 in the cytoplasm of the atrophied muscle cell in short-term organ culture. Capillary density increases in the atrophied soleus muscle (Desplanches et al., 1987). As oxygen delivery is limited by capillary density, oxygen delivery *in vivo* to the atrophied muscle cell should not be less than that in the control (Thomason and Booth, 1990). Thereby, this cytochemical study might suggest the elevated level of H_2O_2 *in vivo* in the cytoplasm of the atrophied muscle cell, which reinforces the preceding prediction.

Source of superoxide anions in the cytoplasm

The next question is what is the source of superoxide anions in the cytoplasm of atrophied muscle. Stripe and Corte (1969) reported that xanthine oxidase (XOD) is localized in the cytoplasm and exists in the two following forms: NAD-dependent XOD (type D) and superoxide-producing XOD (type O). As presented in Figure 10, the XOD activities in atrophy were significantly higher than those in controls, and, especially, the activity of type O increased by as much as about 2.3 times. The substrate levels of XOD, xanthine and hypoxanthine increased in atrophy

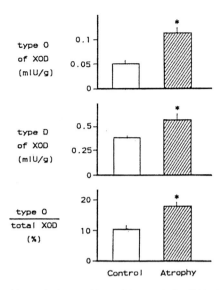

Figure 10. Activities of superoxide-producing xanthine oxidase (type O of XOD) and NAD-dependent xanthine oxidase (type D of XOD) and ratio of type O to total XOD activity in 12-day atrophied and contralateral (control) muscles. Data are mean±SE ($n=6$). * Significant difference at $p<0.05$ compared with control.

(Fig. 11). Actually, levels of urate, the product of the reaction catalysed by XOD, increased in atrophy (Fig. 11). Therefore, superoxide-producing XOD might function more in atrophied muscle, which is believed to be an important source of superoxide anions in the cytoplasm.

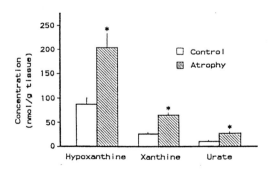

Figure 11. Levels of hypoxanthine, xanthine and urate in 12-day atrophied and contralateral (control) muscles. Data are mean±SE ($n=6$). * Significant difference at $p<0.05$ compared with control.

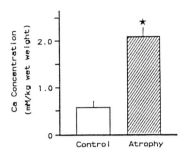

Figure 12. Calcium level in sarcomere (A band) in 8-day atrophied and contralateral (control) muscles. Data are mean±SE ($n = 15$). *Significant difference at $p < 0.05$ compared with control.

Booth and Giannetta (1973) reported increased calcium concentrations in whole tissue of an atrophied gastrocnemius muscle. As extracellular space is a rich source of calcium, unfortunately its intracellular level cannot be estimated by its concentration in whole tissue. To clarify the intracellular calcium level in atrophied muscle cells, it was necessary to perform electron probe X-ray microanalysis (Kondo et al., 1992a). Figure 12 shows the intracellular calcium concentration. The level of intracellular calcium in atrophy was significantly higher than that in the control. On the other hand, the increased ratio of type O:total XOD in atrophy (see Fig. 10) indicates an enhanced conversion of type D to type O. It is known generally that calcium-activated protease participates in this conversion (McCord, 1985). Accordingly, increased intracellular calcium levels in the atrophied muscle cell might bring the conversion about through activation of the protease.

Movement of iron

The generation of superoxide anions and hydrogen peroxides is increased in atrophied muscle cell, as described above. Although they are not very reactive, very reactive radicals – hydroxyl radicals – are generated from them (so-called "Harber–Weiss reaction"). This reaction is generally very slow, but if transition metal ions, such as iron and copper, are present in this reaction system, the reaction is very fast (Halliwell and Gutteridge, 1986). Hence, it is necessary to investigate transition metals in atrophied muscle in order to clarify the metabolism of re-active oxygen species. Moreover, muscular atrophy induced by immobilization causes rapid atrophy of the muscle cell (Nick et al., 1989). Under the condition of a rapid decrease in cell volume and change dynamic in the cell structure, the distribution and balance of metals are expected to be disturbed (Gilbert, 1982; Kondo et al., 1991a).

Figure 13 shows the changes in concentrations of metals in the course of atrophy. The concentration of iron, which is well known to play an important role in the generation of hydroxyl radicals, kept increasing over the 12 days. To clarify this increased iron in whole tissues, the subcellular distribution of iron in atrophy was investigated (Fig. 14). The iron concentrations of

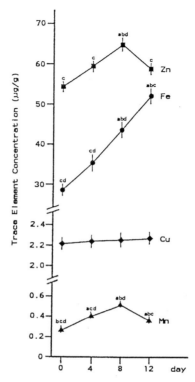

Figure 13. Fe, Zn, Mn, and Cu concentrations in atrophied muscle as function of number of days after start of immobilization. Data are mean±SE ($n=5$). Symbols are explained in Figure 2.

microsomal and supernatant fractions increased in atrophy; in particular, microsomal iron kept increasing throughout the 12 days, and on day 12 was about three times higher than that of the control.

This increase in levels of microsomal iron suggested the possibility that iron-binding proteins appeared in microsomes of atrophied muscle. Hence, iron-binding proteins in the sarcoplasmic reticulum of atrophied soleus muscle were checked; induction of the 54-kDa iron-binding protein was found in atrophy. The level of this 54-kDa protein had already increased on day 4 (Kondo et al., 1994b). The increased microsomal iron might be attached to this protein.

Role of iron

Did the increased iron levels in atrophied muscle play an important role in oxidative stress? To clarify the role of iron, the effect of desferrioxamine – an iron-chelating agent – was investigated (Tab. 1). Desferrioxamine was administered at a rate of 0.21 mmol/day per kg birthweight via an

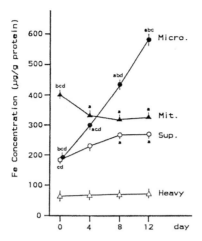

Figure 14. Changes of Fe concentration in subcellular fractions of atrophied muscle as function of number days after start of immobilization. Day 0 is control. Each point is mean±SE ($n = 5$). Heavy, heavy fraction; Mit, mitochondrial fraction; Micro, microsomal fraction; Sup, supernatant fraction. Symbols are explained in Figure 2.

osmotic pump that was implanted subcutaneously on day 4 after immobilization started (Kondo et al., 1992b).

In desferrioxamine injection group (DFX group in Tab. 1), the increases in levels of TBARS and GSSG were suppressed in atrophied muscle. The increased TBARS level strongly suggests

Table 1. Effect of deferrioxamine on the muscle atrophy

	Control		DFX		DFX±Fe	
	C	A	C	A	C	A
Muscle weight (mg)	158.2±3.3[*]	73.6±2.6	144.4±6.2[*]	84.3±4.3[b]	156.9±6.3	68.8±2.4
Degree of atrophy (%)	53.5±1.3		41.6±1.1[ab]		56.1±0.7	
TBARS (nmol MDA/g)	24.6±1.2[*]	38.6±1.5	25.3±1.0	27.7±1.5[ab]	25.8±1.6[*]	38.5±2.0
Total GSH (nmol GSH/g)	2928±118[*]	1805±59	2867±180[*]	1883±85	2753±112[*]	1869±57
GSSG (nmol GSH/g)	51.6±7.3[*]	64.0±4.7	57.6±5.2[*]	33.0±3.5[ab]	56.0±11.5	59.0±5.3
GSSG/total GSH (%)	1.75±0.22[*]	3.60±0.33	2.04±0.20	1.95±0.26[ab]	1.68±0.41[*]	3.21±0.23

Values are means±SE ($n = 5$). Control, double distilled water injection group; DFX, deferrioxamine injection group; DFX+Fe, iron-saturated deferrioxamine injection group. C, contralateral muscle; A, 12-day atrophied muscle. [*]Significant difference at $p < 0.05$ compared with atrophied muscle by paired t-test. [a,b]Significant differences at $p < 0.05$ compared with control and DFX+Fe group, respectively, by Bonferroni method.

acceleration of lipid peroxidation. It is also generally accepted that, under conditions of increased oxidative stress, the level of GSSG is usually increased (Lew et al., 1985). Hence, desferrioxamine is thought to suppress oxidative stress. Iron-saturated desferrioxamine did not have such an effect on oxidative stress (DFX + Fe group in Tab. 1). This indicates that desferrioxamine suppressed the oxidative stress through its iron-chelating action. In other words, this means that iron plays an important role in increasing oxidative stress in atrophied muscle.

On the other hand, the level of TBARS increased only in microsomes during atrophy (Kondo et al., 1992a). Free radicals that are harmful to the biological system are very reactive, and so the distance that they can move is very small. Taking this into consideration, the increased iron levels in the microsome are thought to participate in the generation of free radicals in atrophy. It is generally known that hydroxyl radicals, the most reactive of radicals, are generated from H_2O_2 and superoxide anions (O_2^{-}) in the presence of iron (Halliwell and Gutteridge, 1986). In view of the simultaneous increase in microsomal iron and the increase in H_2O_2 and O_2^{-} in the cytoplasm, it is expected that the generation of hydroxyl radicals could increase in microsomes of atrophied muscles.

The *in vivo* generation of hydroxyl radicals was therefore measured in atrophied muscle (Kondo et al., 1992a). Salicylate was used as a trapping reagent for hydroxyl radicals. When small amounts of salicylate are added to a biological system, the phenolic ring of the salicylate can be attacked by hydroxyl radicals, yielding dihydroxybenzoic acids (DHBs); these can then be

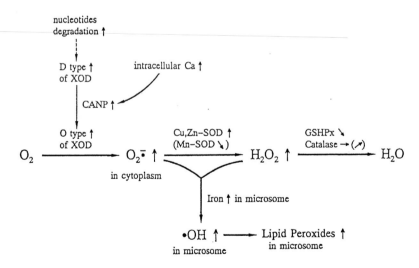

Figure 15. Mechanism of oxidative stress in the skeletal muscle atrophied by immobilization. O_2, molecular oxygen; O_2^{-}, superoxide anion; H_2O_2, hydrogen peroxide; OH', hydroxyl radical; H_2O, water; D-type of XOD, NAD-dependent xanthine oxidase; O-type of XOD, superoxide-producing xanthine oxidase; CANP, calcium activated neutral protease; CuZnSOD, cytoplasmic form of superoxide dismutase; MnSOD, mitochondrial form of superoxide dismutase; GSP, Se-dependent glutathione peroxidase.

determined using high-performance liquid chromatography (HPLC). There was a significant increase in 2,3-DHBs in atrophied muscle. This strongly suggests enhanced generation of hydroxyl radicals *in vivo* in atrophied muscle.

Figure 15 presents the possible mechanism of oxidative stress in atrophy schematically.

Role of oxidative stress in muscular atrophy

Does oxidative stress play a role in the progress of muscular atrophy? To answer this question, the author's group has done an antioxidant injection experiment (Fig. 16). Vitamin E , as an antioxidant, was injected intraperitoneally once a day for 6 days before the period of immobilization and on alternate days during the immobilization period; *dl*-α-tocopherol at a dose of 30 mg/kg birthweight was used (Kondo et al., 1991b).

In the vitamin E group, the TBARS level in atrophied muscle decreased significantly compared with that in the placebo group, which shows that vitamin E injected intraperitoneally served effectively as an antioxidant, lessening the oxidative stress in atrophied muscle. In the vitamin E

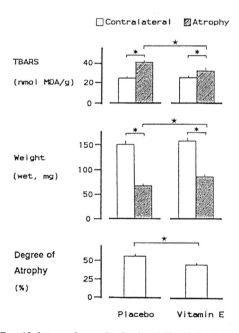

Figure 16. Effect of vitamin E on 12-day muscle atrophy. In vitamin E and placebo groups, TBARS levels, muscle weights and degree of atrophy in atrophied and contralateral muscles are shown. Data are means±SE (*n*=5). *Significant differences at *p*<0.05 between atrophied and contralateral muscles. ★ Indicates significant difference at *p*<0.05 between vitamin E and placebo groups.

group, the muscle weight was significantly heavier, and the degree of atrophy was significantly decreased by about 15% compared with that in the placebo group. The decrease in the degree of atrophy in the vitamin E group suggests that muscular atrophy proceeded mildly with lower oxidative stress. Thus, this indicates that oxidative stress accelerates muscular atrophy.

As shown in Table 1, desferrioxamine lessened the degree of atrophy by about 22%. As desferrioxamine suppressed the increase of oxidative stress in atrophied muscles, this effect is believed to be mediated by the suppression of oxidative stress.

It is not clear how oxidative stress accelerates muscular atrophy, but some possible mechanisms can be proposed:

1. The plasma membrane and sarcoplasmic reticulum are damaged by lipid peroxidation, allowing calcium to accumulate in the cell (Mourelle and Meza, 1991). Ohta et al. (1989) have reported the inhibition of Ca^{2+} ATPase by lipid peroxidation *in vitro*. Increased calcium has been reported to stimulate both phospholipid hydrolysis and non-lysosomal proteolysis (Nicotera et al., 1986; Pascoe and Reed, 1989). An increase of calcium concentration in the sarcomere of atrophied muscle has been shown using electron probe X-ray microanalysis (see earlier). Although it cannot be concluded whether this increase of intracellular calcium was a cause or a effect of increased oxidative stress in muscular atrophy, its initial increase presumably caused the oxidative stress which, in turn, accelerated its increase.

2. Free radical damage to proteins can make the proteins more susceptible to proteolysis (Davis and Goldberg, 1987). The attack of free radicals can directly fragment proteins, followed by an increase in enzymatic proteolysis of the fragments (Davis, 1987; Hunt et al., 1988). If the attacked proteins also have cross-linking and conformational changes, they are more susceptible to proteolysis (Davis, 1987).

3. The damage of lysosomal membranes by free radicals causes leakage of lysosomal protease into the cytoplasm (Mak et al., 1983). Actually, the activities of some lysosomal proteases are reported to increase in muscular atrophy induced by immobilization (Max et al., 1971).

It is also thought that oxidative stress might have some effects on the physiological condition of atrophied muscles. There is a possibility that increased levels of disulphide might be one of the causes of metabolic changes. It remains uncertain what physiological and biochemical functions are altered in atrophied muscle.

Oxidative stress during recovery from muscular atrophy

Oxygen consumption is expected to increase in recovering muscle. Taking into account the catalytic action of increased iron levels in the atrophied muscle, it is possible that oxygen radicals might increase during recovery from muscle atrophy.

Some rats were exsanguinated after 7 days of immobilization (atrophy group). The ankle joints of the other rats were remobilized after 7 days of immobilization and then exsanguinated after 5 days of remobilization (recovery group). The soleus muscles from both hindlimbs (atrophic and contralateral) were collected, and their levels of TBARS, total GSH and GSSG were measured as a parameter of oxidative stress (Kondo et al., 1993c). The levels of TBARS and

GSSG increased significantly in atrophic muscles in the recovery group. These findings prove the enhanced level of oxidative stress during recovery from muscle atrophy.

To clarify the role of oxidative stress during recovery, the effect of vitamin E on recovery from atrophy was examined. Vitamin E was injected intraperitoneally as *dl*-α-tocopherol at a dose of 30 mg/kg birthweight once daily during the remobilization period (Kondo et al., 1993c). In the vitamin E group, the TBARS level in atrophic muscle decreased significantly compared with that in the placebo group; hence, vitamin E injected intraperitoneally was believed to serve effectively as an antioxidant to lessen oxidative stress. Moreover, as the degree of atrophy was significantly decreased in the vitamin E group, it was suggested that recovery proceeded rapidly with lower oxidative stress. In other words, this indicates that oxidative stress slowed the recovery from atrophy.

Conclusion

Atrophied soleus muscles were collected from male Wistar rats (14–16 weeks old), one ankle joint of which had been immobilized in the fully extended position. The degree of atrophy increased rapidly until day 8 and slowly after that. Thiobarbituric acid-reactive substance and GSSG levels increased in the atrophied muscle, indicating enhanced oxidative stress in atrophy.

In atrophy, increased CuZnSOD and decreased MnSOD reflect increased generation of $O_2^{-\cdot}$ in the cytoplasm rather than in the mitochondria. The source of $O_2^{-\cdot}$ in the cytoplasm might be the increased $O_2^{-\cdot}$-producing xanthine oxidase. Enhanced generation of $O_2^{-\cdot}$ and increased CuZnSOD activity in atrophy suggested enhanced generation of H_2O_2 in the cytoplasm. As a result of the unchanged activity of Se-dependent GSP and the unchanged or slightly increased activity of catalase, the ability to degrade H_2O_2 might not increase as much. Hence, H_2O_2 is expected to be increased in atrophy. The cytochemical study supported this expectation.

Increased iron levels were also found, especially in microsomes, in atrophied muscle; its important role in oxidative stress was proved by injection of deferioxamine which suppressed the enhanced oxidative stress in atrophy. The increased iron, could be attached to the 54-kDa protein, is thought to accelerate the Harber–Weiss reaction under the influence of increased $O_2^{-\cdot}$ and H_2O_2 in the cytoplasm. Consequently, the production of hydroxyl radicals, the most aggressive radical, could be elevated in the microsomes of atrophied muscle. This is supported by that observation that only the microsomal fraction showed an increased TBARS level in atrophy.

Vitamin E injections lessened the degree of atrophy, which suggests that muscular atrophy proceeds more slowly with lower oxidative stress, i.e. oxidative stress accelerates muscular atrophy.

Single angle joints of rats were immobilized for 7 days and remobilized for 5 days after the immobilization period. The levels of TBARS and GSSG increased in the recovering muscle, which strongly suggests that enhanced oxidative stress occurred during the recovery from the muscular atrophy. Vitamin E injection accelerates the recovery from atrophy, thus showing that oxidative stress slowed it down.

Acknowledgement
This works were supported by the following funds and companies: Japan Research Foundation for Chronic Disease and Rehabilitation, Foundation for Total Health Promotion, Nakatomi Foundation, Meiji Life Foundation of Health and Welfare Foundation, Osaka Gas Group Welfare Foundation, Scientific Research Fund of the Ministry of Education, Science and Culture of the Government of Japan, Eisai Corporation, and Ciba-Geigy Corporation. I am grateful to Mrs Ruiko Kondo and Shoko Kondo for their kind assistance in the preparation of the manuscript.

References

Appell HJ (1990) Muscular atrophy following immobilization. *Sport Med* 10: 42–58.
Asayama K and Kato K (1990) Oxidative muscular injury and its relevance to hyperthyroidism. *Free Radical Biol Med* 8: 293–303.
Babbs CF, Salaris SC and Turek JJ (1991) Cytochemical studies of hydrogen peroxide generation in postischemic hepatocytes. *Am J Physiol* 260: H128–H129.
Booth FW and Giannetta CL (1973) Effect of hindlimb immobilization upon skeletal muscle calcium in rat. *Calcified Tissue Res* 13: 327–330.
Cooper RR (1972) Alterations during immobilization and regeneration of skeletal muscle in cats. *J Bone Joint Surg Am* 54-A: 919–953.
Davis KJA (1987) Protein damage and degradation by oxygen radicals. *J Biol Chem* 262: 9895–9901.
Davis KJA and Goldberg AL (1987) Oxygen radicals stimulate intracellular proteolysis and lipid peroxidation by independent mehanisms in erthrocytes. *J Biol Chem* 262: 8220–8226.
Desplanches D, Mayet MH, Sempore B and Flandrois R (1987) Structual and functional responses to prolonged hindlimb suspension in rat muscle. *J Appl Physiol* 63: 558–563.
Flohe L and Günzuler WA (1984) Assay of glutathione peroxidase. *Methods Enzymol* 105: 114–121.
GIbert HF (1982) Biological disulfides: the third messenger? *J Biol Chem* 257: 12086–12091.
Habig WH, Pabst MJ and Jakoby WB (1974) Glutathione S-transferase. *J Biol Chem* 249: 7130–7139.
Halliwell B and Gutteridge JMC (1986) Oxygen free radicals and iron in relation to biology and medicine: some problem and concepts. *Arch Biochem Biophys* 246: 501–514.
Hassan HM (1988) Biosynthesis and regulation of superoxide dismutases. *Free Radical Biol Med* 5: 377–385.
Hunt JV, Simpson JA and Dean RT (1988) Hydroperoxide-mediated fragmentation of proteins. *Biochem J* 250: 87–93.
Kondo H, Kimura M and Itokawa Y (1991a) Manganese, zinc, copper and iron concentrations and subcellular distribution in two types of skeletal musce. *Proc Soc Exp Biol Med* 196: 83–88.
Kondo H, Miura M and Itokawa Y (1991b) Oxidative stress in skeletal muscle atrophied by immobilization. *Acta Physiol Scand* 142: 527–528.
Kondo H, Miura M, Nakagaki I, Sasaki S and Itokawa Y (1992a) Trace element movement and oxidative stress in skeletal muscle atrophied by immobilization. *Am J Physiol* 262: E583–E590.
Kondo H, Miura M, Kodama J, Ahmed SM and Itokawa Y (1992b) Role of iron in oxidative stress in skeletal muscle atrophied by immobilization. *Pflügers Arch* 421: 295–297.
Kondo H, Miura M and Itokawa Y (1993a) Antioxidant enzyme systems in skeletal muscle atrophied by immobilization. *Pflügers Arch* 422: 404–406.
Kondo H, Nakagaki I, Sasaki S, Hori S and Itokawa Y (1993b) Mechanism of oxidative stress in skeletal muscle atrophied by immobilization. *Am J Physiol* 265: E839–E844.
Kondo H, Kodama J, Kishibe T and Itokawa Y (1993c) Oxidative stress during recovery from muscle atrophy. *FEBS Lett* 326: 189–191.
Kondo H, Nishino K and Itokawa Y (1994a) Hydroxyl radical generation in skeletal muscle atrophied by immobilization. *FEBS Lett* 349: 169–172.
Kondo H And Itokawa Y (1994b) Oxidative stress in muscular atrophy. *In*: CK Sen, L Packer and O Hänninen (eds): *Exercise and Oxygen Toxicity*. Elsevier, pp 319–342.
Lew H, Pykes S and Quintanilha A (1985) Changes in the glutathione status of plasma, liver and muscle following exhaustive exercise in rats. *FEBS Lett* 185: 262–266.
McCord JM (1985) Oxygen-derived free radicals in postischemic tissue injury. *N Engl J Med* 312: 159–163.
Mak IT, Misra HP and Weglicki WB (1983) Temporal relationship of free radical-induced lipid peroxidation and loss of latent enzyme activity in highly enriched hepatic lysosomes. *J Biol Chem* 258: 13733–13737.
Max SR, Maier RF and Vogelsang L (1971) Lysosomes and disuse atrophy of skeletal muscle. Arch. *Biochem Biophys* 146: 227–232.
Mourelle M and Meza MA (1991) CCl_4-induced lipoperoxidation triggers a lethal defect in the liver plasma membranes. *J Appl Toxicol* 10: 23–27.
Nemeth PM, Meyer D and Kark AP (1980) Effect of denervation and simple disuse on rates of oxidation and on activities of four mitochondrial enzymes in type I muscle. *J Neurochem* 35: 1351–1360.

Nick DK, Beneke WM, Key RM and Timson B (1989) Muscle fibre size and number following immobilization atrophy. *J Anat* 163: 1–5.

Nicotera P, Hartzell P, Baldi C, Svensson S, Bellomo G and Orrenius S (1986) Cystamine induced toxicity in hepatocytes through the elevation of cytosolic Ca^{2+} and the stimulation of a non lysosomal proteolytic system. *J Biol Chem* 261: 14628–14635.

Ohta A, Mohri T and Ohyashiki T (1989) Effect of lipid peroxidation on membrane-bound Ca^{2+}-ATPase activity of the intestinal brush-border membranes. *Biochim Biophys Acta* 984: 151–157.

Pascoe GA and Reed DJ (1989) Cell calcium, vitamin E, and the thiol redox system in cytotoxicity. *Free Radical Biol Med* 6: 209–224.

Stripe F and Corte D (1969) The regulation of rat liver xanthine oxidase. *J Biol Chem* 244: 3855–3863.

Thomason DB and Booth FW (1990) Atrophy of the soleus muscle by hindlimb unweighting. *J Appl Physiol* 68: 1–12.

Oxidative Stress in Skeletal Muscle
A.Z. Reznick et al. (eds)
© 1998 Birkhäuser Verlag Basel/Switzerland

Effect of growth hormone on oxidative stress in immobilized muscles of old animals

I. Roisman, N. Zarzhevsky and A.Z. Reznick

Musculoskeletal Laboratory, Department of Morphological Sciences, The Bruce Rappaport Faculty of Medicine, Technion-Israel Institute of Technology, 31096 Haifa, Israel

Summary. Ageing limbs show spontaneous reduction in both bone mass and skeletal muscle mass, accompanied by reduced functional capacity of these tissues, an increase in bone and joint disorders and muscle weakness. With reduced muscle activity or limb immobilization, there is also a progressive involution of muscle tissue, which has been termed sarcopenia. It has been proposed that growth hormone (GH) could counteract some of these ageing changes observed in bone and muscle.
The authors have developed an aged rat model with limb immobilization, using 24-month-old male Wistar rats whose right hindlimbs were immobilized by external fixation or plaster of Paris for 4 weeks. In some of the immobilized rats, GH was also administered subcutaneously five times weekly for 3 1/2 weeks. At sacrifice, skeletal muscles of the hindlimbs were taken for biochemical, histological and ultrastructural studies.
The results showed that 4 weeks of immobilization caused a large reduction in muscle mass (about 40%). Administration of GH was able to slow down this weight loss quite significantly. Biochemical studies showed that the activity of creatine phosphokinase in the immobilized gastrocnemius muscle was reduced (−18%), while acid phosphatase was greatly elevated (+37%). In addition, parameters of the effects of oxidative stress on muscle proteins and lipids were also examined. Protein oxidation as measured by protein carbonyl assay was 215% higher in immobilized gastrocnemius muscle, which was considerably reduced after GH administration. Also, measuring lipid peroxidation using the thiobarbituric acid reactive substances assay showed a marked increase of 131% in immobilized gastrocnemius muscles compared to controls. Application of GH to immobilized animals checked this increase almost completely.
The above, as well as previous studies, showed that immobilization may be a causative factor in the elevation of oxidative stress in aged skeletal muscle. This oxidative damage is accompanied by other biochemical and morphological changes indicative of muscle damage and degeneration. Administration of GH to immobilized ageing animals can ameliorate some of these muscle changes and considerably reduce the oxidative damage due to immobilization.

Introduction

Skeletal muscle tissue comprises about 40% of the weight of the human body. Between the ages of 20 and 80 years, human beings lose about 30–40% of their skeletal muscle weight. This is observed primarily in the limb muscles, which lose a great deal of their metabolic and physiological capacities in old age. In addition, it is known that old people and animals tend to be less mobile and to suffer more frequently from maladies that cause future immobilization of the limb (Carmeli and Reznick, 1994).

In the past few years, the authors have developed a model for rapid muscle atrophy of old female Wistar rats (25–26 months) immobilized by either casts of plaster of Paris or external fixation (EF) (Carmeli and Reznick, 1992; Carmeli et al., 1993). Indeed, using these techniques, it has been possible to show that muscles of old rats immobilized for 4 weeks lose 40–50% of their muscle mass. Moreover, it has also been possible to show that the EF technique is a more drastic way of immobilization than plaster of Paris (Reznick et al., 1995). The loss of muscle mass was accompanied by biochemical and morphological changes, such as changes in muscle enzymology, oxidative damage to muscle proteins and lipids, and morphological damage seen in muscle fibres (Reznick, 1995; Fares et al., 1996). For example, there was a significant increase in acid phosphatase activity and a decrease in muscle creatine phosphokinase (CPK) activity in immobilized muscles (Carmeli et al., 1993). Administration of rat growth hormone (rGH) to the immobilized old animals was able to slow down some of the immobilization-associated damage (Carmeli and Reznick, 1992; Carmeli et al., 1993).

Recently, the authors initiated a project in which the capacity for recovery from EF of young versus old animals was investigated. Young rats (6 months old) lost up to 60% of their muscle weight after 4 weeks of immobilization. After 4 weeks of recovery, the muscle mass of immobilized limbs did not return to pre-immobilization weights; however, the biochemical and morphological parameters did return to the normal levels of control muscles. This was not the case for old animals (25 months), which recovered more slowly and after 4 weeks of re-mobilization still had not returned biochemically and morphologically to full pre-immobilization capacities.

It was obvious from the above studies that muscles of young rats had more active machinery of protein turnover and were more efficient in rebuilding muscle mass after muscle wasting. Nevertheless, the results described in this chapter and in previous publications indicate that the adverse effects of immobilization on muscles of old rats can be slowed down by exposing these animals to anabolic hormones such as GH.

Immobilization and muscle mass: The effect of GH

In previous studies on female Wistar rats, it was shown that hindleg muscles such as gastrocnemius and plantaris muscles lost about 31–33% of their muscle mass after 4 weeks of immobilization using the technique of plaster of Paris (Carmeli et al., 1993). This loss was quite significant ($p < 0.01$). Weight loss of gastrocnemius and plantaris muscles of animals immobilized and treated with growth hormone was only 15.7% and 14.7%, respectively (Carmeli et al., 1993). Similar observation of muscle weight loss as a result of immobilization was made in many other studies (Booth, 1982).

In Table 1, comparison of body and muscle weights between ageing female and male Wistar rats is shown. The body weights of male rats were 84% greater than those of female rats, whereas increase in weights of major hindleg muscles (i.e. gastrocnemius, quadriceps, plantaris and soleus) was only between 8.5% and 44.2% (Tab. 1). The difference in body weights was the result of excessive adipose tissue acquired by old male animals, whereas female Wistar rats accumulated relatively little fat tissue as they aged. Thus, in view of their very different body composition, it was also of great interest to immobilize male rats by the EF system; Table 2 shows the hindlimb muscle weight loss as a result of immobilization in aged male Wistar rats. After 4 weeks of immobilization, all four muscles lost about 32–43% of their muscle weight. Administration of rGH to these animals for a period of $3\frac{1}{2}$ weeks reduced the weight loss of these muscles to a level 10–19% below that of control muscles (Tab. 2). Thus, GH, which is a strong anabolic hormone, is capable of slowing down muscle weight loss caused by immobilization.

Table 1. Comparison of body and muscle weights in non-immobilized control legs of 26-month-old male and female Wistar rats

	Female rats (n=5)	Male rats (n=6)	Percentage difference between males and females
Body weight (g)	288±40	532±47	+84.0
Gastrocnemius (mg)	1117±116	1437±144	+28.3
Quadriceps (mg)	1629±164	2349±432	+44.2
Plantaris (mg)	211±28	229±49	+8.5
Soleus (mg)	112±10	143±22	+22.6

Values are means±SD.

Table 2. Hindlimb muscle weight (mg) before and after immobilization through either external fixation (EF) or EF followed by administration of rat growth hormone (rGH) in aged male Wistar rats (n=6)

Muscle	4 weeks' EF			4 weeks' EF + rGH administration		
	Control leg	EF leg	% change	Control leg	EF leg	% change
Gastrocnemius	1437.2±144.0	972.2±81.2 $p<0.01$	−32.0±6.5	1524.0±134.4	1320.2±130.5 $p<0.01$	−13.3±4.6
Quadriceps	2349.3±432.0	1344.2±229.0 $p<0.001$	−42.9±7.53	2325.0±122.0	1890.0±107.0 $p<0.001$	−18.7±3.3
Plantaris	229.6±49.2	154.7±39.5 $p<0.01$	−32.7±10.5	269.3±23.4	249.6±26.8 NS	−10.7±3.1
Soleus	143.17±22.6	85.2±10.9 $p<0.001$	−40.6±8.0	157.0±22.9	132.8±17.7 NS	−15.4±3.8

Statistical analysis was performed by paired Student's t-test. Values are means±SD. NS, not significant.

Changes in muscle enzymatic activities resulting from immobilization and GH treatment

Creatine phosphokinase (CPK) is a well-established muscle enzyme that provides the immediate energy (ATP) for muscle contraction. In muscle diseases and muscle atrophy, its level rises in serum and is reduced in muscle tissue, so it is used as a criterion for muscle damage (Sutton et al., 1996). Moreover, its elevated levels in serum indicate the occurrence of skeletal pathologies, as well as cardiac pathologies such as myocardial infarction (Hetland and Dickstein, 1996). Acid phosphatase (AP) is a well-known marker of lysosomal activity in tissues, and its increase in various conditions is indicative of catabolic activity.

Table 3. Changes in creatine phosphokinase (CPK) and acid phosphatase (AP) activities caused by immobilization through external fixation (EF) or EF followed by administration of rat growth hormone (rGH) in aged male Wistar rats ($n = 5$)

Enzyme	4 weeks' EF			4 weeks' EF + rGH administration		
(U/mg protein)	Control leg	EF leg	% change	Control leg	EF leg	% change
CPK	4.35±0.6	3.54±0.5	−18.2±1.3	4.91±1.17	5.17±1.60	+5.3±0.2
		$p < 0.05$			NS	
AP	28.66±2.14	39.43±5.91	+37.6±11.6	28.22±2.26	29.52±2.46	+4.4±2.1
		$p < 0.02$			NS	

Statistical analysis was performed by paired Student's t-test. Values are means ± SD. NS, not significant.

As shown in Table 3, under conditions of immobilization by EF, CPK enzymatic activity was reduced by 18.12% in gastrocnemius muscle of old male Wistar rats immobilized for 4 weeks.

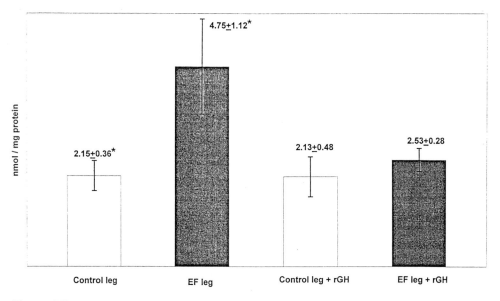

Figure 1. Effect of administration of rat growth hormone (rGH) on protein oxidation (carbonyls) in gastrocnemius muscle from 26-month-old male Wistar rats ($n = 4$) after right hindlimb immobilization: comparison between treated right legs and control left legs.

This decrease in CPK activity was well attenuated, to –5.29%, in gastrocnemius muscles of animals exposed to rGH administration. Acid phosphatase, on the other hand, showed a 37.6% increase in immobilized gastrocnemius muscles, and this elevated acid phosphatase activity was completely abolished in muscles of animals treated with rGH. A very similar trend with both enzymes was also observed in previous studies when old female rats were exposed to rGH (Carmeli et al., 1993).

Effects of immobilization and rGH on levels of protein oxidation and lipid peroxidation

Several procedures for estimation of status of protein oxidation have been developed in the past decade (Levine et al., 1990; and see Reznick et al., this volume). However, the most convenient method for assessing the level of protein oxidation is probably measurement of the level of protein carbonyls using the reagent dinitrophenylhydrazine (DNPH) (Reznick and Packer, 1994). Thus, as shown in Figure 1, the amount of protein carbonyl in gastrocnemius muscle was increased from 2.15 to 4.75 nmol/mg protein when the control contralateral leg was compared

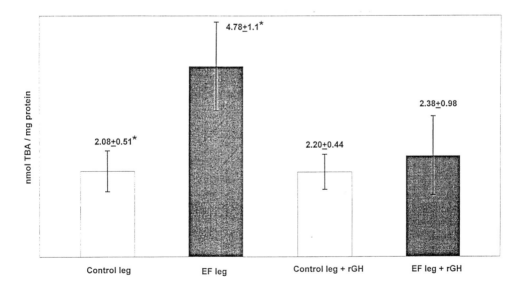

Figure 2. Levels of lipid peroxidation in gastrocnemius muscles of rats subjected to 4 weeks of immobilization by external fixation (EF) and to administration of rat growth hormone (rGH) in treated (right) and control (left) legs.

with immobilized leg (an increase of 215%). This increase was reduced drastically to only 31.2% in animals administered rGH for $3\frac{1}{2}$ weeks (Fig. 1).

Figure 2 shows the levels of lipid peroxidation in gastrocnemius muscles of animals subjected to 4 weeks of EF immobilization. The levels of thiobarbituric acid-reactive substances (TBARS) rose from 2.08±0.51 to 4.78±1.1 nmol TBA/mg protein, an increase of 131% in the immobilized leg. In animals administered rGH, the increase in TBA level was only 8.18%, attesting to the effect of GH in reducing not only protein oxidation, but also lipid peroxidation. Increase in lipid peroxidation in muscles of immobilized legs was also shown in previous studies using the TBARS method (Kondo et al., 1991) or by measuring conjugated dienes (Reznick, 1995). Moreover, the mechanism of oxidative stress in skeletal muscles subjected to immobilization is discussed at length in the chapter by H. Kondo in this volume.

Discussion

Immobilization of muscles leads to rapid muscle atrophy, as a result of significant protein loss (Booth and Seider, 1979; Booth, 1982). As early as 6 h after immobilization, there is a marked decrease in protein synthesis (Booth and Seider, 1979). After the biochemical changes, there are also structural changes indicative of muscle damage (Carmeli et al., 1993). Several reports in the last few years have indicated that this muscle damage observed in immobilization is also accompanied by oxidative stress and damage to proteins as well as to unsaturated lipids (Kondo et al., 1991, 1993). Supplying immobilized animals with antioxidants such as vitamin E could slow down some of the muscle loss caused by immobilization (Kondo et al., 1991). The mechanism of this oxidative stress has been shown to involve calcium-dependent proteases and elevation of free radicals such as OH^{-} and O_2^{-} (Kondo et al., 1993; see also this volume). Nevertheless, it is unclear whether this oxidative stress is a result of or a causative factor in immobilization.

GH secretion has been shown to be impaired in the adult population (Rudman et al., 1981). Administration of GH to adults over 60 years of age has been shown to have positive effects on increasing lean body mass and skin thickness and in reducing lipid content (Rudman et al., 1990). However, other reports on the positive effect of GH on ageing in people and animals have not been that conclusive and remain a controversial matter (Reznick et al., 1996).

Lipid peroxidation and protein oxidation have been shown to increase in ageing tissues and cells (Stadtman, 1992). Administration of antioxidants such as vitamin E and spin traps could reduce the levels of protein oxidation in exercise (Reznick et al., 1992) and in ageing (Carney et al., 1991). In this present chapter and in previous reports, it has been shown that the above parameters of oxidative stress are considerably elevated in muscles damaged by immobilization. Administration of GH could reduce this increase. However, so far it is unclear whether the effect of GH on lipid and protein oxidation is secondary, because GH reduces the general damage to muscles by stimulation of muscle protein synthesis, or whether GH has a direct effect on the status of oxidative stress in skeletal muscles. The above questions should be the subject of future investigations.

Acknowledgment
This work was supported by the generous support of the Krol Foundation, Nutley, NJ, USA.

References

Booth FW (1982) Effect of limb immobilization on skeletal muscle. *J Appl Physiol Respirat Environ Exercise Physiol* 52: 1113–1118.

Booth FW and Seider MJ (1979) Early change in skeletal muscle protein synthesis after limb immobilization of rats. *J Appl Physiol Respirat Environ Exercise Physiol* 47: 974–977.

Carmeli E and Reznick AZ (1992) The effect of growth hormone on skeletal muscles as a model for aging studies. *Gerontology (Israel)* 55: 3–7.

Carmeli E and Reznick AZ (1994) The physiology and biochemistry of skeletal muscle atrophy as a function of age. *Proc Soc Exp Biol Med* 206: 103–113.

Carmeli E, Hochberg Z, Livne E, Lichtenstein I, Kestelboim C, Silbermann M and Reznick AZ (1993) Effect of growth hormone on gastrocnemius muscle of aged rats after immobilization: Biochemistry and morphology. *J Appl Physiol* 75: 1529–1535.

Carney JM, Starke-Reed PE, Oliver CN, Landum RW, Cheng MS, Wu JF and Floyd RA (1991) Reversal of age-related increase in brain protein oxidation, decrease in enzyme activity, and loss in temporal and spatial memory by chronic administration of the spin-trapping compound *N-tert*-butyl-α-phenylnitrone. *Proc Natl Acad Sci USA* 88: 3633–3636.

Fares FA, Gruener N, Carmeli E and Reznick AZ (1996) Growth hormone (GH) retardation of muscle damage due to immobilization in old rats: Possible intervention with a new long-acting recombinant GH. *Ann N Y Acad Sci* 786: 430–443.

Hetland O and Dickstein K (1996) Cardiac markers in the early hours of acute myocardial infarction: Clinical performance of creatine kinase, creatine kinase MB isoenzyme (activity and mass concentration), creatine kinase MM and MB subform ratios, myoglobin and cardiac troponin T. *Scand J Clin Lab Investi* 56: 701–713.

Kondo H, Miura M and Itokawa Y (1991) Oxidative stress in skeletal muscle atrophied by immobilization. *Acta Physiol Scand* 142: 527–528.

Kondo H, Nakagaki I, Sasaki S, Hori S and Itokawa Y (1993) Mechanism of oxidative stress in skeletal muscle atrophied by immobilization. *Am J Physiol* 265: E839–E844.

Levine RL, Garland D, Oliver CN, Amici A, Climent I, Lenz AG, Ahn BW, Shaltiel S and Stadtman ER (1990) Determination of carbonyl content in oxidatively modified proteins. *Methods Enzymol* 186: 464–478.

Reznick AZ (1995) Free radicals and muscle damage due to immobilization of old animals: Effect of growth hormone. *In*: R Cutler, J Bertram and A Mori (eds): *Oxidative Stress and Aging*. Birkhäuser Verlag, Basel, pp 181–188.

Reznick AZ and Packer L (1994) Oxidative damage to proteins: Spectrophotometric method for carbonyl assay. *Methods Enzymol* 233: 357–363.

Reznick AZ, Witt E, Matsumoto M and Packer L (1992) Vitamin E inhibits protein oxidation in skeletal muscle of resting and exercised rats. *Biochem Biophys Res Commun* 189: 801–806.

Reznick AZ, Volpin G, Ben-Ari H, Silbermann M and Stein H (1995) Biochemical and morphological studies on rat skeletal muscles following prolonged immobilization of the knee joint by external fixation and plaster cast: A comparative study. *Eur J Exp Musculoskelet Res* 4: 69–76.

Reznick AZ, Carmeli E and Roisman I (1996) Effects of growth hormone on skeletal muscles of aging systems. *Age* 12: 39–45.

Rudman D, Kutner MH, Rogers CM, Lubin MF, Fleming GA and Bain RP (1981) Impaired growth hormone secretion in the adult population: Relation to age and adiposity. *J Clin Invest* 67: 1361–1369.

Rudman D, Feller AG, Nagraj HS, Gergans GA, Lalitha PY, Goldberg AF, Schlenker RA, Cohn L, Rudman IW and Mattson DE (1990) Effects of human growth hormone in men over 60 years old. *N Engl J Med* 323: 1–6.

Stadtman ER (1992) Protein oxidation and aging. *Science* 257: 1220–1224.

Sutton SC, Evans LA, Rinaldi MT and Norton KA (1996) Predicting injection site muscle damage. II. Evaluation of extended release parenteral formulations in animal models. *Pharmaceut Res* 13: 1514–1518.

Oxidative Stress in Skeletal Muscle
A.Z. Reznick et al. (eds)
© 1998 Birkhäuser Verlag Basel/Switzerland

The diaphragm and oxidative stress

S.K. Powers, J.M. Lawler and H.K. Vincent

Departments of Exercise and Sport Sciences and Physiology, University of Florida, Gainesville, FL 32611, USA
Department of Health and Kinesiology, Texas A&M University, College Station, Texas, USA

Summary. The diaphragm is the principal muscle of inspiration and is the only skeletal muscle considered essential for maintenance of normal ventilation in mammals. It is now clear that increased diaphragmatic contractile activity promotes the production of ROS. Further, experimental evidence demonstrates that in resting muscle, a basal level of ROS are essential for optimal regulation of E-C coupling. In contrast, synthesis and release of NO inhibits E-C coupling and reduces diaphragmatic force production at low stimulation frequencies.
In contracting diaphragm, accumulation of ROS species contributes to muscular fatigue during low frequency stimulation. The molecular identity of those ROS that are associated with the development of diaphragmatic fatigue is unclear. Although several potential mechanisms exist, it seems likely that ROS contribute to diaphragm fatigue by damaging the sarcoplasmic reticulum. This continues to be an exciting area of research.

Introduction

The diaphragm is the principal muscle of inspiration and is the only skeletal muscle considered essential for maintenance of normal ventilation in mammals (Powers et al., 1990a, 1990b). Compared with locomotor skeletal muscles, the diaphragm is unique in that it is chronically active. Indeed, this chronic contractile activity results in a relatively high rate of oxygen consumption and production of reactive oxygen species (ROS). Further, there is growing evidence that increased workloads on the diaphragm, for example, caused by exercise or chronic obstructive lung disease, result in an elevated production of ROS which may contribute to diaphragmatic fatigue. In this chapter, the authors provide a brief synopsis of the current understanding of ROS production in the diaphragm and the role that redox status plays in the regulation of diaphragmatic function in both health and disease. A start is made with an overview of diaphragmatic structure and function.

Overview of diaphragm structure and function

The diaphragm is the most important skeletal muscle involved in mammalian ventilation (DeTroyer and Estenne, 1988). Anatomically, the diaphragm is a musculocutaneous sheet that divides the thoracic and abdominal cavities. The diaphragm is unique anatomically among skeletal muscles in that the muscle fibres radiate from a central tendinous structure (the central tendon) (DeTroyer and Estenne, 1988). Further, the diaphragm can be divided into two functionally distinct parts that have different segmental innervations and embryologic origins: (1) the crural (or vertebral) portion, and (2) the costal portion (Fig. 1). The crural diaphragm originates from the spinal column and the medial and lateral arcuate ligaments and lies in the posterior portion of the thoracic cavity. In humans, the costal diaphragm arises from the inner surfaces and upper margins of the lower six ribs and the sternum (DeTroyer and Estenne, 1988). Both the costal and crural fibres insert into the central tendon of the diaphragm.

As in other skeletal muscles, the diaphragm is composed of functional units (motor units), each comprising a motor neuron and the muscle fibres it innervates. In most mammals, the diaphragm is a mixed muscle that contains both slow and fast fibres. In general, the slow fibres are characterized by a greater expression of type I and IIa myosin heavy chain proteins compared with fast fibres, whereas the fast fibres express more IIa and IIb compared with slow fibres. The

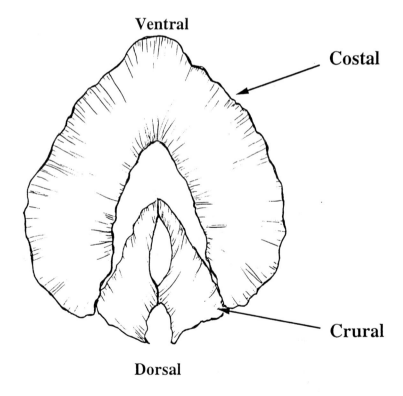

Rodent Costal-Crural Diaphragm
Thoracic View

Ventral

Costal

Crural

Dorsal

Figure 1. Illustration of the anatomical divisions of the rodent diaphragm.

oxidative capacity and fibre type composition of the diaphragm vary across species as a function of body mass (Hodge et al., 1997), i.e. compared with large animals with low metabolic rates, small animals with relatively high metabolic rates possess diaphragms with high oxidative capacities (Hodge et al., 1997).

In a healthy human or animal at rest, the metabolic load placed on the diaphragm is relatively small (about 1% of resting metabolic rate). However, intense muscular exercise (e.g. running) can result in large increases in diaphragmatic activity. In these cases, the work of the diaphragm increases as an exponential function of the metabolic rate (see Milic-Emili, 1991, for a review).

Again, mechanical loading as a result of intense, chronic exercise can also increase the workload and metabolic rate of the diaphragm. Animal studies have shown that increased diaphragmatic recruitment with exercise can increase the oxidative capabilities, as evidenced by increased citrate synthase activity (Powers et al., 1990a, 1994b). Furthermore, preliminary data in the authors' laboratory have revealed that there is a fast to slow shift in the expression of myosin heavy chain proteins in trained animals, such that diaphragms from trained animals express a greater percentage of type I slow myosin and a lower percentage of type IIb fast myosin compared with sedentary controls (unpublished data). This further supports the notion that the diaphragm is capable of adapting both structurally and biochemically to increased metabolic demands.

Several chronic disease states also influence the load placed upon the diaphragm. For example, chronic obstructive lung disease results in an increase in airway resistance which elevates the workload placed on the diaphragm and other inspiratory muscles (Powers et al., 1996). Obesity is a pathology that also influences diaphragmatic work; this condition increases the total mass of the chest wall, and increases both the work of breathing and the metabolic load placed on the diaphragm (Powers et al., 1996).

Production of ROS in the diaphragm

It is now clear that locomotor skeletal muscles contain numerous metabolic pathways that are capable of producing ROS. Similar to locomotor muscles, there is clear evidence that the mammalian diaphragm produces ROS during contractile activity. Indeed, superoxide anions, hydroxyl radicals, and nitric oxide are all produced during diaphragmatic contractions (Reid et al., 1992a; Diaz et al., 1993; Kobicz et al., 1994). Although the origin of ROS in contracting muscle continues to be investigated, it seems likely that oxygen-derived radicals are produced as a by-product of oxidative metabolism within mitochondria (Chance et al., 1979). Other potential sites for production of ROS in the diaphragm include membrane oxidoreductases, the cyclooxygenase pathway of archidonic acid metabolism, and cytosolic endothelial xanthine oxidase (see Halliwell and Gutteridge, 1989, for a review). Here, key evidence is outlined which demonstrates that ROS are produced during diaphragmatic contractile activity.

Although Davies and colleagues (Davies et al., 1982) were the first to detect radical production in locomotor skeletal muscles, Reid et al. (1992a) were the first investigators to report diaphragmatic production of ROS. In these experiments, bundles of diaphragm fibres were loaded *in vitro* with 2',7'-dichlorofluorescin, a fluorochrome that emits at 520 nm when oxidized; the emissions were quantified using a fluoresence microscope. The results indicated that a small but continuous production rate of ROS occurs in the resting (noncontracting) diaphragm. Further, electrically stimulated contraction of diaphragm fibres resulted in a large increase in ROS production. The findings suggested that the rate of diaphragmatic ROS production is dependent upon the metabolic rate of the muscle. These results have been supported by independent experiments using electron spin resonance to detect radicals in the contracting diaphragm *in vivo* (Borzone et al., 1994b).

In an experiment designed to determine which ROS are produced by diaphragm muscle, Reid et al. (1992a) demonstrated that both resting and contracting diaphragm fibres produce superoxide radicals *in vitro*. In this investigation, bundles of diaphragm fibres were incubated in an organ bath containing cytochrome *c* (a standard assay for superoxide). The data revealed that superoxide production occurs in both resting and contracting diaphragm and that the rate of production is greater in contracting fibres. Moreover, the authors concluded that superoxide radicals are capable of crossing the sarcolemma and reaching the interstitial space.

Diaz et al. (1993) first reported that diaphragmatic contractions can also result in the production of hydroxyl radicals. These authors used the salicylate trapping method to detect hydroxyl radicals. This technique is predicated on the fact that salicylate's phenolic ring can be attacked by hyroxyl radicals to form 2,3- or 2,5-dihydroxybenzoic acid. This is a stable product which can be assayed via high-pressure liquid chromatography. In these experiments, hydroxyl radicals could not be detected in resting diaphragm; however, in electrically stimulated diaphragm strips, hydroxyl radical production increased as a linear function of diaphragmatic work as indicated by the time–tension index (i.e. force multiplied by contractile time).

Also, it is now clear that contracting diaphragm produces nitric oxide (Kobzik et al., 1994, 1995). Similar to the findings with superoxide radicals, resting diaphragm produced small but detectable amounts of nitric oxide (NO) whereas contracting diaphragm produced increased amounts of NO. It is of interest that the diaphragm is the first cell type found to express two isoforms of NO synthase (Kobzik et al., 1994, 1995). Indeed, diaphragm myocytes constitutively express both the neuronal-type and the endothelial-type isoforms of NO synthase. The physiological significance of this finding is unclear and is an active area of investigation.

In summary, the diaphragm produces ROS both at rest and during contractile activity. Further, the amount of ROS produced in the diaphragm increases as a function of the total amount of contractile activity (i.e. force multiplied by the time of contraction). Finally, it appears that nitric oxide, superoxide, and hydroxyl radicals are all produced during diaphragmatic contractile activity.

ROS and diaphragmatic contractile properties

Muscular contraction is activated by depolarization of the fibre resulting in the release of calcium from the sarcoplasmic reticulum into the cytoplasm. This free calcium activates the contractile proteins (actin and myosin) to start the contractile process. The series of steps leading from depolarization of the fibre to mechanical shortening is termed "excitation–contraction coupling" (EC coupling). Muscular contraction can be rapidly reversed by removal of cytoplasmic calcium via ATP-dependent transporters which "pump" calcium back into the sarcoplasmic reticulum for storage.

The amount of force produced during muscular contraction can be increased by augmenting the rate of fibre depolarization: an increased rate of fibre depolarization maintains calcium release and therefore elevates muscle force production. Indeed, calcium release and muscle force production can be varied over a finite range by increasing or decreasing the rate of muscle depolarization. This forms the physiological basis for the force–frequency curve (Fig. 2). A pragmatic use of the force–frequency curve is that the position of the curve is indicative of the effectiveness of EC coupling. Specifically, effective EC coupling is indicated by a left shift in the force-frequency curve whereas less effective coupling is indicated by a right shift in the curve.

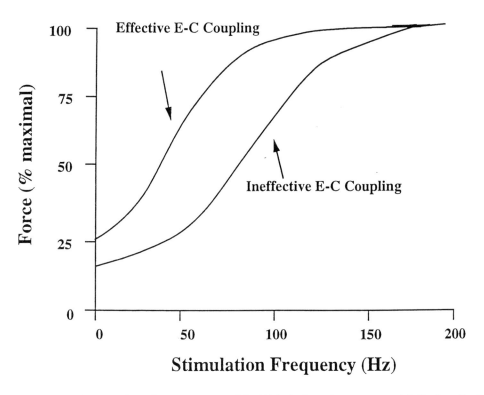

Figure 2. Illustration of the force–frequency curve. A right shift in the force–frequency curve is indicative of ineffective EC coupling whereas a left shift indicates effective EC coupling.

Evidence is accumulating that ROS play a key role in the regulation of EC coupling in the diaphragm and other skeletal muscles. Reid and colleagues (Reid et al., 1993; Reid and Moore, 1994) described a right shift in the force–frequency curve in nonfatigued diaphragm fibre bundles exposed to antioxidants, i.e. introduction of either superoxide dismutase (SOD), catalase, or dimethyl sulfoxide (DMSO) into the fibre bath depressed low-frequency muscle force production of the diaphragm, which results in a right shift in the force–frequency curve (Fig. 3). Consistent with these data, introduction of a low-rate, ROS-generating system (e.g. xanthine oxidase) results in a small increase in low-frequency diaphragmatic force production (Lawler and Hu, 1996; Lawler et al., 1997a). Collectively, these data suggest a potential physiological role of ROS in regulation of diaphragm contractility. Indeed, it appears that ROS produced by resting muscle promote effective EC coupling and are therefore required for optimal contractile function.

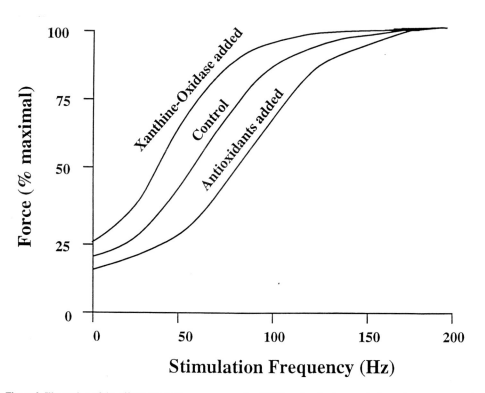

Figure 3. Illustration of the effects of reactive oxygen species (ROS) on the diaphragmatic force-frequency curve in unfatigued muscle. A basal production of ROS is required for optimal contractile function. See text for details. (Data from Reid et al., 1993; Reid, 1996; and Lawler et al., 1997.)

What is the mechanism to explain the influence of ROS on muscle contractile properties? A potential mechanism is by enhancing calcium release from the sarcoplasmic reticulum. In particular, oxidation of sulfhydryl groups on calcium channel proteins results in channel opening and promotes an increased release of calcium into the cytoplasm (Liu and Pessah, 1994; Liu et al., 1994). Further, ROS may also inhibit calcium re-uptake into the sarcoplasmic reticulum. Collectively, these actions result in elevated cytoplasmic calcium levels and enhance muscle force production.

In contrast to the effects of other common ROS, production of NO in muscle impairs EC coupling. Introduction of NO synthase inhibitors results in an increase in low-frequency tension development of the diaphragm and a left shift in the force–frequency curve (Lawler et al., 1997b) (Fig. 4). The mechanism of NO action on the diaphragm remains poorly understood. In smooth muscle, NO exerts second messenger effects via guanosine cyclic 3',5'-monophosphate (cGMP)

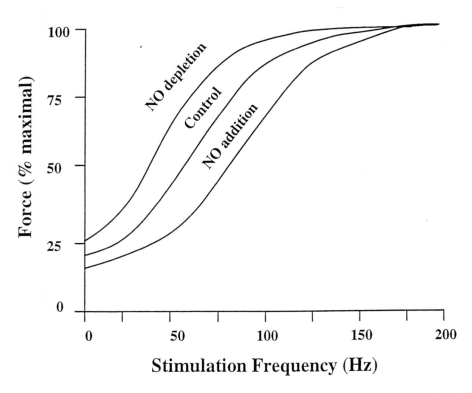

Figure 4. Influence of nitric oxide (NO) on the diaphragmatic force-frequency curve. See text for details. (Data from Reid, 1996 and Lawler et al., 1997.)

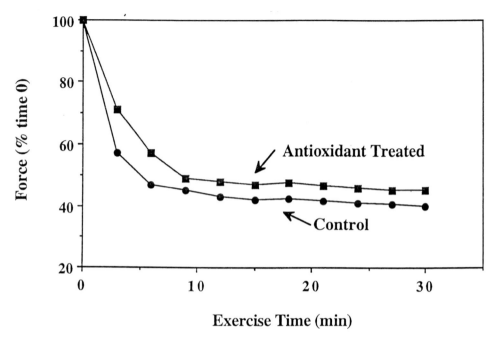

Figure 5. Illustration of the effects of antioxidants on diaphragmatic fatigue development during prolonged contractile periods. (Data from Reid, 1996.)

to inhibit muscular contraction. A similar mechanism exists in the diaphragm but the magnitude of its influence remains unclear (Kobzik et al., 1994). This is currently an active area of investigation.

In review, in resting muscle, the normal production of ROS is essential for optimal regulation of EC coupling. In contrast, synthesis and release of NO inhibits EC coupling and reduce diaphragmatic force production at low frequencies of stimulation.

Radicals and diaphragmatic fatigue

In contrast to the notion that ROS are required for normal muscle function in nonfatigued muscle, there is abundant evidence that, during periods of prolonged contractile activity (e.g. endurance exercise), production of ROS contributes to muscle fatigue. Indeed, numerous studies suggest that ROS production contribute to muscular fatigue. For example, Novelli et al. (1990) and Barclay and Hansel (1991) reported that exogenous antioxidant scavengers reduce the rate of locomotor muscle fatigue both *in vivo* and *in vitro*. These early investigations in locomotor muscles have been followed by a numerous studies investigating the effects of ROS on diaphragmatic fatigue.

Similar to locomotor muscles, much evidence indicates that oxidative stress contributes significantly to the impairment of contractile function in the diaphragm during repetitive contractions. Specifically, there are three primary lines of evidence that collectively establish a cause-and-effect relationship between the production of ROS and diaphragmatic fatigue. First, antioxidant treatment *in vitro* and *in situ* attenuates the rate of fatigue development in the diaphragm (Shindoh et al., 1990; Reid et al., 1992). Second, exposing contracting diaphragm to exogenous ROS decreases muscle force production and promotes fatigue (Lawler et al., 1996, 1997a). Third, diaphragmatic fatigue is accelerated in diaphragms with reduced levels of glutathione (Anzueto et al., 1992; Morales et al., 1993; Borzone et al., 1994a).

Of the above lines of evidence, the strongest experimental support of the notion that ROS actively contribute to diaphragmatic fatigue is the observation that fatigue can be inhibited by exogenous antioxidants. This effect was first observed by Shindoh et al. (1990) who demonstrated that treatment with the antioxidant, *N*-acetylcysteine, reduced the development of rabbit diaphragmatic fatigue *in situ*. The results of this early investigation were followed and confirmed by numerous *in vitro* experiments in rodents incorporating a wide variety of antioxidants (i.e. SOD, catalase, and DMSO) (Fig. 5).

Most of the aforementioned experiments examining the effects of radicals on diaphragmatic contractile performance have used *in vitro* models. Therefore, from a physiological perspective, a key question is: Do oxygen-derived radicals also contribute to diaphragm fatigue during loaded breathing *in vivo*? The answer to this question is yes. Recently, Suspinski et al. (1997) administered *N*-acetylcysteine, a radical scavenger, to decerebrate non-anesthetized rats breathing against a large inspiratory load. Their data clearly showed that this administration of this antioxidant slows the rate of respiratory failure during inspiratory loading.

To date, which ROS are specifically responsible for promoting diaphragmatic fatigue remains unclear. Indeed, Reid et al. (1992a) reported that application of antioxidants specific for superoxide anions (SOD), H_2O_2 (catalase), and hydroxyl radicals (DMSO) attenuate diaphragmatic fatigue with equal potency.

What are the cytotoxic effects of ROS that lead to muscle fatigue? Although a definitive answer is not available, several possibilities exist. It is possible that ROS could down-regulate key bioenergetic enzymes, damage contractile proteins, and/or alter EC coupling by damage to cellular membranes. In this regard, most of the research focus has centered around investigations of ROS-mediated alterations in EC coupling. For example, Nashawati et al. (1993) proposed that ROS disrupt function of diaphragm intracellular membranes and thus impair EC coupling. Other investigators have also hypothesized that disturbances in calcium homoeostasis resulting from ROS-mediated damage to the sarcoplasmic reticulum (SR) could contribute to diaphragmatic fatigue caused by oxidative stress (Brotto and Nosek, 1996; Reid, 1996). Indeed, both Ca^{2+} release (Favero et al., 1995) and Ca^{2+} re-uptake (Byrd, 1992) of SR vesicles may be altered by ROS. Hence, it appears likely that ROS promote fatigue, at least in part, by damage to the sarcoplasmic reticulum resulting in disturbed calcium handling. A search for other possible sites of ROS-mediated damage in the diaphragm is ongoing in several laboratories.

In conclusion, it is clear that ROS contribute to muscular fatigue during low-frequency stimulation in both locomotor muscles and the diaphragm. The molecular identity of which ROS are associated with the development of muscle fatigue remains unclear. Although several potential mechanisms exist, it is postulated that ROS promote diaphragm fatigue by damage to the sarcoplasmic reticulum.

Effect of ROS on diaphragm function: Non-fatigued versus fatigued myocytes

It seems paradoxical that ROS are required for optimal contractile function in the non-fatigued diaphragm, and also that the production of ROS is associated with fatigue in muscles performing repetitive contractions. A potential explanation for these observations has been proposed by Reid (1996). He argues that intracellular levels of oxidants are closely regulated and that alterations of myocyte oxidation status (i.e. increase or decrease) result in compromised contractile function. Further, a logical site in the myocyte where regulation of redox status is important is the sarcoplasmic reticulum (Favero et al., 1993, 1995; Liu and Pessah, 1994; Liu et al., 1994; Reid, 1996). Here, regulatory sulfhydryl groups located on the calcium-release channel and the calcium-dependent ATPase pump could undergo reversible oxidation by various ROS. As mentioned earlier, sulfhydryl oxidation of the calcium-release channel promotes calcium release by opening efflux channels. Sulfhydryl oxidation of the calcium-dependent ATPase pump results in an inhibition of calcium re-uptake into the SR. Conversely, reversal of this oxidation process would result in the closing of calcium-release channels and an increased sarcoplasmic uptake of cytosolic calcium.

Non-fatigued (resting) diaphragm myocytes would normally contain a low level of ROS which are likely to oxidize a small and finite number of regulatory sulfhydryls on the sarcoplasmic reticulum (Reid, 1996). This sulfhydryl oxidation results in an increase in calcium release (i.e. facilitates EC coupling) and an elevation in muscle force production. The application of antioxidants in non-fatigued muscle would reverse this oxidation process and therefore reduce the effectiveness of excitation-contraction coupling.

In contrast, in fatigued diaphragm myocytes, the production of ROS greatly exceeds the rate of production at rest; this results in a prooxidant state which increases the number of oxidized sulfhydryl groups within the sarcoplasmic reticulum (Reid, 1996). The end-result would be a loss of calcium homoeostasis and is a well-recognized component of the fatigue process. In contrast, the addition of antioxidants to muscle before fatiguing exercise would assist in maintaining redox homoeostasis and therefore delay fatigue (Reid, 1996).

In conclusion, it appears that diaphragmatic myocytes are very sensitive to changes in redox status. In resting myocytes, the balance between the production and removal of ROS is important for maintaining redox homoeostasis. Significant perturbations in either direction (i.e. increase or decrease of oxidants) can alter optimum muscle performance in a similar manner to how muscle function can be altered by failure to maintain acid–base status (Reid, 1996). The fundamental understandings of the importance of redox regulation of skeletal muscle myocytes has developed rapidly over the past 5 years and will continue to be an important focus of future research.

Diaphragm and antioxidant enzyme activities

The antioxidant capacity of mammalian organ systems is designed to match the level of tissue O_2 consumption and rate of free radical production (Ji, 1995). For example, tissues with the highest O_2 consumption per day, such as the liver, diaphragm, and heart, have the greatest antioxidant capacity. Furthermore, there is a positive correlation between oxidative potential and antioxidant activities in skeletal muscle (Lawler et al., 1993).

The most widely studied animal model of diaphragmatic antioxidant enzyme capacity is the rat. When contrasting diaphragmatic biochemical properties with those of other skeletal muscles, plantaris muscle is often used as a comparison. Indeed, plantaris is a locomotor muscle with a mixed fibre type profile similar to that of the diaphragm. Figure 6 provides a comparison of the oxidative capacity (represented by citrate synthase activity) and the antioxidant enzyme activity (represented by glutathione peroxidase) in the costal diaphragm and plantaris muscles of young adult and senescent rats (Powers et al., 1992). Note that, in young animals, the oxidative potential and the antioxidant capacity of the diaphragm are more than 100% higher in the diaphragm compared with plantaris muscle. This observation is not surprising given that the diaphragm is chronically active whereas plantaris muscle is recruited infrequently during periods of locomotion. This chronic diaphragmatic activity requires a constant production of ATP via oxidative phosphorylation; this elevated rate of electron transport via the cytochrome chain is associated with an increased production of superoxide radicals. Hence, it is not surprising that the diaphragm is well equipped with antioxidant enzymes to protect this important muscle from oxidative damage.

Ageing and diaphragmatic antioxidant status

Also, note that, although there is an age-related decrease in oxidative capacity in plantaris muscle, this decline in oxidative potential is not observed in the diaphragm (Fig. 6). Furthermore, ageing tends to result in an increase in the activity of glutathione peroxidase in both the diaphragm and plantaris muscles. This observation has been interpreted as an indication that senescent skeletal muscle produces more ROS than muscles from young animals (Powers et al., 1992).

Effects of chronic and acute exercise on diaphragamtic antioxidant enzymes

Endurance exercise and antioxidant enzyme activity

It is well known that the antioxidant defence systems of many mammalian tissues are plastic and capable of adaptation in response to increased exposure to ROS. For example, irradiation of the mouse heart results in an increased expression of the manganese-dependent isoform of SOD (Oberley et al., 1987). As aerobic metabolism is increased with exercise, the production of ROS in the diaphragm is increased as well (Reid et al., 1992). Therefore, exercise should stress the antioxidant capacity of active tissues such as the diaphragm. It follows that endurance training would augment antioxidant enzyme activities in contracting skeletal muscle as an adaptive re-

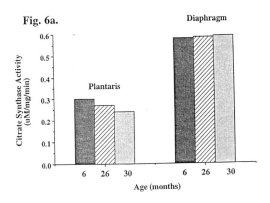

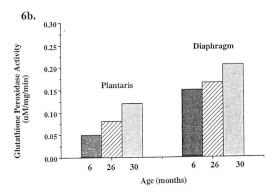

Figure 6. Effects of age on citrate synthase activity and glutathione peroxidase activity in the rat diaphragm. (Data from Powers et al., 1992.)

sponse to oxidative stress with exercise. Indeed, this is the case. A brief overview of the effects of endurance training on antioxidant enzyme activity in the diaphragm follows.

Superoxide dismutase (SOD)
Numerous studies have reported training-induced increases in total SOD activity in locomotor muscles (Higuchi, 1985; Hammeren et al., 1992; Criswell et al., 1993; Powers et al., 1994a, b). To date, only one investigation has examined the effects of endurance training on SOD activity of the diaphragm (Powers et al., 1994b). Ten weeks of endurance treadmill training increased SOD activity in the costal diaphragm. It was interesting that the exercise induced diaphragmatic SOD activity was independent of exercise intensity and duration (see Powers and Criswell, 1996, for a complete review).

Glutathione peroxidase (GPX)

Similar to SOD activity, endurance exercise training promotes an increase in GPX activity in the diaphragm (Powers et al., 1994). Further, upregulation of diaphragm GPX with endurance training is relatively independent of intensity and duration (Powers et al., 1994a).

Catalase (CAT)

The effects of regular exercise on CAT activity in the diaphragm is unknown. Given the lack of exercise effect on the heart, and the inconsistent findings in locomotor skeletal muscle, it would seem logical that the impact of exercise training on CAT activity would be minimal.

Acute exercise and antioxidant enzyme activity

It well established that acute exercise results in an increase in the activities of SOD and GPX in locomotor muscles (Ji et al., 1988; Lawler et al., 1993). Similarly, Lawler et al. (1994b) has shown that 40 minutes of treadmill exercise (about 70% $VO_{2\,max}$ or maximum oxygen consumption) resulted in a small up-regulation of diaphragm SOD, with no influence on GPX activity. In general, the magnitude of the antioxidant enzyme upregulation with an acute bout of exercise is substantially smaller than seen with endurance training. This rapid increase in antioxidant enzyme activity occurs too quickly to be a result simply of new protein synthesis (Ji and Fu, 1992), and suggests the involvement of post-translational modification or positive allosteric modification. The precise molecular mechanisms underlying exercise-induced upregulation of antioxidant enzymes remain to be elucidated.

Acknowledgements
Work in the laboratory of Scott K. Powers has been supported by the American Heart Association-Florida and the American Lung Association-Florida. Work in John M. Lawler's laboratory has been supported by the American Lung Association.

References

Anzueto A, Andrade FH, Maxwell LC, Levine SM, Lawrence RA, Gibbons GA and Jenkinson SR (1992) Resistive breathing activates the glutathione redox cycle and impairs performance of the rat diaphragm. *J Appl Physiol* 72: 529–534.

Barclay JK and Hansel M (1991) Free radicals may contribute to oxidative skeletal muscle fatigue. *Can J Physiol Pharmacol* 69: 279–284.

Borzone G, Julian MW, Merola AJ and Clanton TL (1994a) Loss of diaphragm glutathione is associated with respiratory failure induced by resistive breathing. *J Appl Physiol* 76: 2825–2831.

Borzone G, Zhao B, Merola AJ, Berliner L and Clanton TL (1994b) Detection of free radicals by electron spin resonance in the rat diaphragm after resistive breathing. *J Appl Physiol* 77: 812–818.

Brotto MAP and Nosek TM (1996) Hydrogen peroxide disrupts Ca^{2+} release from sarcoplasmic reticulum of rat skeletal muscle fibers. *J Appl Physiol* 81: 731–737.

Byrd SK (1992) Alteration in the sarcoplasmic reticulum: a possible link to exercise-induced muscle damage. *Med Sci Sports Exerc* 24: 531–536.

Chance B, Sies H and Boveris A (1979) Hydroperoxide metabolism in mammalian organisms. *Physiol Rev* 59: 527–605.

Criswell DS, Powers S, Dodd S, Lawler J, Edwards W, Renshler K and Grinton S (1993) Cellular oxidative and antioxidant response in skeletal muscle to interval and continuous exercise training. *Med Sci Sports Exerc* 25: 1135–1140.
Davies KJA, Quintanilha AT, Brooks GA and Packer L (1982) Free radicals and tissue damage produced by exercise. *Biochim Biophys Acta* 107: 1198–1205.
DeTroyer A and Estenne M (1988) Functional anatomy of the respiratory muscles. *Clinics in Chest Medicine* 9: 175–194.
Diaz PT, She Z-W, Davis WB and Clanton TL (1993) Hydroxylation of salicylate by the *in vitro* diaphragm: evidence for hydroxyl radical production during fatigue. *J Appl Physiol* 75: 540–545.
Favero TG, Zable AC, Bowman MB, Thompson A and Abramson JJ (1995) Metabolic end products inhibit sarcoplasmic reticulum Ca^{2+} release and [^{3}H]ryanodine binding. *J Appl Physiol* 78: 1665–1672.
Favero TG, Henry R and Abramson JJ (1993) Hydrogen peroxide stimulated calcium release from rat skeletal muscle sarcoplasmic reticulum. *Biophys J* 64: A303.
Halliwell B and Gutteridge J (1989) *Free Radicals in Biology and Medicine*. Clarendon Press, Oxford, England.
Hammeren J, Powers S, Lawler J, Criswell D, Martin D, Lowenthal D and Pollock M (1992) Exercise training-induced alterations in skeletal muscle oxidative and antioxidant enzyme activity in senescent rats. *Int J Sport Med* 13: 412–416.
Higuchi M, Cartier L-J, Chen M and Holloszy JO (1985) Superoxide dismutase and catalase in skeletal muscle: adaptive response to exercise. *J Gerontol* 40: 281–286.
Hodge K, Powers S, Coombes J, Fletcher L, Dodd S and Martin D (1997) Bioenergetic characteristics of the costal and crural diaphragm in mammals. *Resp Physiol; in press*.
Ji L (1995) Exercise and oxidative stress: role of cellular antioxidant systems. *Exerc Sport Sci Rev* 23: 135–166.
Ji LL and Fu R (1992) Responses of the glutathione system and antioxidant enzymes to exhaustive exercise and hydroperoxide. *J Appl Physiol* 2: 549–554.
Ji LL, Stratman FW and Lardy HA (1988b) Antioxidant enzyme systems in rat liver and skeletal muscle. *Arch Biochem Biophys* 263: 150–160.
Kobzik L, Reid MB, Bredt DS and Stamler JS (1994) Nitric oxide in skeletal muscle. *Nature* 372: 546–548.
Kobzik L, Stringer B, Balligand J, Reid M and Stamler J (1995) Endothelial type nitric oxide synthase (ec-NOS) in skeletal muscle fibers: mitochondrial relationships. *Biochem Biophys Res Commun* 211: 375–381.
Lawler JM and Hu Z (1996) Effect of superoxide dismutase on xanthine oxidase-induced depression of contractility in the fatigued diaphragm. *The Physiologist* 39: A15.
Lawler JM, Powers SK, Visser T, Van Dijk H, Kordus M and Ji LL (1993) Acute exercise and skeletal muscle antioxidant and metabolic enzymes: effects of fiber-type and age. *Am J Physiol* 265: R1344–R1350.
Lawler JM, Cline CC and Barnes WS (1994a) Effect of xanthine oxidase-induced free radical stress on K^{+} contractures in the rat diaphragm. *Med Sci Sports Exerc* 26: A193.
Lawler JM, Powers SK, Van Dijk H, Visser T, Kordus M and Ji LL (1994b) Metabolic and antioxidant enzyme activities in the diaphragm: effects of acute exercise. *Resp Physiol* 96: 139–149.
Lawler JM, Cline CC, Hu Z and Coast JR (1997a) Effect of oxidant challenge on contractile function of the aging diaphragm. *Am J Physiol* 272: E201–207.
Lawler JM, Hu Z, Demaree S and Reid M (1997b) Does nitric oxide influence xanthine oxidase-induced potentiation of low-frequency tension in the unfatigued diaphragm? *Med Sci Sports Exerc* 29: S27.
Liu G and Pessah IN (1994) Molecular interaction between ryanodine receptor and glycoprotein triadin involves redox cycling of functionally important hyperreactive sulfhydryls. *J Biol Chem* 269: 33028–33034.
Liu G, Abramson J, Zable A and Pessah I (1994) Direct evidence for the existence and functional role of hyperreactive sulfhydrals on the ryanodine receptor-triadin complex selectively labeled by the courmarin maleimide 7-diethylamino-3-(4'-maleimidylphenyl)-4-methylcoumarin. *Mol Pharmacol* 45: 189–200.
Milic-Emili J (1991) Work of breathing. *In*: R Crystal and JB West (eds): *The Lung: Scientific Foundations*. Raven Press, New York, pp 1065–1075.
Morales CF, Anzueto A, Andrade F, Levine SM, Maxwell LC, Lawrence RA and Jenkinson SG (1993) Diethylmaleate produces diaphragmatic impairment after resistive breathing. *J Appl Physiol* 75: 2406–2411.
Morales CF, Anzueto A, Andrade F, Brassard J, Levine SM, Maxwell LC, Lawrence RA and Jenkinson SG (1994) Buthione sulfoxamine treatment impairs rat diaphragm function. *Am J Respir Crit Care Med* 149: 915–919.
Nashawati E, DiMarco A and Supinski G (1993) Effects produced by infusion of a free radical-generating solution into the diaphragm. *Am Rev Respir Dis* 147: 60–65.
Novelli GP, Bracciotti G and Falsini S (1990) Spin trappers and Vitamin E prolong endurance to muscle fatigue in mice. *Free Radical Biol Med* 8: 8–13.
Oberley L, St Clair D, Autor A and Oberley T (1987) Increase in manganese superoxide dismutase activity in the mouse heart after irradiation. *Arch Biochem Biophys* 254: 69–80.
Powers S and Criswell D (1996) Adaptive strategies of respiratory muscles in response to endurance exercise. *Med Sci Sports Exerc* 28: 1115–1122.
Powers S, Lawler J, Criswell D, Dodd S, Grinton S, Bagby G and Silverman H (1990a) Endurance training-induced cellular adaptations in respiratory muscles. *J.Appl. Physiol.* 68: 2114–2118.

Powers S, Lawler J, Criswell D, Silverman H, Forster HV, Grinton S and Harkins D (1990b) Regional metabolic differences in the rat diaphragm. *J Appl Physiol* 69: 648–650.

Powers S, Lawler J, Criswell D, Lieu F and Dodd S (1992) Alterations in diaphragmatic oxidative and antioxidant enzymes in the senescent Fischer-344 rat. *J Appl Physiol* 72: 2317–2321.

Powers S, Criswell D, Lawler J, Ji L, Martin D, Herb R and Dudley G (1994a) Influence of exercise and fiber type on antioxidant enzyme activity in rat skeletal muscle. *Am J Physiol* 266: R375–R380.

Powers S, Criswell D, Lawler J, Martin D, Ji L, Herb R and Dudley G (1994b) Regional training-induced alterations in diaphragmatic oxidative and antioxidant enzymes. *Resp Physiol* 95: 227–237.

Powers S, Farkas G, Demirel H, Coombes J, Fletcher L, Hughes M, Hodge K, Dodd S and Schlenker E (1996) Effects of aging and obesity on respiratory muscle phenotype in Zucker rats. *J. Appl.Physiol.* 81: 1347–1354.

Reid MB (1996) Reactive oxygen and nitric oxide in skeletal muscle. *News Physiol Sci* 11: 114–119.

Reid MB, Haack KE, Franchek FM, Valberg PA, Kobzik L and West MS (1992) Reactive oxygen in skeletal muscle. I. Intracellular oxidant kinetics and fatigue *in vitro. J Appl Physiol* 73: 1797–1804.

Reid MB, Khawli FA and Moore MR (1993) Reactive oxygen in skeletal muscle. III. Contractility of unfatigued muscle. *J Appl Physiol* 75: 1081–1087.

Shindoh C, DiMarco A, Thomas A, Manubag P and Supinski G (1990) Effect of *N*-acetylcysteine on diaphragm fatigue. *J Appl Physiol* 68: 2107–2113.

Suspinski G, Stofan D, Ciufo R and Dimarco A (1997) *N*-Acetylcysteine administration alters the response to inspiratory loading in oxygen-supplemented rats. *J Appl Physiol* 82: 1119–1125.

Oxidative damage after ischemia/reperfusion in skeletal muscle

A. Hochman

Department of Biochemistry, George S. Wise Faculty of Life Sciences, Tel Aviv University, Tel Aviv, Israel 69978

Summary. Skeletal muscle ischemia may result from intrinsic events which cause vascular insufficiency, or extrinsic interventions such as surgery and transplantation. Due to the low resting energy demands and large intracellular stores of available energy the ischemia-induced biochemical and physiological perturbations may be prominent, but are reversible most of the time. On the other hand during reperfusion there are additional, significant structural and functional changes which are irreversible in many cases, and in combination with the ischemic damages, may result in skeletal muscle necrosis. Injuries inflicted to skeletal muscle following ischemia–reperfusion may also affect other organs, and have also been implicated in the development of final multiple system organ failure which can cause death. A growing body of experimental data, in tissue cultures, isolated organs, animal models in humans point to the involvement of reactive oxygen metabolites (ROS), superoxide, hydrogen peroxide and hydroxyl radicals in the etiology of ischemia–reperfusion damages ROS may be generated during reperfusion by three main metabolic pathways: 1) oxidation of xanthine and hypoxanthine by xanthine oxidase which is activated by ischemia; 2) activation and influx of activated neutrophils into the ischemic tissue and 3) arachidonic acid metabolism and metabolites. Oxidative damage was identified as depletion of internal antioxidants, altered levels of antioxidant enzymes, lipid peroxidation, conjugated dienes, TBARS and changes in mitochondrial structure and function. Furthermore, in most studies it was shown that inhibition of ROS production as well as administration of external antioxidants during ischemia or reperfusion reduced structural and functional damages.

Introduction

Ischemia occurs when there is lack of oxygenation of cells, tissues or organs. It is a common clinical event produced by generalized reduction in cardiac output as well as intrinsic or extrinsic vascular obstruction. It occurs during surgery, with transplantation and under conditions of shock, hypovolaemia and trauma, such as crush injury. Limb ischemia may occur as a result of vascular insufficiency, including chronic atherosclerotic occlusive disease, arterial embolism, vascular trauma and acute arterial thrombosis. It is also inflicted electively in orthopaedic procedures that use a tourniquet to provide a bloodless organ. Peripheral tissue ischemia is relatively well tolerated, compared with organs such as the brain, liver and heart (Soussi et al., 1990). Experimental findings as well as clinical experience have shown that functioning of skeletal muscle recovers after several hours of ischemia without persistent damage (Harris et al., 1986).

The relative resistance of skeletal muscle to normothermic ischemia is related to its low resting energy demands and large intracellular stores of available energy. Ischemia-induced decrease in skeletal muscle membrane potential was restored during reperfusion, provided that the duration of the ischemia was not too long (Oredsson et al., 1993). The energy charge is less sensitive to reperfusion injury, and was restored to its control value after longer periods of ischemia. Under certain conditions skeletal muscle can undergo up to 8 h of circulatory interruption with no irreversible damage to the myocytes (Paul et al., 1990). Prolonged ischemia to skeletal muscle results in alterations in adenine nucleotide metabolism. Adenosine triphosphate (ATP) continues to be used for cellular functions, and there is a progressive degradation of compounds rich in energy (Aldman et al., 1987). Aerobic metabolism shifts to anaerobic glycolysis, resulting in decreased production of ATP and intracellular accumulation of lactate which results in acidosis (Enger et al., 1978).

When the infrarenal aorta of rats was clamped for 90 min, muscle resting transmembrane potential difference (E_m) was significantly depolarized, creatine phosphate was depleted and ATP levels maintained (Yokota et al., 1989). On reperfusion, persistent depolarization of resting E_m was observed despite restoration of the muscle creatine phosphate content. When partial skeletal muscle ischemia–reperfusion occurred in dogs and rats, 1–3 h caused cell membrane depolarization (Perry and Fantini, 1989). Intracellular levels of ATP remained normal, suggesting that direct membrane injury rather than electrogenic pump failure occurred. A 2-h ischemic insult in a canine gracilis muscle model resulted in minimal ultrastructural damage and complete regeneration of intramuscular phosphagens and glycogen on reperfusion (Harris et al., 1986). However, a 7-h ischemic insult caused profound injury at the ultrastructural level, with inability to restore intramuscular phosphagens and glycogen on reperfusion. This severe muscle injury correlated with a decline in ATP levels to below 5 μmol/g dry weight on reperfusion. Acute arterial occlusion for 4 and 5 h resulted in complete depletion of creatine phosphate and decrease in ATP which was more significant after 5 h (Lindsay et al., 1990). Necrosis, which was assessed after 48 h, was much higher in the 5-h ischemic group. In accordance, it was shown that supplementation of intracellular ATP stores had a protective effect. In a model of 3.5-h ischemia of rectus femoris muscle of rabbits, a mixture of phosphoenolpyruvate (PEP) and ATP was infused intra-arterially at the end of ischemia (Hickey et al., 1995). At 24 h the viability of the PEP/ATP-infused muscles was significantly greater than that of controls, ATP stores were significantly higher and water content significantly lower. At 24 h and 4 days, muscles infused with PEP/ATP showed less necrosis.

Ischemia-induced biochemical events may be prominent but in many cases are reversible, although during reperfusion functional changes are also significant, but not all are reversible. The combination of cellular damage during the period of ischemia and its exacerbation during reperfusion may result in the production of skeletal muscle necrosis. These multifactorial damages are manifested at the structural level as microvascular dysfunction, which is characterized by an increased membrane permeability to plasma proteins (Diana and Laughlin, 1974; Korthuis et al., 1985; Korthuis et al., 1989; Sexton et al., 1990), tissue oedema (Strock and Majno, 1969a, b; Sexton et al., 1990), compartment syndrome (Mubarek and Hargens, 1983) contractile dysfunction (Gardner et al., 1984), mitochondrial, sarcolemmal and myofibrillar disruption, increased lipid peroxidation, reduction in transmembrane potential and vascular damage. Furthermore, some muscles may fail to reperfuse upon reinstitution of flow, thus demonstrating the "no-reflow" phenomenon (Menger et al., 1992a, b). These events, termed "reflow paradox" are characterized by leukocyte–endothelium interaction and further increase in microvascular permeability.

Injuries inflicted on skeletal muscle after ischemia–reperfusion may also affect other organs. Reperfusion injury after ischemia to skeletal muscle has also been implicated in the development of a severe form of circulatory shock, tissue and organ oxidative stress, and final multi-system organ failure (MSOF) which results in death of the animals within 24 h of tourniquet release (Ward et al., 1992; Ward et al., 1995). The damage observed in hind-limb muscle tissue after reperfusion does not, by itself, account for the final systemic and lethal MSOF. Therefore, it was suggested that organ failure has its genesis in a primary perfusion abnormality in the hindlimbs, which is followed by secondary hypoperfusion of other organs, such as the liver, as has been shown to be the case in several septic shock models (Ward et al., 1995). Ischemia and reperfusion of the lower torso may also lead to leukotriene- and neutrophil-dependent lung injury characterized by lung polymorphonuclear leukocyte (PMNL) sequestration, increased permeability and non-cardiogenic oedema.

A conventional view has attributed ischemia–reperfusion injuries to ischemia itself, but a growing body of experimental data indicates that a variable, but often substantial, proportion of the injury is caused by reactive oxygen species (ROS) such as superoxide (O_2^{-}), hydrogen peroxide (H_2O_2), and hydroxyl radicals (OH') which are generated at the time of reperfusion. This mechanism was first identified and characterized in a model of vascular occlusion in cat small intestine (McCord (1985), but similar mechanisms have been subsequently confirmed in the liver, pancreas, heart, stomach, kidney, brain, skin, skeletal muscle and lung. Various sources of ROS generated during reperfusion have been elucidated (Bulkley, 1987; Kloner et al., 1989):

1. During ischemia, the NAD^+-dependent xanthine dehydrogenase is converted to xanthine oxidase which produces, upon reperfusion, O_2^{-} and H_2O_2.
2. Activation and influx of activated neutrophils into ischemic tissue.
3. ROS are generated in arachidonate metabolism.

Research to elucidate the contribution of ROS to skeletal muscle injuries after ischemia–reperfusion was performed with both animal models and humans. In animal models they used tourniquet hindlimb ischemia–reperfusion, compression beneath pneumatic tourniquets, as well as muscle flaps, and limb transplantation. In humans these studies were performed during surgery or disease. The studies focus on the effects of the duration of ischemia and reperfusion, assessment of damages and their reversibility, measurement of ROS production and their sources, remote organ injuries, prevention, and treatment and transplantation. Direct measurement of steady-state concentrations of O_2^{-}, H_2O_2 and OH' *in vivo* is inapplicable as a result of very short half-lives. Consequently, measurement of ROS production during ischemia–reperfusion had to rely on indirect methods, which include spin traps and other reagents that interact with them *in situ*, as well as assessment of oxidative damage in the tissue. Another approach to investigate the involvement of ROS in these injuries was through studies that used supplementation of antioxidants in an attempt to reduce ischemia–reperfusion injuries or which measure tissue antioxidants as a marker of oxidative stress. Studies on the effects of supplementation of antioxidants also serve to formulate approaches to therapy, and include: free radical scavengers such as superoxide dismutase (SOD), catalase or glutathione (GSH), xanthine oxidase inhibitors, such as allopurinol, and leukopenic agents and inhibitors of PMNL chemotaxis and accumulation.

Assessment of ROS production

A spin-trapping technique, along with electron spin resonance ESR spectroscopy, have been used to study an experimental model of rat muscle of pedicled rectus femoris flap (De-Santis and Pinelli, 1994). No ESR signal could be detected either before the ischemic period or after only 15 minutes of ischemia. α-Phenyl-N-*tert*-butyl nitrone (PBN) radical adducts were detected after 30, 60, 120 and 180 min of ischemia. A similar signal was detected when PBN was injected during reperfusion 10 min after the ischemic periods. Measurements of H_2O_2 were performed on patients undergoing knee surgery (Mathru et al., 1996). Tourniquet-induced limb exsanguination was induced for about 2 h, and blood samples were collected during a 2-h period after tourniquet release. At 5 min reperfusion, in local blood, H_2O_2 concentrations peaked at about 500% and, from 20 to 120 min, it returned to pre-tourniquet levels. Systemic blood H_2O_2 was not increased during the study.

Oxidative damages

The malondialdehyde assay method has been used for many years to assess lipid peroxidation which, in turn, served as a an indicator of cellular oxidative stress. Recently, it has been shown that this assay, now designated "thiobarbituric acid-reactive substances" (TBARS), is not specific for peroxidized lipid. However, as it may serve as a parameter to quantify peroxidative products of many cellular macromolecules, it is still a useful assay as a general indicator of oxidative damage. Malondialdehyde levels were higher than those in controls in the hindlimbs of rabbits subjected to tourniquet ischemia and showed direct correlation with duration of ischemia (Concannon et al., 1992). When assayed in the plasma, malondialdehyde showed no variation during the ischemia phase, but a significant increase 1 min after release of the tourniquet (Chopineau et al., 1994). In rat cremaster muscle prepared as a tourniquet ischemia model and subjected to 2 h of ischemia followed by 1 h of reperfusion, lipid peroxidation increased sixfold in the reperfusion group (Lee et al., 1995). A control experiment revealed that 3 h of anaesthesia of rats caused a significant rise of TBARs concentration in muscle compared with normal controls with no anaesthesia (Nylander et al., 1989). Furthermore, an increase of similar magnitude was seen after 3 h of ischemia, with or without reperfusion.

Several other studies used another parameter to evaluate oxidative damages – the "conjugated dienes"; these are more closely correlated with lipid peroxidation. In canine gracilis muscle a 2-h ischemic insult resulted in minimal ultrastructural damage and complete normalization of lipid oxidation products assessed as free fatty acid-conjugated dienes (Harris et al., 1986). In contrast, a 7-h ischemic insult caused profound injury at the ultrastructural level with an inability to restore intramuscular phosphagens and glycogen on reperfusion. This severe muscle injury correlated with a 2.5-fold increase in lipid oxidation products. In a bilateral canine gracilis muscle model, in which one muscle of each pair was exposed to 3 h and the other to 5 h of normothermic ischemia. (Lindsay et al., 1988), there was no significant increase in the level of conjugated dienes during ischemia. However, significant increases were detected during the period of reperfusion, and hydroxy-conjugated diene isomers 18:2 and 20:4 were positively identified in reperfusion biopsies on gas chromatograph–mass spectroscopy. Tourniquet ischemia and reperfusion of rat skeletal muscle affected mitochondrial structure and function (Soussi et al., 1990). Four hours of ischemia caused cardiolipin peroxidation and a decrease in the V_{max} of cytochrome *c* oxidase. Tissue reperfusion which followed resulted in a more dramatic decrease in both V_{max} (38% of the control) of cytochrome *c* oxidase and cardiolipin. Two hours of ischemia, with or without subsequent reperfusion, had no effect on either parameters. The Michaelis constant, K_m, of cytochrome c oxidase remained unchanged under all conditions. These findings suggest that mitochondrial damages during ischemia may result in dysfunction and increase in ROS production during reperfusion.

Xanthine dehydrogenase/xanthine oxidase

ROS generated by xanthine oxidase (XO) are considered to play a central role in the pathogenesis of reperfusion injuries. It is based on a mechanism that implicates breakdown of ATP, and accumulation of its products, xanthine and hypoxanthine, during the ischemic period and a concomitant conversion of xanthine dehydrogenase (XDH) to xanthine oxidase (XO) (Fig. 1). Upon reperfusion, XO oxidizes xanthine and hypoxanthine to uric acid, using molecular oxygen as an electron acceptor, and reducing it to O_2^{-} and H_2O_2. Studies on the involvement of xanthine oxidase in reperfusion injuries were conducted in animal models and humans, by two

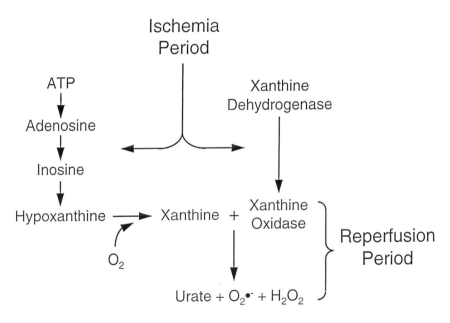

Figure 1. Generation of reactive oxygen species O_2^{-} and H_2O_2 in the reperfusion period of ischemia–reperfusion.

approaches – direct measurements of activities of XDH and XO and treatment with their inhibitors, such as allopurinol, in an attempt to attenuate the injuries. Six hours of tourniquet ischemia in rat hindlimbs resulted in a severalfold increase in intracellular hypoxanthine (Idstrom et al., 1990), indicating enhanced breakdown of adenine nucleotides. Uric acid formation was observed only after reperfusion, in accordance with the activation of XO activity. After 4 h of rat hindlimb ischemia followed by reperfusion, plasma XO activity rose threefold over pre-ischemia levels (Punch et al., 1992), whereas xanthine dehydrogenase activity did not change. Pre-treatment with the XO inhibitor allopurinol reduced XO activity to negligible levels and significantly attenuated conjugated diene levels, creatine phosphokinase (CPK) levels and albumin extravasation. Similar decrease of albumin extravasation was also attained by pre-treating animals with the antioxidant superoxide dismutase (SOD), together with catalase, dimethylthiourea, and dimethylsulphoxide. Three hours of ischemia and 4 h of reperfusion of rat gastrocnemius muscles caused a decrease in its contractile function (McCutchan et al., 1990). Xanthine oxidase activity was suppressed by feeding animals tungsten or treating them with allopurinol. The reperfused muscles showed increased production of H_2O_2, but both the injuries and H_2O_2 production were lower in rats with decreased XO activities (Smith et al., 1989b; McCutchan et al., 1990).

Most studies with humans confirmed the activation of XO after ischemia reperfusion of the lower torso. Xanthine oxidase activity in the venous effluent, after reperfusion of human rectus abdominis muscle following microvascular transfer, correlated with duration of ischemia and biochemical markers of cellular injury (Wilkins et al., 1993). Effluent blood from extremities of patients undergoing surgical treatment which involves application of a tourniquet, showed increases in the plasma levels of XO and its product, uric acid upon reperfusion (Friedl et al., 1990, 1991). Tourniquet-induced limb exsanguination was induced for about 2 h in patients undergoing knee surgery (Mathru et al., 1996). After tourniquet release there was a 500% and 520% increase, respectively, in H_2O_2 concentrations and XO activity, as well as in xanthine and uric acid in local blood. Xanthine oxidase activity also increased in systemic blood, but not the concentration of H_2O_2, indicating a balanced oxidant scavenging in the systemic circulation. Ischemia of 6–12 h in a rabbit model of femoral artery occlusion, followed by 2 h of reperfusion, were associated with increases in the hydrostatic pressure of anterior muscle compartment (Perler et al., 1990). Ablation of free radicals generated from XO with allopurinol or oxypurinol, either by scavenging superoxide radical at reperfusion with SOD, or by blocking secondary hydroxyl radical formation with desferrioxamine, significantly ameliorated the rise in compartment pressure. A rat cremaster model was used to study "no-reflow" in skeletal muscle, by application of 5 h of ischemia, followed by 45 min of reperfusion (O'Farrell et al., 1995). Animals treated with allopurinol had a significantly higher percentage twitch contraction and a lower CPK release than untreated saline controls. The effectiveness of allopurinol and SOD was also evaluated in preventing reperfusion injury in a rat limb replantation model (Concannon et al., 1991). Treatment of the animals with SOD together with allopurinol, immediately before reperfusion resulted in higher limb survival relative to each regent separately.

Polymorphonuclear leukocytes

White blood cells, platelets, vascular endothelial cells and complement proteins may participate in reperfusion injury. Xanthine oxidase-derived oxidants, produced at the time of reperfusion, initiate the formation and release of proinflammatory agents, which subsequently attract and activate neutrophils. Activated PMNLs adhere to vascular endothelium in a mechanism promoted by neutrophil adhesion molecule. They pass through endothelial intercellular junctions, and migrate distally into the interstitial spaces of the skeletal muscle tissue, where they release cytotoxic oxidants and/or non-oxidative toxins (e.g. proteases) that contribute to tissue destruction (Carden and Korthuis, 1989; Pang, 1990; Formigli et al., 1995). The leukocytes were also shown to induce injury directly by adhering to endothelium and obliterating capillaries, especially during low-flow states that may result in the "no-reflow" paradox (Menger et al., 1992a,b). Cytokines may also contribute to ischemia–reperfusion injuries by regulating endothelial adhesion molecules (Seekamp et al., 1993). The involvement of neutrophils in reperfusion injuries was studied by direct estimation of their concentration in muscle tissue and levels of activation in the blood, as well as by depletion and manipulation of their recruitment and adhesion. PMNLs isolated from muscle venous effluent of gracilis muscle subjected to ischemia–reperfusion showed increased rates of O_2^- production (Cambria et al., 1991) and their degree of activation was directly related to the extent of muscle infarction. In rat hindlimb ischemia, 4 but not 3 h resulted in the "no-reflow" phenomenon (Sirsjo et al., 1990a). There was no significant accumulation of PMNLs in skeletal muscle and there was no correlation between the number of PMNLs in postischemic muscle and restricted blood flow. However, a statistically significant positive correlation between number of PMNLs and the amount of oedema was shown after 30 min of reperfusion. The activity of myeloperoxidase (MPO) was evaluated as an index of the time course of PMNL accumulation in a tourniquet model, in which the hindleg was made ischemic for 1.5, 3 or 5 h (Sirsjo et al., 1990b). A significant increase was first seen after 5 h of reperfusion with the peak at 24 h. After 74 h of reperfusion, the MPO activity had almost returned to control levels. Prolonging the ischemia from 3 to 5 h did not cause any further significant MPO increase. Amputated rabbit hindlimbs were subjected to 4 h of ischemia followed by 2 h of reperfusion with oxygenated Krebs' buffer, or with PMNL-supplemented buffer (Oredsson et al., 1995). PMNLs aggravated histological changes seen after reperfusion, which was prevented by SOD and catalase. These free radical scavengers also prevented accumulation of PMNLs, suggesting that oxygen-derived free radicals are engaged in the interaction between PMNLs and microvascular endothelium.

Pre-treatment of animals with ATP–MgCl$_2$ before 4 h of occlusive ischemia, followed by 1 h of reperfusion, significantly attenuated superoxide production by activated neutrophils and increase in skeletal muscle vascular resistance and permeability (Korthuis et al., 1988a). Reperfusion of isolated canine gracilis muscles with whole blood was associated with a dramatic increase in vascular permeability and resistance, whereas the use of leukocyte-depleted blood avoided these injuries (Korthuis et al., 1988b). Wistar rats were rendered neutropenic by administering 750 rad of whole-body radiation before the application of tourniquet hindlimb ischemia and 1 h of reperfusion (Belkin et al., 1989). Evaluation of ischemic damages by a quantitative spectrophotometric assay of triphenyltetrazolium chloride reduction revealed that neutropenia had a protective effect. Furthermore, treatment with SOD and catalase prevented the damage, indicating that neutrophils exerted their damaging effect by formation of oxygen-derived free radicals. Inhibition of neutrophil recruitment and activation during reperfusion by the platelet-activating factor (PAF) receptor antagonist, WEB 2170, significantly improved the survival of rabbit skeletal muscle (Lepore et al., 1995). This treatment also reduced tissue lipid peroxide levels and release into the blood of the enzyme CPK, but it did not affect oedema in muscles.

Antineutrophil serum and neutrophil elastase inhibitor Elafin reduced neutrophil recruitment during reperfusion of gastrocnemius muscle and muscle viability was preserved (Crinnion et al., 1994). Reperfusion oedema still occurred, however, suggesting that altered endothelial permeability is mediated by factors other than neutrophils. Ischemia–reperfusion elicited marked enhancement of leukocyte rolling during initial reperfusion and a twentyfold increase of leukocyte adherence which lasted for the entire postischemic reperfusion period (Menger et al., 1992b). SOD and allopurinol were effective in attenuating leukocyte rolling and adherence, and the resulting microvascular leakage. In a model of canine skeletal muscle, prevention of neutrophil adherence with monoclonal antibody IB4 directed against the neutrophil CD11/CD18 glycoprotein adherence complex, or neutrophil depletion with a specific polyclonal antineutrophil serum, prevented the increase in vascular permeability and resistance over control values (Carden et al., 1990). The role of endothelial E-selectin in the genesis of the ischemia–reperfusion syndrome was studied in patients undergoing reconstructive vascular surgery (Formigli et al., 1995). Immunohistochemistry revealed a strong positive reaction for E-selectin on the venular endothelium during ischemia and reperfusion, which matched neutrophil accumulation in skeletal muscle tissue; this in turn, correlated with tissue damage at reperfusion.

Arachidonic acid metabolites

During revascularization of skeletal muscle, lipid mediators derived from arachidonic acid are released and were shown to have a role in the pathogenesis of reperfusion injury. These metabolites include prostaglandins, such as PGE_1, PGE_2 and PGI_2 (prostacyclin), thromboxane (TxA_2), and leukotrienes (LTs) (Fig. 2), whose normal function is to control blood flow and capillary permeability in skeletal muscle. Research to explore the involvement of these agents in reperfusion injuries included assessment of their blood levels and treatment with inhibitors and antagonists. A study of ischemia–reperfusion in rabbit hindlimbs showed that oxygen free radicals are essential for ischemia-induced synthesis of LTB_4 by PMNLs which in turn, mediated their diapedesis (Goldman et al., 1992). Plasma derived from rabbit hindlimbs after 3 h of tourniquet ischemia and 10 minutes of reperfusion showed an increased LTB_4 level which was lower than the levels found in ischemic plasma, derived from neutropenic animals. Introduction of ischemic plasma in abraded skin chambers placed on the dorsum of normal rabbits, led after 3 h to PMNL diapedesis that was associated with a further increase in LTB_4 levels. A correlation was found between LTB4 levels in ischemic plasma and PMNL accumulations in blister fluid. Intravenous pre-treatment of rabbits used in the blister chamber bioassay with the LT receptor antagonist FPL-55712 attenuated diapedesis induced by ischemic and ischemic–neutropenic plasma (Goldman et al., 1992). Pre-treatment with SOD and catalase, or allopurinol, prevented

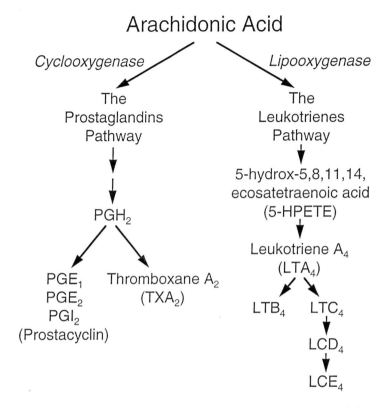

Figure 2. Schematic representation of arachidonic acid metabolites involved in ischemia–reperfusion.

ischemic plasma-induced LTB_4 synthesis, as well as ischemic plasma-induced diapedesis. Plasma TxB_2 levels increased at 5 minutes of reperfusion after 4 h of bilateral tourniquet ischemia of hindlimbs (Paterson et al., 1989). Treatment of neutrophils with phorbol myristate acetate (PMA) led to a 91% increase in neutrophil H_2O_2 production before ischemia, and 5 minutes after reperfusion there was an enhanced response to PMA. Pre-treatment of animals with the thromboxane synthase inhibitor OKY 046 prevented ischemia-induced thromboxane generation, neutrophil H_2O_2 production and this enhanced response to PMA stimulation Treatment of rats with the thromboxane receptor antagonist SQ 29,548 did not affect the increase in plasma thromboxane levels after ischemia, but was as effective as OKY 046 in preventing the ischemia-induced increase in neutrophil H_2O_2 production and the enhanced response to PMA stimulation. The protective efficacy of a TxA_2 receptor antagonist, GR32191 and an LTB_4 receptor antagonist, C41930, was evaluated in a rat hindlimb model of 6 h of ischemia and 4 h of reperfusion (Homer-Vanniasinkam and Gough, 1994). Ischemia itself did not result in muscle oedema or necrosis but both occurred after reperfusion. The inhibitors preserved muscle viability and pre-vented oedema. A similar protective effect was found with cinnamophilin, a novel dual inhibitor of thromboxane synthase and TxA_2 receptor (Cheng, and Chang, 1995).

Administration of external antioxidants

A central approach to studying the contribution of ROS to reperfusion injuries was to assess the protective effects of antioxidants. It included pre-treatment with external antioxidants and estima-tion or manipulation of tissue levels of internal antioxidants. In most studies free radical scaven-gers, or agents that prevent their production, applied during ischemia or reperfusion were found to reduce structural and functional damage. The antioxidant studied the most is the superoxide scav-enging enzyme SOD. Membrane depolarization, resulting from 1–3 h of partial skeletal muscle ischemia–reperfusion in dogs and rats, can be prevented by administration of SOD and catalase, or by neutrophil depletion (Perry and Fantini, 1989). SOD infused during removal of the aortic clamp prevented the continued decrease in E_m in rats subjected to 60 min of infrarenal aortic occlusion (Perry and Fantini, 1989). After 60 min of no-flow ischemia in a rat spinotrapezius muscle preparation, the maximum tetanic force fell significantly during 90 min of reperfusion compared with control, non-ischemic muscles (Suzuki et al., 1995). Treatment with either dimethylthiourea (DMTU) or SOD decreased cell death but only dimethylthiourea attenuated the fall in force. A comparison was made between the effects of intravenous SOD and catalase or DMTU on reperfusion injuries in rats (Hardy et al., 1992). The SOD and catalase or DMTU were given intravenously 30 min before tourniquet release and continued throughout the period of reperfusion. The SOD and catalase reversed the low reflow, whereas DMTU had no effect on perfusion at either 10 or 240 min. The results suggested that, although superoxide radicals are harmful during postischaemic reperfusion, hydroxyl radicals may be beneficial. However, in another study, that of a replantation model, it was found that SOD provided protection after 5 h of ischemia, as evidenced by normal strength and histological appearance, but it was ineffective after 8 h (Feller et al., 1989). Dimethylsulphoxide (DMSO), on the other hand, had no beneficial effects after 5 h of ischemia, although after 8 h the treated muscle had significantly better function than the untreated one. Pre-treatment with SOD and catalase also maintained higher rates of Ca^{2+} uptake by the SR of skeletal muscle from postischemic, reperfused rats (Lee et al., 1987).

Various studies compared the efficacy of different types and preparations of SOD in reducing reperfusion injuries. Studies of SOD isozymes from different sources were studied in the rat tourniquet poditis model which showed that, although all the enzymes have the same specific enzymic activity, they function differently (Jadot and Michelson, 1987). Both bovine CuSOD and *Escherichia coli* MnSOD were very effective, whereas yeast CuSOD and the homologous rat CuSOD showed no activity. In addition, liposomal encapsulation of bovine CuSOD greatly en-hances biological efficacy, by providing a slow release mechanism for the enzyme. These results may explain the differences in findings regarding the use of SOD as a drug for the treatment of ischemic injury.

The rat cremaster model was used to study the effect of recombinant human manganese SOD (rhMnSOD) on "no-reflow" in skeletal muscle (O'Farrell et al., 1995). The specific advantage of this form of enzyme over other SOD types is a much longer plasma half-life (5–7 h) which allows better equilibration between extra- and intracellular compartments. Compared with untreated saline control animals, those treated with rhMnSOD had a significantly higher percentage area of blood reflow, a greater percentage tetanic and twitch contractile strength and less CPK release, after 5 h of cremaster ischemia. A significant difference in the survival rate was found in animals treated with a modified form of m-polyethylene glycol-SOD (mPEG-SOD) after $4\frac{1}{2}$ h of complete warm ischemia (Giardino et al., 1995). Six hours of ischemia and 24 h of reperfusion caused death of all the rats in the control group, whereas 55% of the animals treated with liposomal SOD survived for 24 h or more, and two completely recovered (Aoki et al., 1992). Whereas most studies reported beneficial effects of SOD, there were several others that could not confirm it. Application of an intra-arterial bolus of SOD immediately before reperfusion, and its infusion during the first hour of reperfusion, did not affect oxygen consumption, lactate clearance and blood flow after canine hindlimb ischemia–reperfusion (Hoch et al., 1991). Bolus administration of SOD and catalase at the start of reperfusion offered no significant improvement in metabolic or contractile function in an *in vivo*, autoperfused, canine hindlimb model (Long et al., 1989). On the other hand, mannitol, a well-known hydroxyl radical scavenger, significantly reduced muscle damage and improved neuromuscular contractile function compared with controls. In other experiments, it was found that lazaroid analogues U-74389F and U-74389FG could provide protection (Potter et al., 1994; Hoballah et al., 1995). U-74389F improved microvascular perfusion after 3 h of no-flow ischemia and 90 min of reperfusion, whereas SOD reduced it (Potter et al.,1994).

Treatment of rats with cobra venom factor reduced acute lung injury caused by transient hindlimb ischemia (Leff et al., 1991). The protection was attributed to reperfusion-induced erythrocyte haemolysis which caused an increase in plasma catalase and consequently in its H_2O_2-scavenging activity. Recently, it has been shown that coenzyme Q_{10} could play a protective role against injuries induced by ischemia–reperfusion by attenuating the degree of peroxidative damage. Patients undergoing elective vascular surgery for abdominal aortic aneurysm or obstructive aortoiliac disease were treated with coenzyme Q_{10} for 7 days before surgery (Chello et al., 1996). The concentrations of malondialdehyde, conjugated dienes, CPK and lactate dehydrogenase in plasma samples from both arterial and inferior vena cavae in patients who received coenzyme Q_{10} were significantly lower than in the placebo group. Decrease of plasma malondialdehyde concentrations correlated positively with a decrease in both CPK and lactate dehydrogenase levels. Coenzyme Q_{10} was also shown to provide protection in an isolated gracilis muscle of a dog (Choudhury et al., 1991).

Ascorbate was explored as a potential protectant (Lagerwall et al., 1994). Treatment of rat skeletal muscle with ascorbate at the time of reperfusion had an immediate and positive effect on the recovery of high-energy phosphates, and pH. It was suggested that ascorbate served as an antioxidant, but that it also had other ancillary effects, mainly its provision of additional buffer capacity. Several inhibitors of lipid peroxidation were studied in ischemia–reperfusion models. U-74500A enhanced gastrocnemius muscle blood flow throughout reperfusion with complete muscle salvage (Homer-Vanniasinkam et al., 1993) and the lazaroid analogue U-74389F improved microvascular perfusion after 3 h of no-flow ischemia and 90 min of reperfusion, although it was not as effective as SOD (Potter et al., 1994). In contrast, the antioxidant U-

74006F did not have any protective effect on peak tension, rate of force production, contraction time, half-relaxation time or resistance to fatigue (Mohler et al., 1996). Fructose-1,6-diphosphate or adenosine, administered at the onset of reperfusion in isolated canine gracilis muscle and mouse cremaster muscle, attenuated postischemic microvascular barrier dysfunction (Akimitsu et al., 1995). It was suggested that these reagents function by a mechanism that is related to their ability to inhibit leukocyte adhesion and migration. The involvement of iron-catalysed formation of hydroxyl radicals in ischemia–reperfusion injuries was studied by pre-treatment with the iron chelator desferrioxamine and the iron-binding protein apotransferrin (Smith et al., 1989c; Fantini and Yoshioka, 1993). Both reagents attenuated the increased vascular permeability produced by ischemia–reperfusion, whereas iron-loaded desferrioxamine afforded no protection. Acetyl-carnitine was also found to ameliorate ischemia–reperfusion-induced damage of human skeletal muscle (Adembri et al., 1994). Patients undergoing aortic reconstructive surgery were given acetylcarnitine before the induction of ischemia. During ischemia and reperfusion, complement activation recruited numerous granulocytes into the muscle tissue, but, contrary to the untreated samples, the ability of these cells to generate O_2^{-} remained at low levels and was comparable to that of ischemia

Efect of ischemia–reperfusion on levels of endogenous antioxidants

Studies of endogenous antioxidants showed that tissue damage is sometimes a secondary event after depletion of cellular antioxidants. Reperfusion of ischemic skeletal muscle was associated with 50% decrease in reduced glutathione (GSH) content and a dramatic increase in tissue neutrophil content, as reflected by a 26-fold increase over control in tissue MPO activity after 1 h of reperfusion. Tissue CAT and SOD activities were unaffected by ischemia–reperfusion (Smith et al., 1989a). In another series of experiments, it was shown that 60 min of ischemia followed by 60 min of reperfusion resulted in a decrease in glutathione peroxidase activity, increase in catalase and no effect on SOD (Laughlin et al., 1990). Four hours of tourniquet ischemia in rats resulted in an increase in GSH, GSSG, and TBARS which was similar in white and red muscle, whereas the GSSG:GSH ratio remained unchanged in muscle (Purucker et al., 1991). However, during the subsequent reperfusion period, GSH decreased within 2 h by 39% and 89% in white and red muscle, respectively. No recovery from the depletion was observed after up to 12 h of reperfusion. The GSH decrease was paralleled by a marked increase of the GSSG:GSH ratio of 150% in white and 450% in red muscle, and followed by about a 150% increase in TBARS in both muscle types.

Treatment of the animals during the reperfusion period with antioxidants did not prevent the GSH decrease, but were effective in reducing the GSSG:GSH ratio to near normal and the TBARS increase by about 50%. A way to increase internal antioxidants is through exercise training. Exercised Dolly rats had increased oxidative capacity and increased glutathione peroxidase activity, lower catalase and similar SOD activities compared with the same muscles of sedentary animals (Laughlin et al., 1990). Subjecting the rats to 60 min of ischemia and 60 min of reperfusion resulted in increased catalase activities, which was higher in the trained animals. It also caused a decrease in glutathione peroxidase and little or no effect on SOD activities in muscles from both trained and untrained animals. In a model of ischemia of the tibialis anterior muscle of the rat hindlimb, 4 h of ischemia resulted in a slight increase in GSH levels compared with these in controls (Sirsjo et al., 1996). After 1 h of reperfusion the levels of GSH decreased by 50% compared with those in control and remained at that level after 5 h of reperfusion. This depletion was not seen after 2 h of ischemia.

Remote organ

Ischemia in rat hindlimbs followed by reperfusion results in local as well as remote organ injuries which were shown to affect the lung, liver and gut. Lower torso ischemia and reperfusion led to lung injury characterized by lung polymorphonuclear neutrophils (PMNL) sequestration, increased permeability and oedema, which is dependent on activated PMNL adherence, and release of cytotoxic products. Three hours of ischemia in rabbit bilateral hindlimbs followed by 10 min of reperfusion resulted in increased plasma LTB_4 levels, reduction in circulating white blood cells and sequestration of PMNLs in the hindlimbs (Welbourn et al., 1992). Four hours after release of the tourniquets, there was sequestration of PMNLs in the lungs, as increase in the protein content of bronchoalveolar lavage (BAL) fluid and lung oedema. In another study it was shown that complement activation, which occurred during tourniquet ischemia, mediated permeability changes in the ischemic muscle and the lungs during reperfusion (Lindsay et al., 1992). Lung microvascular permeability was increased twofold after 30 min of reperfusion (Punch et al., 1991). Pulmonary injury, as well as pulmonary neutrophil sequestration, was blocked with DMSO, DMTU, allopurinol, indomethacin and SOD plus catalase. The prostaglandin inhibitor, indomethacin, reduced lung permeability, but did not attenuate the neutrophil sequestration within the pulmonary parenchyma. Four hours of bilateral hindlimb tourniquet ischemia followed by 4 h of reperfusion in rats resulted in PMNL sequestration in lungs and increases in LTB_4 levels of BAL fluid, and in permeability and oedema (Welbourn et al., 1991).

   Treatment of the animals with the specific elastase inhibitor methoxysuccinyl-L-Ala-L-Ala-L-Pro-L-Val-chloromethylketone (MAAPV) reduced BAL fluid and the lung wet weight:dry weight ratio, but did not affect lung neutrophil sequestration and rise of LTB_4 in BAL fluid. These findings indicate that lung injury is affected by both ROS and elastase. One minute after tourniquet release in sheep that underwent 2 h of bilateral hindlimb ischemia, the mean pulmonary artery pressure increased whereas the pulmonary artery wedge pressure was unchanged (Klausner et al., 1989). At 30 minutes of reperfusion there was an increase in lung–lymph TxB_2 levels and lung microvascular permeability. The white blood cell count fell during the first hour of reperfusion and there was marked leukosequestration in the blood. Pre-treatment with SOD and catalase, with both conjugated to polyethylene glycol (PEG) blunted the rise in mean pulmonary artery pressure and prevented the increase in plasma and lymph TxB_2 lymph flow and lymph protein clearance. Five hours of tourniquet ischemia in rat hindlimbs followed by reperfusion caused a severe form of circulatory shock, characterized by hypotension and death within 24 h of tourniquet release (Ward et al., 1992). This multi-system organ failure cannot be explained as a direct effect of the oxidative damage to muscle tissue, but correlated with oxidative stress symptoms in the liver. The findings of a decrease in hepatic tissue thiol levels and an increase in

TBARS indicated that ROS are involved in liver damage. Furthermore, allopurinol and a SOD–catalase–DMSO combination offered protection after 2 h of tourniquet release, whereas only allopurinol was able to prevent ischemia-induced hepatic thiol loss. The small intestines are also affected by hindlimb ischemia–reperfusion (Corson et al., 1992). After reperfusion, horseradish peroxidase permeability had not developed by 15 min, but was present in all animals by 2 h.

Ischemia reperfusion injuries are necessarily mediated by ROS

Even though there are numerous studies that substantiate the role of ROS in skeletal muscle reperfusion injuries, there are several that could not confirm it. For example, canine gracilis muscle made ischemic for 4, and administration of allopurinol, SOD or mannitol during reperfusion (Faust et al., 1988). After 60 min of reperfusion there was no improvement in muscle O_2 consumption ($MVO_{2\,max}$) or tissue oedema, whereas ATP was significantly depressed with allopurinol and SOD treatment. Perfusion of capillaries and postcapillary venules of striated muscle were assessed after 4 h of ischemia (Menger et al., 1992). Ischemia–reperfusion was characterized by a significant reduction in functional capillary density to 35% of baseline values during initial reperfusion, with incomplete recovery after 24 h. Treatment with either SOD or allopurinol resulted in maintenance of capillary density of 60% baseline. From the fact that, in SOD- and allopurinol-treated animals, 40% of the capillaries were still found to be non-perfused, it was concluded that mechanisms other than oxygen radicals play an important role in the development of postischemic "reflow".

Conclusion

Until the start of the 1980s it was generally accepted than ischemia–reperfusion injuries were an outcome of the hypoxic insult. A growing body of experimental data, which has accumulated since then, shows beyond doubt that ROS contribute to the injuries, but there are still several unresolved issues:

1. What proportion of the damage is caused by ROS?
2. Which ROS-generating system is important in skeletal muscle ischemia–reperfusion?
3. Which of the ROS is causing the damages?
4. To what extent are antioxidants effective in protecting the affected tissue?

It is apparent that, as a result of the multifactorial nature of the system there are no unequivocal answers to these questions. The outcome of skeletal muscle ischemia–reperfusion injuries is dependent on the duration of the ischemia, the type of muscle (its level of oxidative capacity), the age of the organism and the tissue levels of antioxidants; these, in turn, are dependent on nutrition as well as the history of physical activity. Research on ischemia–reperfusion focused mainly on heart and brain, as a result of its important contribution to mortality and morbidity in Western culture. However, skeletal ischemia–reperfusion can result in compartment syndrome, circulatory shock and rhabdomyolysis, with consequent oedema and muscle infarction, and may necessitate amputation. Furthermore its complications include remote organ injuries, mainly to the lungs, liver, kidneys and gut, and can lead to kidney failure, and death caused by MSOF. Further studies on the contribution of ROS to ischemia–reperfusion injuries and their clinical implications may open the way to the formulation of treatments, both for various pathologies and in preparation for medical interventions such as surgery and transplantation.

References

Adembri C, Domenici LL, Formigli L, Brunelleschi S, Ferrari E and Novelli GP (1994) Ischemia–reperfusion of human skeletal muscle during aortoiliac surgery: effects of acetylcarnitine. *Histol Histopathol* 9: 683–690.

Akimitsu T, White JA, Carden DL, Gute DC and Korthuis RJ (1995) Fructose-1,6-diphosphate or adenosine attenuate leukocyte adherence in postischemic skeletal muscle. *Am J Physiol* 269: H1748–H1751.

Aldman A, Lewis DH and Larsson J (1987) Muscle metabolic changes in induced subtotal ischemia of the leg in a pig model. *Acta Chir Scand* 153: 345–351.

Aoki Y, Nata M, Odaira T and Sagisaka K (1992) Suppression of ischemia-reperfusion injury by liposomal superoxide dismutase in rats subjected to tourniquet shock. *Int J Legal Med* 105: 5–9.

Belkin M, LaMorte WL, Wright JG and Hobson RW (1989) The role of leukocytes in the pathophysiology of skeletal muscle ischemic injury. *J Vasc Surg* 10: 14–18; 18–19.

Bulkley GB (1987) Free radical-mediated reperfusion injury: a selective review. *Brit J Cancer* 8: 66–73.

Cambria RA, Anderson RJ, Dikdan G, Teehan EP, Hernandez-Maldonado JJ and Hobson RW (1991) Leukocyte activation in ischemia-reperfusion injury of skeletal muscle. *J Surg Res* 51: 13–17.

Carden DL and Korthuis RJ (1989) Mechanisms of postischemic vascular dysfunction in skeletal muscle: implications for therapeutic intervention. *Microcirc Endothel Lymphat* 5: 277–298.

Carden DL, Smith JK and Korthuis RJ (1990) Neutrophil-mediated microvascular dysfunction in postischemic canine skeletal muscle. Role of granulocyte adherence. *Circ Res* 66: 1436–1444.

Chello M, Mastroroberto P, Romano R, Castaldo P, Bevacqua E and Marchese AR (1996) Protection by coenzyme Q10 of tissue reperfusion injury during abdominal aortic cross-clamping. *J Cardiovasc Surg Torino* 37: 229–235.

Cheng HT and Chang H (1995) Reduction of reperfusion injury in rat skeletal muscle following administration of cinnamophilin, a novel dual inhibitor of thromboxane synthase and thromboxane A2 receptor. *Thorac Cardiovasc Surg* 43: 73–76.

Chopineau J, Sommier MF and Sautou V (1994) Evaluation of free radical production in an ischemia-reperfusion model in the rabbit using a tourniquet. *J Pharm Pharmacol* 46: 519–520.

Choudhury NA, Sakaguchi S, Koyano K, Matin AF and Muro H (1991) Free radical injury in skeletal muscle ischemia and reperfusion. *J Surg Res* 51: 392–398.

Concannon MJ, Dooley TW and Puckett CL (1991) Improved survival in a replantation model containing ischemic muscle. *Microsurgery* 12: 18–22.

Concannon MJ, Kester CG, Welsh CF and Puckett CL (1992) Patterns of free-radical production after tourniquet ischemia: implications for the hand surgeon. *Plast Reconstr Surg* 89: 846–852.

Corson RJ, Paterson IS, O'Dwyer ST, Rowland P, Kirkman E, Little RA and McCollum CN (1992) Lower limb ischemia and reperfusion alters gut permeability. *Eur J Vasc Surg* 6: 158–163.

Crinnion JN, Homer-Vanniasinkam S, Hatton R, Parkin SM and Gough MJ (1994) Role of neutrophil depletion and elastase inhibition in modifying skeletal muscle reperfusion injury. *Cardiovasc Surg* 2: 749–753.

De-Santis G and Pinelli M (1994) Microsurgical model of ischemia reperfusion in rat muscle: evidence of free radical injury by spin trapping. *Microsurgery* 15: 655–659.

Diana JN and Laughlin MH (1974) Effect of ischemia on capillary pressure and equivalent pore radius in capillaries of the isolated dog hindlimb. *Circ Res* 35: 77–101.

Enger EA, Jennische E, Medegard A and Haljamamae H (1978) Cellular restitution after 3 h of complete tourniquet ischemia. *Eur Surg Res* 10: 230–239.

Fantini GA and Yoshioka T (1993) Deferoxamine prevents lipid peroxidation and attenuates reoxygenation injury in postischemic skeletal muscle. *Am J Physiol* 264: H1958–H1959.

Faust KB, Chiantella V, Vinten-Johansen J and Meredith JH (1988) Oxygen-derived free radical scavengers and skeletal muscle ischemic/reperfusion injury. *Am Surg* 54: 709–719.

Feller AM, Roth AC, Russell RC, Eagleton B, Suchy H and Debs N (1989) Experimental evaluation of oxygen free radical scavengers in the prevention of reperfusion injury to skeletal muscle. *Ann Plast Surg* 22: 321–331.

Formigli L, Manneschi LI, Adembri C, Orlandini SZ, Pratesi C and Novelli GP (1995) Expression of E-selectin in ischemic and reperfused human skeletal muscle. *Ultrastruct Pathol* 19: 193–200.

Friedl HP, Smith DJ, Till GO, Thomson PD, Louis DS and Ward PA (1990) Ischemia-reperfusion in humans. Appearance of xanthine oxidase activity. *Am J Pathol* 136: 491–495.

Friedl HP, Till GO, Trentz O and Ward PA (1991) Role of oxygen radicals in tourniquet-related ischemia-reperfusion injury of human patients. *Klin Wochenschr* 69: 1109–1112.

Gardner VO, Caiozzo VJ, Long ST, Stoffel J, McMaster WC and Prietto CA (1984) Contractile properties of slow and fast muscle following tourniquet ischemia. *Am J Sport Med* 12: 417–423.

Giardino R, Capelli S, Fini M, Giavaresi G, Orienti L, Veronese FM, Caliceti P and Rocca M (1995) Biopolymeric modification of superoxide dismutase (mPEG-SOD) to prevent muscular ischemia-reperfusion damage. *Int J Artif Organs* 18: 167–172.

Goldman G, Welbourn R, Klausner JM, Valeri CR, Shepro D and Hechtman HB (1992) Oxygen free radicals are required for ischemia-induced leukotriene B4 synthesis and diapedesis. *Surgery* 111: 287–293.

Hardy SC, Homer-Vanniasinkam S and Gough MJ (1992) Effect of free radical scavenging on skeletal muscle blood flow during postischaemic reperfusion. *Brit J Surg* 79: 1289–1292.

Harris K, Walker PM, Mickle DA, Harding R, Gatley R, Wilson G J, Kuzon B, McKee N And Romaschin AD (1986) Metabolic response of skeletal muscle to ischemia *Am J Physiol* 250: H213–H220.

Hickey MJ, Knight KR, Hurley JV and Lepore DA (1995) Phosphoenolpyruvate/adenosine triphosphate enhances post-ischemic survival of skeletal muscle. *J Reconstr Microsurg* 11: 415–422.

Hoballah JJ, Mohan CR, Sharp WJ, Kresowik TF and Corson JD (1995) Lazaroid U74389G attenuates skeletal muscle reperfusion injury in a canine model. *Transplant Proc* 28: 2836–2839.

Hoch JR, Stevens RP, Keller MP and Silver D (1991) Recovery of neuromuscular function during reperfusion of the ischemic extremity: effect of mannitol and superoxide dismutase. *Surgery* 110: 656–663.

Homer-Vanniasinkam S and Gough MJ (1994) Role of lipid mediators in the pathogenesis of skeletal muscle infarction and oedema during reperfusion after ischaemia. *Brit J Surg* 81: 1500–1503.

Homer-Vanniasinkam S, Hardy SC and Gough MJ (1993) Reversal of the post-ischaemic changes in skeletal muscle blood flow and viability by a novel inhibitor of lipid peroxidation. *Eur J Vasc Surg* 7: 41–45.

Idstrom JP, Soussi B, Elander A And Bylund-Fellenius AC (1990) Purine metabolism after *in vivo* ischemia and reperfusion in rat skeletal muscle. *Am J Physiol* 258: H1668–H1673.

Jadot G and Michelson AM (1987) Comparative anti-inflammatory activity of different superoxide dismutases and liposomal SOD in ischemia. *Free Radical Res Commun* 3: 389–394.

Klausner JM, Paterson IS, Kobzik L, Valeri CR, Shepro D and Hechtman HB (1989) Oxygen free radicals mediate ischemia-induced lung injury. *Surgery* 105: 192–199.

Kloner RA, Przyklenk M and Whittaker P (1989) Deleterious effects of oxygen radicals in ischemia/reperfusion. Resolved and unresolved issues. *Circulation* 80: 1115–1127.

Korthuis RJ, Granger DN, Townsley MI and Taylor AE (1985) The role of oxygen-derived free radicals in ischemia-induced increases in canine skeletal muscle microvascular permeability. *Circ Res* 57: 599–609.

Korthuis RJ, Grisham MB, Zimmerman BJ, Granger DN and Taylor AE (1988a) Vascular injury in dogs during ischemia-reperfusion: improvement with ATP-MgCl2 pretreatment. *Am J Physiol* 254: H702–H708.

Korthuis RJ, Grisham MB and Granger DN (1988b) Leukocyte depletion attenuates vascular injury in postischemic skeletal muscle. *Am J Physiol* 254: H828–H827.

Korthuis RJ, Smith JK and Garden DL (1989) Hypoxic reperfusion attenuates postischemic microvascular injury. *Am J Physiol* 256: H315–H319.

Lagerwall K, Daneryd P, Schersten T and Soussi B (1994) *In vivo* 31P nuclear magnetic resonance evidence of the salvage effect of ascorbate on the postischemic reperfused rat skeletal muscle. *Life Sci* 56: 389–397.

Laughlin M, H, Simpson T, Sexton WL, Brown OR, Smith JK and Korthuis RJ (1990) Skeletal muscle oxidative capacity, antioxidant enzymes, and exercise training. *J Appl Physiol* 68: 2337–2343.

Lee KR, Cronenwett JL, Shlafer M, Corpron C and Zelenock GB (1987) Effect of superoxide dismutase plus catalase on Ca^{2+} transport in ischemic and reperfused skeletal muscle. *J Surg Res* 42: 24–32.

Lee YH, Wei FC, Lee J, Su MS and Chang YC (1995) Effect of postischemic reperfusion on microcirculation and lipid metabolism of skeletal muscle. *Microsurgery* 16: 522–527.

Leff JA, Kennedy DA, Terada LS, Emmett M, McCutchan HJ, Walden DL and Repine JE (1991) Reperfusion of ischemic skeletal muscle causes erythrocyte hemolysis and decreases subsequent oxidant-mediated lung injury. *J Lab Clin Med* 118: 352–358.

Lepore DA, Knight KR, Stewart AG, Riccio M and Morrison WA (1995) Platelet-activating factor (PAF) receptor antagonism by WEB 2170 improves the survival of ischaemic skeletal muscle. *Ann Acad Med Singapore* 24 (suppl): 63–67.

Lindsay T, Walker PM, Mickle DA and Romaschin AD (1988) Measurement of hydroxy-conjugated dienes after ischemia-reperfusion in canine skeletal muscle. *Am J Physiol* 254: H578–H583.

Lindsay TF, Liauw S, Romaschin AD and Walker PM (1990) The effect of ischemia/reperfusion on adenine nucleotide metabolism and xanthine oxidase production in skeletal muscle. *J Vasc Surg* 12: 8–15.

Lindsay TF, Hill J, Ortiz F, Rudolph A, Valeri CR, Hechtman HB and Moore FD Jr (1992) Blockade of complement activation prevents local and pulmonary albumin leak after lower torso ischemia-reperfusion. *Ann Surg* 216: 677–683.

Long JW Jr, Laster JL, Stevens RP, Silver WP and Silver D (1989) Contractile and metabolic function following an ischemia-reperfusion injury in skeletal muscle: influence of oxygen free radical scavengers. *Microcirc Endothel Lymph* 5: 351–363.

McCord JM (1985) Oxygen-derived free radicals in postischemic tissue injury. *N Engl J Med* 312: 159–163.

McCutchan HJ, Schwappach JR, Enquist EG, Walden DL, Terada LS, Reiss OK, Leff JA and Repine JE (1990) Xanthine oxidase-derived H_2O_2 contributes to reperfusion injury of ischemic skeletal muscle. *Am J Physiol* 258: H1415–H1419.

Mathru M, Dries DJ, Barnes L, Tonino P, Sukhani R and Rooney MW (1996) Tourniquet-induced exsanguination in patients requiring lower limb surgery. An ischemia-reperfusion model of oxidant and antioxidant metabolism. *Anesthesiology* 84: 14–22.

Menger MD, Steiner D and Messmer K (1992a) Microvascular ischemia-reperfusion injury in striated muscle: significance of "no reflow". *Am J Physiol* 263: H1892–H1900.

Menger, M.D., Pelikan, S., Steiner, D. and Messmer, K. (1992b) Microvascular ischemia-reperfusion injury in striated muscle: significance of "reflow paradox". *Am J Physiol* 263: H1901–H1906.

Mohler LR, Pedowitz RA, Ohara WM, Oyama BK, Lopez MA and Gershuni DH (1996) Effects of an antioxidant in a rabbit model of tourniquet-induced skeletal muscle ischemia-reperfusion injury. *J Surg Res* 60: 23–28.

Mubarek SJ and Hargens AR (1983) Acute compartment syndromes. Surg. Clin. North Am. 63: 539–565.

Nylander G, Otamiri T, Lewis DH and Larsson J (1989) Lipid peroxidation products in postischemic skeletal muscle and after treatment with hyperbaric oxygen. *Scand J Plast Reconstr Surg Hand Surg* 23: 97–103.

O'Farrell D, Chen LE, Seaber AV, Murrell GA and Urbaniak JR (1995) Efficacy of recombinant human manganese superoxide dismutase compared to allopurinol in protection of ischemic skeletal muscle against "no-reflow". *J Reconstr Microsurg* 11: 207–214.

Oredsson S, Arlock P, Plate G and Qvarfordt P (1993) Metabolic and electrophysiological changes in rabbit skeletal muscle during ischaemia and reperfusion. *J Surg* 159: 3–8.

Oredsson S, Qvarfordt P and Plate G (1995) Polymorphonuclear leukocytes increase reperfusion injury in skeletal muscle. *Int Angiol* 14: 80–88.

Pang CY (1990) Ischemia-induced reperfusion injury in muscle flaps: pathogenesis and. major source of free radicals. *J Reconstr Microsurg* 6: 77–83.

Paterson IS, Klausner JM, Goldman G, Kobzik L, Welbourn R, Valeri CR, Shepro D and Hechtman HB (1989) Thromboxane mediates the ischemia-induced neutrophil oxidative burst. *Surgery* 106: 224–229.

Paul J, Bekker AY and Duran WN (1990) Calcium entry block prevents leakage of macromolecules induced by ischemia-reperfusion in skeletal muscle. *Circ Res* 66: 1636–1642.

Perler BA, Tohmeh AG and Bulkley GB (1990) Inhibition of the compartment syndrome by the ablation of free radical-mediated reperfusion injury. *Surgery* 108: 40–47.

Perry MO and Fantini G (1987) Ischemia: profile of an enemy. Reperfusion injury of skeletal muscle. *J Vasc Surg* 6: 231–234.

Perry MO and Fantini G (1989) Ischemia-reperfusion and cell membrane dysfunction. *Microcirc Endothel Lymph* 5: 241–258.

Potter RF, Ellis CG, Tyml K and Groom AC (1994) Effect of superoxide dismutase and 21-aminosteroids (lazaroids) on microvascular perfusion following ischemia-reperfusion in skeletal muscle. *Int J Microcirc Clin Exp* 14: 313–318.

Punch J, Rees R, Cashmer B, Oldham K, Wilkins E and Smith DJ Jr (1991) Acute lung injury following reperfusion after ischemia in the hind limbs of rats. *J Trauma* 31: 760–767.

Punch J, Rees R, Cashmer B, Wilkins E, Smith DJ and Till G (1992) Xanthine oxidase: its role in the no-reflow phenomenon. *Surgery* 111: 169–176.

Purucker E, Egri L, Hamar H, Augustin AJ and Lutz J (1991) Differences in glutathione status and lipid peroxidation of red and white muscles: alterations following ischemia and reperfusion. *Res Exp Med Berl* 19: 209–17.

Seekamp A, Warren JS, Remick DG, Till GO and Ward PA (1993) Requirements for tumor necrosis factor-alpha and interleukin-1 in limb ischemia/reperfusion injury and associated lung injury. *Am J Pathol* 143: 453–463.

Sexton W L, Korthuis RJ and Laughlin M, H (1990) Ischemia reperfusion injury in isolated rat hindquarters. *J Appl Physiol* 68: 387–392.

Sirsjo A, Soderkvist P, Gustafsson U, Lewis DH and Nylander G (1990a) The relationship between blood flow, development of edema and leukocyte accumulation in post-ischemic rat skeletal muscle. *Microcirc Endothel Lymph* 6: 21–34.

Sirsjo A, Lewis DH and Nylander G (1990b) The accumulation of polymorphonuclear leukocytes in post-ischemic skeletal muscle in the rat, measured by quantitating tissue myeloperoxidase. *Int J Microcirc Clin Exp* 9: 163–173.

Sirsjo A, Kagedal B, Arstrand K, Lewis DH, Nylander G and Gidlof A (1996) Altered glutathione levels in ischemic and postischemic skeletal muscle: difference between severe and moderate ischemic insult. *J Trauma* 41: 123–128.

Smith JK, Grisham MB, Granger DN And Korthuis RJ (1989a) Free radical defense mechanisms and neutrophil infiltration in postischemic skeletal muscle. *Am J Physiol* 256: H789–H793.

Smith JK, Carden DL and Korthuis RJ (1989b) Role of xanthine oxidase in postischemic microvascular injury in skeletal muscle. *Am J Physiol* 257: H1782–H1789.

Smith JK, Carden DL, Grisham MB, Granger DN and Korthuis RJ (1989c) Role of iron in postischemic microvascular injury. *Am J Physiol* 256: H1472–H1477.

Soussi B, Idstrom JP, Schersten T and Bylund-Fellenius AC (1990) Cytochrome c oxidase and cardiolipin alterations in response to skeletal muscle ischaemia and reperfusion. *Acta Physiol Scand* 138: 107–114.

Strock PE and Majno G (1969a) Vascular responses to tourniquet ischemia. *Surg Gynecol Obstet* 129: 309–318.

Strock PE and Majno G (1969b) microvascular changes in acutely ischemic rat muscle. *Surg Gynecol Obstet* 129: 1213–1224.

Suzuki H, Poole DC, Zweifach BW and Schmid-Schonbein GW (1995) Temporal correlation between maximum tetanic force and cell death in postischemic rat skeletal muscle. *J Clin Invest* 96: 2892–2897.

Ward PH, Maldonado M and Vivaldi E (1992) Oxygen-derived free radicals mediate liver damage in rats subjected to tourniquet shock. *Free Radical Res Commun* 17: 313–325.

Ward PH, Maldonado M, Roa J, Manriquez V and Vivaldi E (1995) Ibuprofen protects rat livers from oxygen-derived free radical-mediated injury after tourniquet shock. *Free Radical Res Commun* 22: 561–569.

Welbourn CR, Goldman G, Paterson IS, Valeri CR, Shepro D and Hechtman HB (1991) Neutrophil elastase and oxygen radicals: synergism in lung injury after hindlimb ischemia. *Am J Physiol* 260: H1852–H1856.

Welbourn R, Goldman G, Kobzik L, Paterson IS, Valeri CR, Shepro D and Hechtman HB (1992) Role of neutrophil adherence receptors (CD 18) in lung permeability following lower torso ischemia. *Circ Res* 71: 82–86.

Wilkins EG, Rees RS, Smith D, Cashmer B, Punch J, Till GO and Smith DJ Jr (1993) Identification of xanthine oxidase activity following reperfusion in human tissue. *Ann Plast Surg* 31: 60–65.

Oxidative damage in rat skeletal muscle after excessive L-tryptophan and atherogenic diets

E. Livne[1], N. Ronen[1], S. Mokady[2], A.Z. Reznick[1] and B. Gross[3]

[1]Division of Morphological Sciences, Bruce Rappaport Faculty of Medicine – Technion-Israel Institute of Technology, Haifa, Israel
[2]Faculty of Food Engineering and Biotechnology – Technion-Israel Institute of Technology, Haifa, Israel
[3]Department of Neurology, Carmel Medical Center, Haifa, Israel

Summary. Abnormalities in metabolism of L-tryptophan have been reported to be associated with tissue fibrosis and inflammation and were also documented in a number of other syndromes such as carcinoid syndrome, scleroderma, rheumatoid arthritis and eosinophilia myalgia syndrome (EMS). It has been constantly reported that the clinical features of EMS associated with excessive L-tryptophan consumption included myalgia, rash, edema, arthralgia, fatigue and cough. Pathological study of skeletal muscle showed infiltration of eosinophils and inflammatory cells along with fibrosis in muscle fascia. Some cell functions depend also on fatty acids composition which were shown to influence cytokine production and to modulate the immune system. In damaged skeletal muscle such activities can enhance generation of free radicals. The purpose of the present study was to investigate the combined effects of an atherogenic diet enriched with L-tryptophan on lipid peroxidation and protein oxidation in skeletal muscle of rats and furthermore to test which of the tryptophan pathways was involved in the induction of cell proliferation and activation of inflammatory reaction in skeletal muscle. Female CD-1 rats were fed for 3 weeks on a control or atherogenic diet, and on the same diets supplemented with 0.4%–2.0% L-tryptophan. On the 3rd week of feeding, half of the animals that were fed on the control diet and half of the animals that were fed on a diet supplemented with L-tryptophan were injected with 2 doses of para chlorophenyl alanine (p-CPA), (300 mg/kg body weight, i.p.) followed by 3 doses (100 mg/kg body weight) on every alternate day. Results indicated that treatment with p-CPA, an inhibitor of tryptophan hydroxylase, lead to a reduction of serotonin levels in blood. An increased amount of connective tissue and cell infiltration was observed in gastrocnemia muscle from rats fed L-tryptophan alone or in combination with p-CPA, as well as in rats fed atherogenic diet. In these experimantal groups no changes were observed by autoradiography in ³H-thymidine incorporation into DNA. Indirect immunohistochemistry revealed that in animals treated with L-tryptophan alone, or in combination with p-CPA, an increased TNFα positive reaction, indicative of macrophages, was observed in gastrocnemius muscle. In addition, an increased amount of eosinophils was observed in myofascia adjacent to blood vessels in skeletal muscle of animals following excessive L-tryptophan consumption.
The lipid peroxidation level was significantly ($p < 0.01$) increased in skeletal muscle of animals fed 0.4% L-tryptophan, but not in animals fed an atherogenic diet or an atherogenic diet with 0.4% L-tryptophan. In contrast, protein oxidation levels were significantly ($p < 0.01$) increased in skeletal muscle of animals fed an atherogenic diet and 0.4% L-tryptophan, or the control diet with 0.4% L-tryptophan, with no effect observed in animals fed the atherogenic diet alone. No significant changes in lipid peroxidation were detected in skeletal muscle following injections of p-CPA, or in animals fed 1%, and 2% L-tryptophan.
It is concluded that L-tryptophan induced inflammatory processes and tissue fibrosis in skeletal muscle. These effects were further augmented through the kynurenine pathway following p-CPA treatment. Lipid peroxidation as well as protein oxidation levels were induced in skeletal muscle following consumption of L-tryptophan, but were not elevated following consumption of the atherogenic diet alone. Protein oxidation appeared to be induced by atherogenic diet in combination with L-tryptophan, whereas no such effect on lipid peroxidation levels following atherogenic diet was observed in this tissue. Atherogenic diet in combination with L-tryptophan induced protein oxidation in this tissue, with no such effect on lipid peroxidation.

Introduction

L-Tryptophan, an essential amino acid, is absorbed from food ingested by animals and humans. About 90% of the tryptophan intake is incorporated into proteins, 9% serves as substrate for tryptophan catabolism in the kynurenine pathway, producing nicotinic acid, xanthurenic acid and picolinic acid, and about 1% is used as a substrate for serotonin, an important neurotransmitter in the central nervous system (Harper et al., 1984) (Fig. 1). Over the past decade, it was reported that diets containing excessive L-tryptophan caused a multisystemic syndrome, known as eosinophilia myalgia syndrome (EMS) (Duffy, 1992). Abnormalities in metabolism of L-tryptophan have been also reported in other syndromes, such as carcinoid syndrome (Harrison et al., 1986; Cooper et al., 1991), scleroderma-like syndrome (Bruce et al., 1990; Smith et al., 1990; Varge et al., 1990), eosinophilic fasciitis (Freis et al., 1973), idiopathic scleroderma (Stenerg et al., 1980) and systemic sclerosis. Inflammatory responses were elicited in all these syndromes, together with leukocyte infiltration and fibroblast proliferation, which have been consistently reported in histological sections from human skeletal muscle. Patients with EMS were characterized by the presence of activated T lymphocytes in blood from affected tissues during the clinically active, progressive phase of EMS (Kita et al., 1995). Even though an epidemic study indicated that a contaminant 1,1-ethylidenebis-L-tryptophan (EBT) was involved in EMS (Love et al., 1993), abnormalities in the metabolism of tryptophan were also reported to be associated with tissue fibrosis and inflammation. Animal studies were controversial, but there have been reports that either female Lewis rats (LEW/N) fed on tryptophan which are implicated in the EMS epidemic or LEW/N rats fed synthesized EBT had developed myofascial fibrosis similar to that observed in patients with EMS (Silver et al., 1994).

It has been demonstrated previously that some cell and tissue functions depend on fatty acid composition and that specific types of dietary lipids can alter the function of the immune system by affecting membrane fluidity of macrophages and other cytokine-secreting cells. Such cells are one of the main sources of oxidizing agents in cells and tissues (Robinson, 1991; Newsholm et al., 1993). Mokady et al. (1990) reported that dietary tryptophan was also found to increase platelet aggregation; there was a synergistic effect between atherogenic diet and tryptophan on induction of uptake of plasma cholesterol by macrophages (Aviram et al., 1991). In addition, there have been reports that dietary tryptophan supplementation increased plasma lipid peroxidation in control and hypercholesterolaemic animals (Newsholm et al., 1993). Similarly, excessive dietary tryptophan enhanced the esterification rate of macrophage cholesterol after cell incubation with plasma obtained from control and hypercholesterolaemic animals.

In human patients, interferon-γ (IFN-γ) appears to be the principal mediator of indoleamine-2,3-dioxygenase induction in macrophages (Meyer et al., 1995). The immune system is a target for many toxins, including those obtained through degradation of tryptophan such as kynurenine, quinolinic acid and picolinic acid (Leonhardt et al., 1986; Freese et al., 1990; Varesio et al., 1990; Heyes et al., 1992; Saito et al., 1993). Previous reports revealed that excessive dietary tryptophan also resulted in the appearance of numerous tumour necrosis factor α (TNFα)-positive cells in skeletal muscle of rats fed 0.4%–1% tryptophan for 3 weeks (Ronen et al., 1996). Such induction is indicative of macrophage activation. Thus, an interaction between tryptophan and dietary lipid may be involved in induction of damage to cell membranes as well as activation of macrophages and other cytokine-producing cells. Rudzite and Jurika (1991) have also reported on the interaction between lipid metabolism and kynurenine, one of the metabolites of tryptophan.

To answer the question of which possible environmental factors elicited such a dramatic response to consumption of tryptophan-rich diets as observed in EMS, the effects of possible interactions of tryptophan, its metabolites or saturated lipid-enriched diets in rat skeletal muscle were tested.

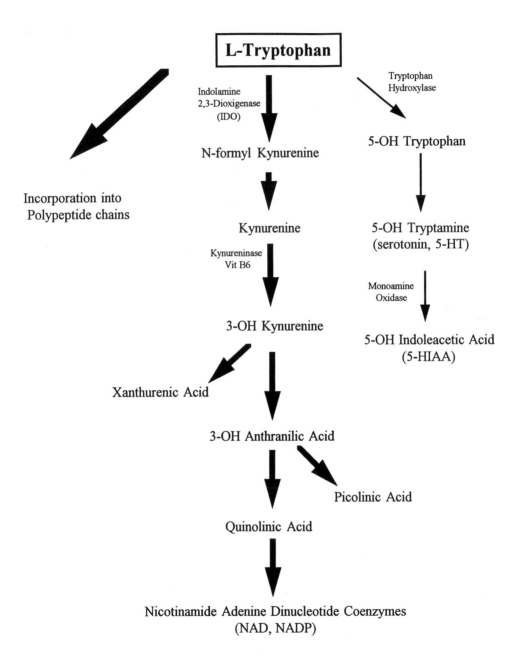

Figure 1. The biochemical pathways of tryptophan metabolism.

Materials and methods

Three-month-old female rats (Charles River CD-1) were divided into 10 groups (six animals each), fed for 3 weeks on a 20% protein diet derived from casein and supplemented with either 8% soybean oil (control) or 8% coconut oil, 1% cholesterol and 0.5% cholic acid (atherogenic diet) (Anonymous, 1977). These basic diets were further supplemented with 0.4%, 1.0% or 2.0% tryptophan (Sigma, St Louis, MO, USA). At the end of the second week, half of the animals of each group were injected with p-chlorophenylalanine (p-CPA), an inhibitor of serotonin synthesis (300 mg/kg i.p., 24 h apart, acute treatment), followed by three additional injections of 100 mg/kg on every alternate day (subchronic treatment). All animals were kept in conditions of controlled temperature (24 °C), maintained on a 12–12 h light–dark cycle. Food and water were supplied *ad libitum*. Body weights were determined weekly and food intake was calculated on every alternate day. At the end of the feeding period, animals were sacrificed and blood was collected into a disodium EDTA solution for measuring serotonin levels. Plasma serotonin levels were determined using (N-ω-methyl-5-hydroxytryptamine oxalate or NMET) as internal standard for HPLC. Gastrocnemius muscles were individually dissected and quickly frozen in 2-methylbutane isopentane bathed in liquid nitrogen (−180°C) and kept at −70 °C until use. Additional gastrocnemius muscle samples were fixed in 3% glutaraldehyde in 0.1 M cacodylate buffer and embedded flat in Epon 812. Ultrathin sections (50–70 nm) were used for photographing on a Jeol SX100 electron microscope operating at 80 kV. Other similar samples of gastrocnemius muscle were fixed in 10% neutral phosphate-buffered formaldehyde pH 7.4, sectioned (6 μm) and stained with haematoxylin and eosin (H&E) for general morphology observations, or with May–Grünwald–Giemsa for detection of mast cells and eosinophils. Other samples of gastro-cnemius muscles were collected for immunocytochemistry; immersed in 2-methylbutan-isopentan bathed in liquid nitrogen (−180 °C) and kept at −70 °C until used. Frozen sections (8–10 μm) were used for immunohistochemical demonstration of TNFα. The primary antibody employed was polyclonal rabbit anti-rat TNFα (1:50–1:100) (Zymed Laboratories, South San Francisco, CA). The second antibody was biotinylated, Strepavidin–peroxidase was then added and the presence of peroxidase was revealed by adding a mixture of substrate–chromogen solution, for visualization of TNFα-positive sites. In addition, three animals from each group were injected (i.p.) with 2 μCi/gr body weight of [^{3}H]-thymidine (specific activity of 40–50 Ci/mmole; Nuclear Research Center, Beer Sheva, Israel). Skeletal muscle samples were than dissected, fixed and embedded in paraplast as previously described. Unstained paraffin sections were dipped in NTB II photographic emulsion (Kodak, Rochester, N.Y., USA) and kept in light-tight boxes (4°C for 3 weeks) for demonstration of autoradiography. Sections were then developed and stained lightly with eosin for light microscope evaluation of radioisotope labelling.

   Lipid peroxidation was measured in gastrocnemius muscle homogenates using the method of Pryor and Castle (1984).The thiobarbituric acid (TBA)-reactive substances (TBARS) assay was used to measure malondialdehyde (MDA). 2,2'-Azobis-dimethylvaleronitrile (AMVN) was used as the oxidizing agent and the pink reaction colour developed was read spectrophotometrically at 532 nm. Data were expressed as nanomoles of MDA per milligram of tissue. The level of protein oxidation was determined in the supernatant of muscle samples that were minced gently in 0.1 M phosphate buffer pH 7.4, containing 0.1% digetonin, using spectrophotometric assay for the reduction of dinitrophenylhydrazine (DNPH) to dinitrophenylhydrazone (Reznick and Packer, 1994). For all measurements statistical analysis was performed using unpaired the Student's t-test and results were reported as mean±SEM. The Student's t-test was performed together with analysis of variance (ANOVA) for all groups; a p value of less than 0.05 was considered as the significance level.

Results

Injection of p-CPA into the animals resulted in significant ($p < 0.001$) reduction in serotonin content in blood (Tab. 1). A similar significant ($p < 0.001$) reduction was observed in the contents of serotonin (5-HT) and 5-hydroxyindoleacetic acid (5-HIAA) in the midbrain region of p-CPA-treated animals, as well as in animals fed 1% tryptophan and treated with p-CPA, but not in animals fed 1% tryptophan alone (Tab. 1). An increased infiltration of cells was observed in tissue sections of gastrocnemius muscles from animals treated with 1% tryptophan compared with the control group (Figs 2a,b), and in animals treated with 1% tryptophan and injected with p-CPA (Fig. 2c). In animals fed an atherogenic diet, accumulation of lipocytes adjacent to blood vessels was also observed (Fig. 2d). Autoradiograms of gastrocnemius muscle from control and treated animals indicated that there was no increase in the number of [^{3}H]-thymidine-labelled cells in animals treated with either L-tryptophan or p-CPA (Fig. 3). The nature of these infiltrating cells was tested immunohistochemically using TNFα antibody. It was shown that, compared with control tissue (Fig. 4a), gastrocnemius muscle tissue from L-tryptophan-treated-animal, appeared to contain large numbers of cells that were positive for TNFα (Fig. 4b). Similar results were obtained in gastrocnemius muscle from animals fed 1% tryptophan and injected with p-CPA (not shown). Gastrocnemius muscle from animals treated with 1% tryptophan also revealed increased numbers of eosinophils and mast cells in connective tissue adjacent to blood vessels (Fig. 5). The ultrastructured appearance of skeletal muscle from non-treated animal revealed organized myo-

Table 1. Serotonin (5-HT) contents in the serum and midbrain and 5-hydroxyindole acetic acid (5-HIAA) contents in the midbrain of rats fed 1% tryptophan and injected with p-CPA

Diet	5-HT (ng/ml serum)	5-HT Midbrain (pg/mg tissue)	5-HIAA in midbrain (pg/mg tissue)
Control ($n=6$)	116.32±48.79	4.11±0.49	4.07±0.78
1% Tryptophan ($n=6$)	609.00±90.96[*] (+419.26%)	4.42±0.29 (+7.54%)	3.46±0.33 (−14.99%)
Control + p-CPA ($n=6$)	24.00±7.50[*] (−79.37%)	0.24±0.06[*] (−94.39%)	0.28±0.07[*] (−93.12%)
1% Tryptophan +p-CPA ($n=6$)	186.00±70.00[*] (+59.90%)	0.20±0.03[*] (−95.13%)	0.20±0.03[*] (−95.09%)

[*] $p < 0.001$ vs. control; % versus control. Values in parantheses are percentages versus control.

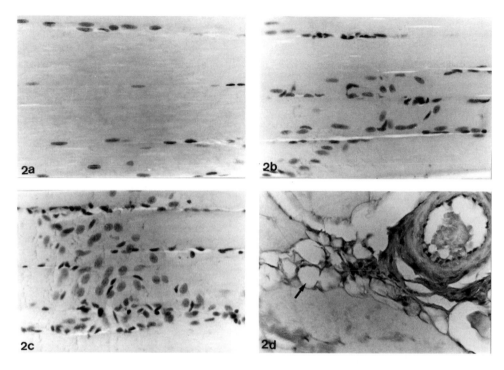

Figure 2. The morphological appearance of gastrocnemius muscle from (a) control animal, (b) animal treated with 1% tryptophan, (c) animal treated with 1% tryptophan + p-CPA and (d) animal fed an atherogenic diet. Note excessive accumulation of nuclei among muscle fibres in (b, c) and numerous lipocytes (arrow) next to a blood vessel in (d). Stained with haematoxylin and eosin (H&E); magnification ×90.

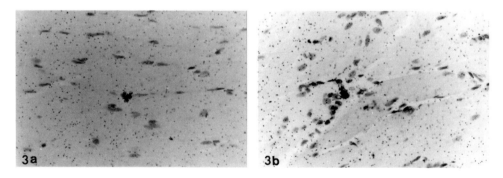

Figure 3. An autoradiogram of gastrocnemius muscle after [^{3}H]-thymidine incorporation obtained from a (a) control animal and (b) an animal fed 1% L-tryptophan. Magnification ×360.

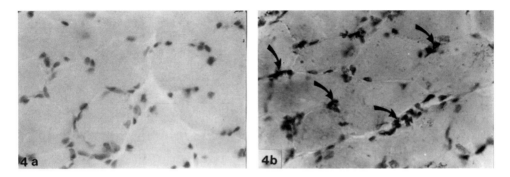

Figure 4. Immunohistochemistry performed on frozen sections of gastrocnemius muscle for demonstrating the localization of TNFα in gastrocnemius muscle from (a) control animal and (b) animal fed 1% tryptophan. Note numerous positively labelled cells in (b) (arrow). Magnification ×360.

fibrils and sarcomeres (Fig. 6a). In skeletal muscle from animals fed an atherogenic diet and treated with p-CPA, there was marked damage to myofibrils along with the appearance of swollen T-tubes in the triad regions (Fig. 6b). A significant increase in lipid peroxidation ($p<0.01$) was observed after administration of 0.4% tryptophan (Fig. 7). No similar effects were observed, however, after consumption of an atherogenic diet alone or in combination with 0.4% L-tryptophan and lipid peroxidation levels were similar to the levels in the control group.

Protein oxidation appeared to be significantly increased by almost +200% ($p<0.01$) in skeletal muscle after consumption of atherogenic diet and 0.4% tryptophan. There was also a significant (90%) increase in protein oxidation ($p<0.05$) after 0.4% L-tryptophan, but not

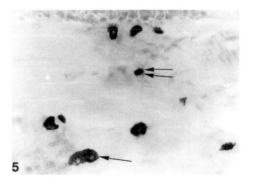

Figure 5. The morphological appearance of gastrocnemius muscle from an animal fed a 1% tryptophan-rich diet and stained with Giemsa–Romanowsky. Note mast cell (arrow) and eosinophils (two arrows). Magnification ×720.

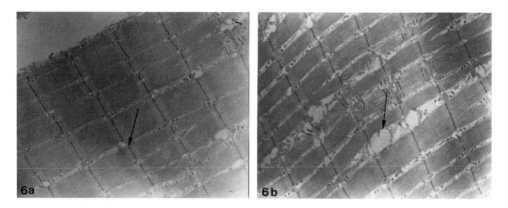

Figure 6. An electron micrograph of gastrocnemius muscle obtained from (a) control animal and (b) animal fed a 1% tryptophan diet and p-CPA. Typical organization of myofibrils and sarcomeres is seen as well as organized T-tubes (arrow). In (b) disorganization of myofibrils and dilated T-tubes in the triad region (arrow) are observed. Magnification × 10 000.

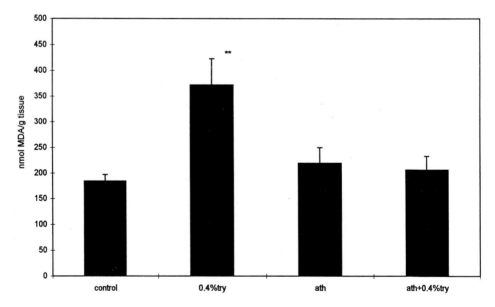

Figure 7. Lipid peroxidation levels (nmol MDA/g tissue) in gastrocnemius muscle obtained from animals ($n = 6$) fed on control diet, a diet containing 0.4% (Try), an atherogenic diet (Ath) alone, or in combination with L-tryptophan.

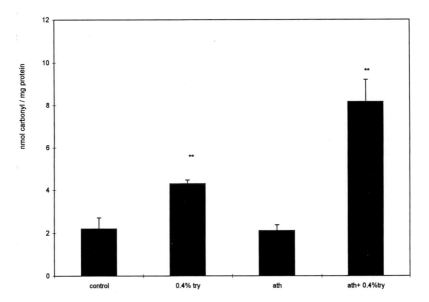

Figure 8. Protein oxidation levels (nmol carbonyl/mg protein) in gastrocnemius muscle obtained from animals ($n = 6$) fed on control diet, on a diet containing 0.4% tryptophan (Try), atherogenic diet (Ath) alone or in combination with L-tryptophan.

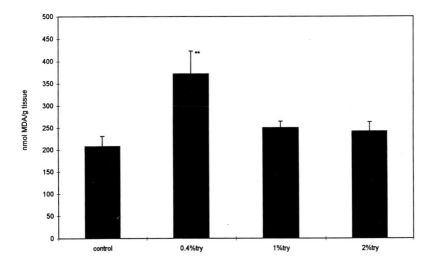

Figure 9. Lipid peroxidation levels (nmol MDA/mg tissue) in gastrocnemius muscle obtained from animals ($n = 6$) fed 0.4%, 1% or 2% L-tryptophan.

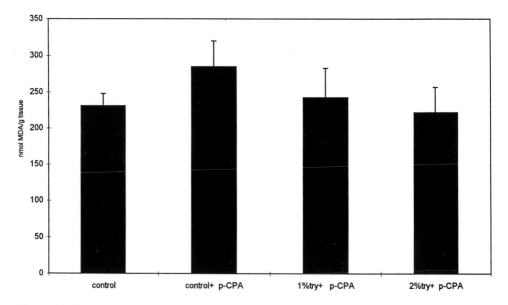

Figure 10. Lipid peroxidation levels (nmoles MDA/g tissue) in gastrocnemius muscle from animals ($n=6$) fed on a control diet, a diet containing 1% or 2% tryptophan (Try) and p-CPA.

after feeding with atherogenic diet alone (Fig. 8). Comparison between the effects of different levels of tryptophan consumption on MDA levels, revealed that 0.4% tryptophan had the greatest inductive effect, whereas the 1% and 2% levels were no different from the control (Fig. 9). Injection of p-CPA caused a slight increase in lipid peroxidation in skeletal muscle, but had no significant effect when compared with controls (Fig. 10). The possible metabolic pathways resulting from excessive L-tryptophan and atherogenic diets, in relation to macrophage activation and lipid peroxidation, are summarized in Figure 11.

Discussion

Abnormalities in tryptophan metabolism were reported to be associated with disease in which skeletal muscle was damaged (Mokady et al., 1990). Excessive L-tryptophan consumption induced inflammatory processes in muscle. In the present study, endogenous serotonin stores were reduced by blocking serotonin synthesis with p-CPA. It was shown that this treatment inhibited the serotonin pathway in serum and brain, and most probably activated the kynurenine pathway (Richard et al., 1990). This activation appeared to result in tissue fibrosis and inflammatory processes in rat muscle. The kynurenine pathway is known to be activated by two enzymes that degrade L-tryptophan by catalysing the oxidative cleavage of the pyrrole ring to give N-formyl-L-kynurenine: (1) the hepatic enzyme tryptophan-2,3-dioxygenase (TDO) which is

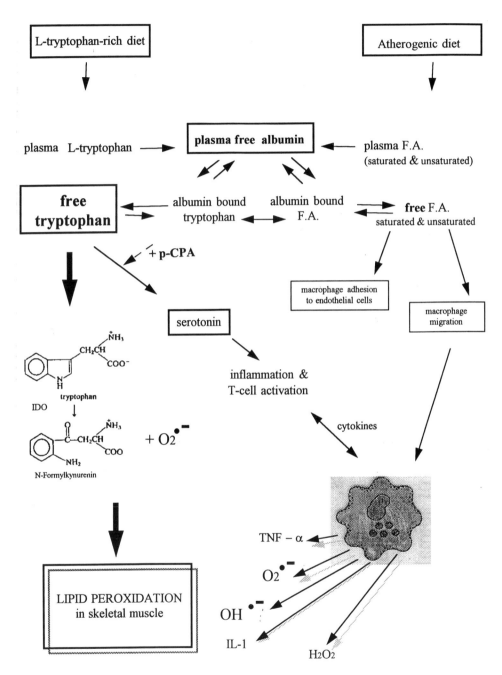

Figure 11. The possible metabolic routes of L-tryptophan and atherogenic diets in relation to macrophage activation and lipid peroxidation. (F.A. = fatty acid; ◄– – – = inhibition).

inducible by cortisol or tryptophan and is involved in regulation of tryotophan levels and the initiation of NAD biosynthesis; and (2) indoleamine 2,3-dioxygenase (IDO) which also degrades tryptophan and is a distinct enzyme found in a number of extrahepatic tissues, such as mono-cytes, macrophages, fibroblasts and other cell types, and is induced by interferons or by condi-tions that stimulate their immune-mediated release. The hepatic enzyme TDO is constitutively expressed and has less affinity for tryptophan compared with IDO (Meyer et al., 1995).

It has been reported that products of L-tryptophan stimulate peripheral blood mononuclear cells (PBMCs) which have been cultured to produce granulocyte–macrophage colony-stimulating factor (GM-CSF). This response was caused by endotoxin contamination of the L-tryptophan products and not by a specific L-tryptophan contaminant. Fractionation of PBMCs indicated that the cells producing the cytokine are mainly monocytes (Kita et al., 1995). The mechanisms that link immune stimulation to an increase in kynurenine pathway metabolism, including the produc-tion of cytokines, such as INF-γ and TNFα, from monocytes, macrophages or lymphocytes, are not yet known. However, IDO activity in the brain and systemic tissues is induced by INF-γ or by tryptophan itself. It has been observed that there is significant correlation between the concen-trations of L-kynurenine, quinolinic acid (Leonhardt et al., 1986), and kynurinic acid (Freese et al., 1990) in the central nervous system (CNS), and quantitative measures of the severity of neurological dysfunction, the severity of brain injury and the concentrations of immune markers in a broad spectrum of inflammatory conditions caused by infectious organisms and autoimmune processes. This observation supports a role for these metabolites in the aetiology of inflammatory neurological diseases.

Systemic dysfunction as a result of amplification of the kynurenine pathway, are not been re-ported yet. Systemic effects on muscle were found in the present study, and results similar to those reported in work on the CNS were obtained after inhibition of the serotonin pathway and amplification of the kynurenine pathway (Richard et al., 1990); this had caused inflammation and induced production of the cytokine TNFα. Positive TNFα cells observed in muscle are usually indicative of macrophages. In the authors' work, TNFα-positive cells were induced in muscle treated with tryptophan or p-CPA, thus inhibiting the serotonin pathway and inducing the kynu-renine pathway. Picolinic acid, which is produced in this pathway, is also a co-inducer of macro-phage activation (Varesio et al., 1990; Heyes et al., 1992). The mechanisms by which picolinic acid activates macrophages are not known yet. In the present work it has been shown that, by injection of p-CPA, the serotonin pathway was inhibited and consequently the kynurenine path-way was amplified (Tab. 1). This process also resulted in enhanced proliferation of muscle cells, as shown by histology. Autoradiography was performed to ascertain whether these cells origin-ated from induced proliferation of existing cells in the tissue. However, in muscle tissue, no significant increase of [^{3}H]-thymidine-labelled cells was observed but DNA content was shown to increase. The conclusions are that the increased DNA content in muscle resulted from infiltration of cells into the muscle (Ronen et al., 1996).

In the present study it was also shown that 0.4% L-tryptophan induced lipid peroxidation levels in skeletal muscle. Malondialdehyde is a biomarker of lipid peroxidation and is used for quantification of oxidative stress (Reznick at el., 1994). Increased lipid peroxidation, which is a marker for free radical production, may indicate that tryptophan did indeed induce tissue damage, as seen in the muscle. When free radical generation exceeds the capacity of the antioxidant defence systems, the result is oxidative stress. Production of free radicals by phagocytes is useful in killing foreign organisms, but it can also harm the phagocyte itself and the surrounding tissues. Lipid peroxidation was not increased with higher levels of tryptophan or after injection of p-PCA, because the oxidative metabolites of tryptophan groups have antioxidant properties by the virtue of their active hydroxyl groups. Indoleamine-2,3-dioxygenase cleaves the pyrrole ring of tryptophan and indoleamines by using the superoxide anion radical as a co-factor and substrate in its catalytic process (Halliwell, 1991). Thus, high levels of metabolites of the kynurenine pathway can act as scavengers of the hydroxyl radicals (Saito et al., 1993a; Paguirigan et al., 1994).

One of the targets of free radical attack in human cells is the membrane lipids. Free polyunsaturated fatty acids (PUFAs) and other fatty acids that are incorporated into the membrane phospholipid are readily attacked by free radicals, thereby being oxidized to lipid peroxides (Reznick and Packer, 1993). Lipid peroxidation of cell membranes results in decreased membrane fluidity, inability to maintain ionic gradients, cellular swelling and tissue inflammation (Parkhause et al., 1995).

The ultrastructural appearance of treated muscle revealed swelling and deformation of intracellular membranes, which may indicate that this damage can be caused by free radicals. Among the various ways for increasing free radical production, is the increase of $O_2^{-\cdot}$ generation in the mitochondrial electron transport chain, or the increase of metal-catalysed free radical production as a result of mechanical and morphological damage to muscle (Watson, 1994). The hydroperoxides produced in this process can induce stimulation of cell proliferation and cause induction of cytokine secretion. The authors' study demonstrated that excessive dietary tryptophan, when consumed with saturated fatty acids and cholesterol, could mediate inflammation, tissue damage and fibrosis in skeletal muscles, as seen in EMS (Esterbauer, 1993; Varge et al., 1990). However, a combination of atherogenic diet and tryptophan decreased lipid peroxidation probably as a result of high levels of saturated fatty acids incorporated into muscle membranes. At the same time, protein oxidation was increased in muscle from animals fed atherogenic diet and L-tryptophan. This may result from the fact that muscle tissue contains a high concentration of proteins, which may contribute to increased oxidative stress in this tissue.

The metabolic routes of tryptophan and atherogenic diets in relation to macrophage activation, free radical production and peroxidative reactions are summarized in Figure 11. In animals fed normally, the fraction of total plasma tryptophan bound to albumin is about 85–90%. This equilibrium can be shifted under conditions that raise plasma non-esterified fatty acid (NEFA) concentrations, such as during fasting or stress (Peter, 1991). This is because NEFAs compete with tryptophan for binding sites on the albumin molecule, so when the concentration of NEFAs rises, tryptophan is displaced from the albumin molecule, increasing the concentration of free tryptophan. It thus appears that, in a diet of tryptophan and the fatty acids, the actual concentrations of unbound tryptophan in the plasma could be much higher than the calculated values (Cristen et al., 1990; Halliwell, 1991). It is concluded that the L-tryptophan-induced inflammatory processes and tissue fibrosis were probably enhanced via the kynurenine pathway. An atherogenic diet induced similar morphological effects. However, tissue oxidative stress was induced by 0.4% tryptophan, but not higher levels of tryptophan. An atherogenic diet induced increased protein oxidation but not lipid peroxidation. Finally, it appears that oxidative stress and free radicals are not causative in the pathological process, but their increase is a secondary phenomenon that is influenced by a balance between oxidative stress and local antioxidant defence mechanisms.

Acknowlegment
This work was supported by Technion V.P.R. Grant No. 181-681, Israel Ministry of Health Grant No. 184-207 and The Krol Foundation Tiftain, NJ, USA, grant No. 184-198.

References

Anonymous (1977) Report of the American Institute of Nutrition Ad hoc Committee on Standards for Nutritional Studies. *J Nutr* 107: 1340–1348.
Aviram M, Cogan U and Mokady S (1991) Excessive dietary tryptophan enhances plasma lipid peroxidation in rats. *Atherosclerosis* 88: 29–34.
Bruce VP, Werth AH, Rook CR, O'Connor HR and Schumacher (1990) L-Tryptophan ingestion assoiated with eosinophilic fasci but not progressive systemic sclerosis. *Ann Intern Med* 112: 758–762.
Cooper JR, Bloom FE and Roth RH (1991) Serotonin (5-hydroxytryptamine) and histamin in carcinoid syndrom. *In: The Biochemical Basis of Neuropharmacology*, 6th ed. Oxford University Press, New York, pp 280–338.
Cristen S, Peterson E and Stocker R (1990) Antioxidant activities of some tryptophan metabolites: Possible implication for inflammatory diseases. *Proc Natl Acad Sci USA* 87: 2506–2510.
Duffy J (1992) The lessons of eosinophilia-myalgia syndrome. *Hospital Practic* 30: 65–90.
Esterbauer H (1993) Cytotxicity and genotoxicity of lipid-oxidation products. (Review). *Amer J Clin Nutr* 57 (suppl): 799S–786S.
Freese J, Kenton BA, Swartz J and Mattew (1990) Kynurenine metabolites of tryptophan: Implications for neurologic diseases. *Neurology* 40: 691–695.
Freis JF, Lingren JA and Bull MJ (1973) Scleroderma-like lesions and carcinoid syndrome. *Arch Intern Med* 131: 550–552.
Halliwell B (1991) *Lipid Peroxidation Free-Radical Reactions and Human Disease*. Upjohn Company Calamazoo, ISBN Press, Michigan.
Harper HA, Rodwell VW and Mayers PA (1984) Conversion of amino acids to specialized products. *In: Review of Physiological Chemistry*, 17th edn. pp 384, 415, 434.
Harrison TR, Petesdorf RG, Adams PD, Braunwald E, Isselbacher KJ and Wilson JD (1986) *Harrison's Principles of Internal Medicine*, 10th edn. McGraw Hill, New York, pp 825–829.
Heyes MP, Saito K and Crowley JS (1992) Quinolinic acid and kynurenine pathway metabolism in inflammatory and non-inflammatory neurologic disease. *Brain* 115: 1249–1273.
Kita H, Mayer AN, Weyand CM, Goronzy JJ, Weiler DA, Lundy SK, Abrams JS and Gleich GJ (1995) Eosino-phil-active cytokine from mononuclear cells cultured with L-tryptophan products: An unexpected consequence of endoxin contamination. *J Allergy Clin Immunol* 6: 1261–1267.
Leonhardt AB, Neal JW, Kuln JA, Schwarz M and Plimmer JR (1986) Quinolinic acid: An endgenus metabolite that produces axon-sparing lesins in rat brain. *Science* 219: 316–318.
Love AL, Rader JZ, Confford LJ, Raybourne RB, Principato MA, Page SW, Truckess MW, Smith MJ, Dugan EM and Turner ML (1993) Pathological and immunological effects of ingesting L-tryptophan and 1,1'-ethylidenebis (L-tryptophan) in lewis rats. *J Clin Invest* 91: 804–811.
Meyer KC, Arend RA, Kalayoglu MV, Rosenthal NS, Byrne GI and Brown RR (1995) Tryptophan metabolism in chronic inflammatory lung disease. *J Lab Clin Med* 126: 530–540.
Mokady S, Cogan U and Aviram M (1990) Dietary tryptophan enhances platelet aggregation in rats. *J Nutr Sci Vitaminol* 36(suppl): S177–S180.
Newsholm EA, Calder P and Yaqoob P (1993) The regulatory informational and immunomodulatory roles of fat fuels. *Am Clin Nutr* 57 (suppl): 738S–751S.
Paguirigan AM, Byrne GI, Becht S and Carlin JM (1994) Cytokine-Mediated indoleamine 2,3-dioxygenase induc-tion in respone to *Chlamydia* infection in human macrophage cultures. *Infection Immunity* 62: 4: 1131–1136.
Parkhause WS, Willis PE, Zahang J (1995) Hepatic lipid peroxidation and antioxidant enzyme responses to long-term voluntary physical activity and aging. *Age* 18: 11–17.
Peter JC (1991) Tryptophan nutrition and metabolism: an overview. *Adv Exp Biol* 294: 345–358.
Pryor W, Castle L (1984) Chemical methods for the detection of lipid hydroperoxides. *Methods Enzymol* 105: 293–299.
Reznick AZ and Packer L (1993) Free Radicals and antioxidants in musclar and neurological diseases and dis-orders. *In: G Poli, E Albano and MU Dianzani (eds): Free Radicals: From basic Science to Medicine*. Birkhäuser Verlag, Basel, pp 425–437.

Reznick AZ and Packer L (1994) Oxidative damage to proteins: spectrophotometric method for carbonyl assay, *Methods Enzymol* 233: 357–363.

Richard F, Sanne JL, Borde O, Weissman D, Ehret M, Cash C, Maitre M and Pujol JF (1990) Variation of tryptophan-5-hydroxylase concentration in rat raphe dorsalis nucleus after *p*-chlorophenylalanine administration. A model to study the turnover of enzymatic protein. *Brain Res* 536: 41–45.

Robinson DR (1991) Alleviation of autoimmune disease by dietary lipids containing omega-3 fatty acid. *Rheum Dis Clin North Am* 17: 213–222.

Ronen N, Gross B, Ben-Shachar D and Livne E (1996) The effects of induced kynurenin pathway on immunohistochemical changes in rat muscle following excessive L-tryptophan. *Adv Exp Med Biol* 398: 177–182.

Rudzite E and Jurika E (1991) Kynurenine and lipid metabolism. *Adv Exp Med Biol* 294: 463–466.

Saito K, Crowley JS, Markry SP and Heyes MP (1993a) A mechanism for increased quinolinic acid formation following acute systemic immune stimulation. *J Biol Chem* 268: 15496–15503.

Saito K, Chen CY, Masana M, Crowley JS, Markey SP and Heyes MP (1993b) 4-Chloro-3-hydroxyanthranilate, 6-chlorotryptophan and norharmane attenuate quinolinic acid formation by interferon-gamma-stimulated monocytes (THP-1 cells). *Biochem J* 291: 4–11.

Silver RM, Ludwicka A, Hampton M, Ohba T, Bingel S, Smith T, Russell AH, Maize J and Heyes PH (1994) A murine model of the EMS induced by 1'1"-ethylidenebis L-tryptophan. *J Clin Invest* 93: 1473–1480.

Smith SA, Roelofs RI and Gertner E (1990) Microangiopathy in eosinophilia–myalgia syndrome. *J Rheumatol* 17: 1544–1550.

Stenerg EM, Vav Woert MH, Young SN and Magnussen IB (1980) Development of a scleroderma-like illness during therapy with L-5-hydroxytryptophan and carbidopa. *N Engl J Med* 303: 782–787.

Varesio L, Calyton M, Ruffman E and Radzioch D (1990) Picolonic acid, a catabolite of tryptophan, as the second signal in the activation of interferon-γ-primed macrophges. *J Immunol* 145: 4265–4271.

Varge J, Peltonen J, Uitto S and Jimenez (1990) Development of diffuse fasciitis with eosinophilia during L-tryptophan treatment: Demonstration of elevated type I collagen gene expression in affected tissues. *Ann Intern Med* 112: 344–351.

Watson RR (1994) Age-related membrane alterations: modulation by dietary restriction. In: *Handbook of Nutrition in the Aged*, 2nd edn. CRC Press, Boca Raton, FL, pp 113–131.

Oxidative stress and muscle wasting of cachexia

M. Buck and M. Chojkier

Department of Medicine, Veterans Affairs Medical Center, and Center for Molecular Genetics, University of California, San Diego, CA 92161, USA

Summary. Cachexia is a frequent feature of patients afflicted with chronic diseases, including AIDS, cancer and inflammatory disorders. In cachexia, muscle wasting accounts for most of the weight loss, which may occur independently of the decreased food intake or malabsorption of nutrients. Moreover, there is evidence to suggest that tumor necrosis factor α (TNFα), may mediate, perhaps in concert with other cytokines, the muscle wasting of cachexia. However, in spite of the relevance of this issue to human diseases, the molecular mechanisms by which TNFα, or other mediators, induce muscle wasting remain to be determined. Therefore, in an attempt to elucidate this question, we evaluated the biological cascade leading to muscle wasting in a murine model of cachexia induced by chronically elevated serum TNFα. We found that TNFα induces oxidative stress and nitric oxide synthase (NOS) in skeletal muscle, leading to decreased myosin creatinine phosphokinase (MCK) expression and binding activities. The impaired MCK-E box binding activities resulted from abnormal myogenin/Jun-D complexes, and were normalized by the addition of Jun-D, DTT or Ref-1, a nuclear redox protein. Treatment of skeletal muscle cells with a phorbol ester, a superoxide-generating system, a NO donor, or a Jun-D antisense oligonucleotide decreased Jun-D activity and transcription from the MCK-E box, which were prevented by antioxidants, a scavenger of reducing equivalents, a NOS inhibitor, and/or overexpression of Jun-D. More importantly, the decreased body weight, muscle wasting and skeletal muscle molecular abnormalities of cachexia in the murine model were prevented by treatment with the antioxidants *d*-α-tocopherol or BW755c, or the NOS inhibitor nitro-L-arginine, which act at a post-receptor level.

Introduction

Cachexia is a frequent feature of patients afflicted with chronic diseases (Beutler, 1992; Grunfeld and Feingold, 1992; Tracey and Cerami, 1993), including AIDS (when associated with secondary infections or tumors), cancer, and inflammatory disorders, such as arthritis and those affecting the intestine, kidney, or lung (Voth et al., 1990; Beutler, 1992; Grunfeld and Feingold, 1992; Tracey, 1992; Roubenoff et al., 1994). The syndrome of cachexia is characterized by weight loss, decreased albumin synthesis, anemia, abnormal wound healing, and impaired immunity (Tracey, 1992). In cachexia, muscle wasting accounts for most of the weight loss, which may occur independently of the decreased food intake or malabsorption of nutrients (Tracey et al., 1990; Spiegelman and Hotamisligil, 1993), which is sometimes associated with chronic diseases (Grunfeld and Feingold, 1992). Moreover, there is evidence to suggest that tumor necrosis factor α (TNFα), a product of monocytes and macrophages (Akira et al., 1990), may mediate, perhaps in concert with other cytokines, the muscle wasting of cachexia (Beutler and Cerami, 1986; Fong et al., 1989; Grunfeld and Feingold, 1992; Strassmann et al., 1992; Spiegelman and Hotamisligil, 1993). Chronic increases in serum TNFα produced by tumor cells (Oliff et al., 1987; Brenner et al., 1990; Tracey et al., 1990; Costelli et al., 1993) or by expression of a TNFα transgene (Cheng et al., 1992) can induce muscle wasting. Cachexia in tumor models can be ameliorated with antibodies against TNFα (Sherry et al., 1989; Yoneda et al., 1991; Costelli et al., 1993). In addition, TNFα induces many other manifestations of cachexia (Tracey et al., 1988; Brenner et al., 1990;

Yoneda et al., 1991). However, in spite of the relevance of this issue to human diseases, the molecular mechanisms by which TNFα or other mediators induce muscle wasting remain to be determined (Beutler, 1992; Spiegelman and Hotamisligil, 1993; Tracey and Cerami, 1993). Therefore, in an attempt to elucidate this question, the authors have evaluated the biological cascade leading to muscle wasting (Buck and Chojkier, 1996) in a murine model of cachexia induced by chronically elevated serum TNFα (Oliff et al., 1987; Brenner et al., 1990; Tracey et al., 1990).

Results and discussion

As the animal model of cachexia developed by Oliff and co-workers (Oliff et al., 1987) closely follows the main features observed in patients with this syndrome, i.e. muscle wasting, decreased albumin synthesis, anemia, and impaired wound healing (Oliff et al., 1987; Brenner et al., 1990; Tracey et al., 1990; Buck et al., 1996), it provides a valuable system for analyzing the molecular mechanisms responsible for these abnormalities. At the onset of weight loss, TNFα serum levels were only moderately increased (100–300 pg/ml) as described previously (Brenner et al., 1990). At the time of sacrifice, the TNFα mice had high serum TNFα levels; however, similar values have been found in patients with trauma or infectious, parasitic, and neoplastic diseases (Scuderi et al., 1986; Waage et al., 1987; Grau et al., 1989; Goodman et al., 1990). Although a hypercatabolic state has been suggested as a possible explanation for the weight loss in cachexia (Starnes et al., 1988; Grunfeld and Feingold, 1992; Costelli et al., 1993; Roubenoff et al., 1994), the physiological and molecular disturbances leading to this syndrome have remained elusive (Beutler and Cerami, 1986; Spiegelman and Hotamisligil, 1993; Tracey and Cerami, 1993).

Recently the authors have demonstrated that the antioxidants d-α-tocopherol and BW755c and the nitric oxide synthase (NOS) inhibitor nitro-L-arginine prevented weight loss and muscle wasting in this animal model of cachexia (Fig. 1) (Buck and Chojkier, 1996). As with the findings in patients with cachexia, the weight loss in cachectic animals occurs, at least in the early stages, independently of decreased food intake and without changes in total body water content. The induction of an oxidative pathway in the skeletal muscle of TNFα animals was indicated by the presence of malondialdehyde (MDA)–protein adducts, which result from the oxidation of polyunsaturated fatty acids (Houglum et al., 1990; Chaudhary et al., 1994; Holvoet et al., 1995). These findings are in agreement with evidence that TNFα stimulates oxidative stress in many cells and tissues (Wong et al., 1989). In addition, NOS expression was markedly stimulated in the skeletal muscle of TNFα mice, and this effect is prevented by treating these animals with d-α-tocopherol or BW755c, suggesting that NO may mediate the muscle wasting and dedifferentiation induced by oxidative stress (Buck and Chojkier, 1996).

To ascertain that the dramatic effects of d-α-tocopherol and BW755c were not the spurious result of decreased synthesis of TNFα, the authors analyzed the influence of antioxidants on TNFα cells. Antioxidants did not affect either TNFα cell viability or the secretion of biologically active TNFα by these cells. Moreover, the TNFα cell tumor size, as well as the serum levels of TNFα were similar in TNFα animals whether or not they were treated with d-α-tocopherol, BW755c, or nitro-L-arginine. Finally, the end-organ biological effects of TNFα, such as anemia and decreased collagen $a_1(I)$ gene expression, remained abnormal in TNFα animals treated with

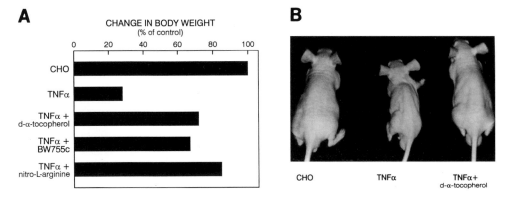

Figure 1. Muscle wasting of cachexia is prevented by antioxidants and nitro-L-arginine. (A) Change in body weight: athymic nude male mice were injected intramuscularly with either CHO cells (CHO, control group; $n=6$) or TNFα-secreting cells (Brenner et al., 1990). Some of the TNFα mice received no treatment (TNFα group; $n=6$), a diet supplemented with d-α-tocopherol ($n=6$) or were treated with either BW755c ($n=4$) or nitro-L-arginine ($n=6$). Values are percentages of change in body weight during the 30-day experiment, with the value for CHO animals set at 100. $p<0.05$ for all treatments compared with TNFα mice. (B) Representative examples of CHO, TNFα and TNFα + d-α-tocopherol groups. TNFα mice treated with either BW755c or nitro-L-arginine had an appearance similar to the TNFα + d-α-tocopherol mouse shown. (Reproduced by permission of Oxford University Press from Buck and Chojkier, 1996.)

d-α-tocopherol, BW755c, or nitro-L-arginine. Collectively, these results indicate that a TNFα receptor or postreceptor pathway(s) in skeletal muscle is the target of antioxidants. This pathway may include activation of other cytokines such as interleukins IL-1b and IL-6 (Dinarello et al., 1986; Akira et al., 1990), which in turn could induce muscle wasting (Flores et al., 1989; Fong et al., 1989; Strassmann et al., 1992; Spiegelman and Hotamisligil, 1993). Also, a 24-kD proteoglycan induces cachexia in rodents and may contribute to the muscle wasting of cachectic cancer patients (Todorov et al., 1996).

In cachectic animals, the skeletal muscle fibrils are depleted of myosin, a major structural protein, and their organization is disrupted. This decrease in myosin protein content in TNFα mice was prevented by treatment with d-α-tocopherol, BW755c, or nitro-L-arginine, indicating that oxidative pathways and activation of NOS are critical in the development of muscle wasting in cachectic animals. As muscle kreatine kinase (MCK) plays an important role in skeletal muscle differentiated functions (Kirchberger, 1991), it has been used as an indicator of muscle-specific gene expression (Li, Chambard et al., 1992; Tapscott et al., 1993; Kaushal et al., 1994; Skapek et al., 1995). The expression of the MCK gene was markedly reduced in the skeletal muscle of cachectic animals, indicating a pre-translational down-regulation of this gene. In agreement with previous reports in muscle cell lines (Brennan and Olson, 1990), myogenin is a major component of the MCK-E box-binding activities of nuclear extracts from normal skeletal muscle. Although the MCK-E box-binding activities were substantially decreased in nuclear extracts from cachectic skeletal muscle (Fig. 2a), the expression of myogenin was essentially unchanged (Fig. 3a). In addition, phosphatase treatment of skeletal muscle nuclear extracts from cachectic animals did not modify their binding affinity to the MCK-E box, suggesting that phosphorylation of myogenin or

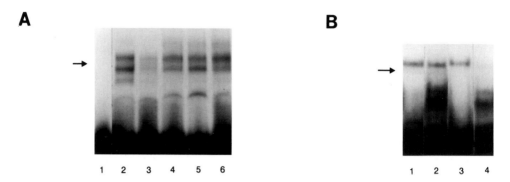

Figure 2. MCK-E box binding activities are decreased in skeletal muscle from cachectic animals. Mobility shift analysis of skeletal muscle nuclear extracts. Equal amounts of nuclear protein were incubated with ^{32}P-labelled MCK-E box oligonucleotide (1 ng). The DNA–protein complexes were resolved by electrophoresis on a 6% nondenaturing polyacrylamide gel. The position of the bound DNA is indicated by arrows. Some samples were incubated with specific antibodies or unlabeled oligonucleotide as indicated by + signs. The lanes are shown below in parentheses. (A) Representative samples of CHO (2); TNFα (3); TNFα/d-α-tocopherol (4); TNFα/BW755c (5); and TNFα/nitro-L-arginine (6). On lane 1, the probe was processed without nuclear extracts. (B) Representative samples of CHO (1); CHO/TNFα mix (2); CHO/TNFα mix + myogenin antibodies (3); and CHO/TNFα mix + myogenin and Jun-D antibodies (4). (Reproduced by permission of Oxford University Press from Buck and Chojkier, 1996.)

other nuclear factors, which could inhibit its binding to cognate DNA sequences (Li et al., 1992a), was not responsible for the impaired DNA-binding affinity. Addition of skeletal muscle nuclear extracts from cachectic animals to those from normal animals did not inhibit the binding of the latter to the MCK-E box (Fig. 2b). These data strongly suggested that the lack of an activator, rather than the presence of an inhibitor, is responsible for the decreased MCK-binding activities in skeletal muscle from cachectic animals.

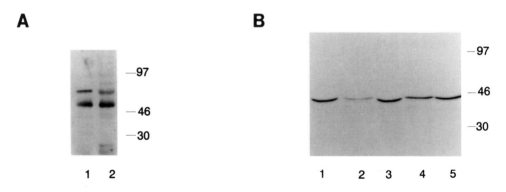

Figure 3. Expression of myogenin and Jun-D in skeletal muscle of cachectic mice. (A) Western blot was performed using monoclonal antibodies against myogenin in extracts of skeletal muscle from CHO (1) and TNFα (2) mice. The mobility of molecular weight standards is shown. (B) Western blot using antibodies against Jun-D (amino acids 327–341) in extracts of skeletal muscle from CHO (1); TNFα (2); TNFα/d-α-tocopherol (3); TNFα/BW755c (4); and TNFα/nitro-L-arginine (5) mice. (Reproduced by permission of Oxford University Press from Buck and Chojkier, 1996.)

Using cognate *cis*-elements and specific antibodies for various transcription factors, the authors demonstrated that Jun-D, but not Jun-B or C-Jun, is a major contributor to the MCK-binding activities of normal skeletal muscle. This finding allowed the authors to establish that Jun-D-binding activity and Jun-D protein were decreased in skeletal muscle nuclear extracts from cachectic animals (Fig. 3b). The addition of recombinant Jun-D to these nuclear extracts, in concentrations similar to those found in nuclear extracts from normal skeletal muscle, normalized the MCK-binding activities (Fig. 4a). The nuclear redox factor Ref-1 is sensitive to the redox state of the cell (Xanthoudakis and Curran, 1992), functions as a DNA repair enzyme (Xanthoudakis et al., 1992), stimulates the DNA-binding activity of several transcription factors, and may itself be under post-translational regulation. In this context, it was found that, although Ref-1 expression is induced in the skeletal muscle of cachectic animals (unpublished data), it is apparent that this mechanism is insufficient to protect Jun-D from the effects of oxidation. More importantly, addition of recombinant Ref-1 or dithiothreitol normalized the MCK-E box-binding activities of nuclear extracts from TNFα mice (Fig. 4b), suggesting an oxidative modification of a critical binding factor such as Jun-D. It is interesting that the Jun-D-reconstituted MCK/nuclear extract complex from skeletal muscle of cachectic animals now contained myogenin, indicating a direct or indirect physical association between Jun-D and myogenin. The precise nature of the eventual interaction between myogenin-E12 or other heterodimers (Murre et al., 1989; Brennan and Olson, 1990) and Jun-D remains to be determined. It is relevant to future therapeutic approaches in the treatment of cachexia that the antioxidants *d*-α-tocopherol and BW755c, or the NOS inhibitor, nitro-L-arginine (Kilbourn et al., 1990), normalized the following skeletal muscle abnormalities in TNFα mice: (1) myosin expression; (2) MCK expression; (3) MCK-binding activity (see Fig. 2a); and (4) Jun-D-binding activity (Buck and Chojkier, 1996).

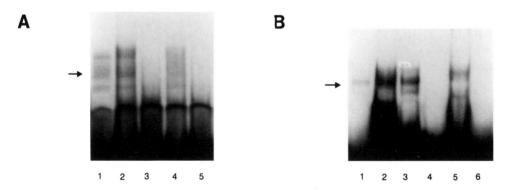

Figure 4. Jun-D and Ref-1 normalize the MCK-E box binding activities of skeletal muscle from cachectic animals. Mobility shift analysis of skeletal muscle nuclear extracts. Samples were incubated with ^{32}P-labeled MCK-E box oligonucleotide, and processed as described in Figure 2. The position of the bound DNA is indicated by arrows. Some samples were incubated with specific antibodies, recombinant protein, or dithiothreitol (DTT) as indicated by the + signs. (A) Representative samples of TNFα without additions (1); or with addition of recombinant Jun-D (2); recombinant Jun-D and myogenin antibodies (3); recombinant Jun-D and Jun-D antibodies (4); and recombinant Jun-D, myogenin and Jun-D antibodies (5). (B) Representative samples of TNFα without additions (1); or with the addition of DTT (2); recombinant Ref-1 (3); DTT, myogenin and Jun-D antibodies (4); recombinant Jun-D (5); and oxidized recombinant Jun-D (6). (Reproduced by permission of Oxford University Press from Buck and Chojkier, 1996.)

Although c-Jun, by virtue of its interactions with Myo-D and myogenin, may repress skeletal muscle differentiated functions such as MCK gene expression (Bengal et al., 1992; Li et al., 1992b), the authors found that c-Jun activity was unchanged from control levels in the nuclei of skeletal muscle from cachectic animals.

As proteolysis has been reported to be induced in skeletal muscle by a single injection of TNFα (Flores et al., 1989), and it has also been incriminated as a mechanism of muscle wasting during fasting (Wing and Goldberg, 1993), we analyzed whether protein degradation could contribute to the muscle dedifferentiation of cachexia. The synthesis and nuclear translocation of NF-κB is the result of proteolysis by the ubiquitin–proteasome pathway of the NF-κB precursor (Palombella et al., 1994) and the NF-κB inhibitor, IκB (Palombella et al., 1994; Traenckner et al., 1994), respectively; because NF-κB activation can be induced by TNFα and oxidative stress

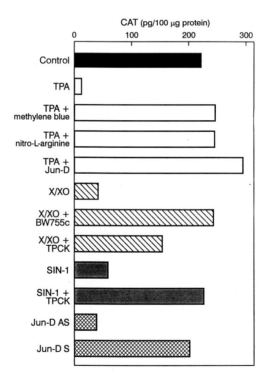

Figure 5. Oxidative stress inhibits transcription from the MCK-E box in mouse skeletal muscle cells. Mouse skeletal muscle C2C12 cells were transfected with 4RTK-CAT, an ε box reporter vector, alone or with RSV-Jun-D and treated with: 12-O-tetradecanoyl-phorbol-13-acetate (TPA), methylene blue, nitro-L-arginine, xanthine (X)/xanthine oxidase (XO), BW755c, SIN-1; Jun-D oligonucleotide antisense (AS); and Jun-D oligonucleotide sense (S), as indicated. Forty-eight hours later the cells were collected and catalase activity was determined. The results represent averages of three independent experiments. $p < 0.05$ for TPA, X/XO, SIN-1 and Jun-D oligonucleotide antisense. (Reproduced by permission of Oxford University Press from Buck and Chojkier, 1996.)

(Henkel et al., 1993; Palombella et al., 1994), the NF-κB activity in skeletal muscle was assessed. NF-κB-binding activities were comparable in skeletal muscle nuclear extracts from control and cachectic animals (data not shown). Although these data exclude the participation of the NF-κB pathway in the muscle wasting of cachectic animals, it is conceivable that proteolysis of a critical regulatory pathway participates in the cascade, leading to muscle wasting in cachexia. In this context, degradation of Jun-D is induced by serum in fibroblasts and it may play a role in modulating the proliferation of these cells (Pfarr et al., 1994). In addition, Treier and co-workers (Treier et al., 1994) have reported that another AP-1 factor, c-Jun, undergoes proteolysis via the ubiquitin system, and that Jun-D also reacts with ubiquitin in fibroblasts. The down-regulation of Jun-D expression in the skeletal muscle of cachectic animals was determined to occur at the translational or post-translational level, given that Jun-D mRNA was not affected.

Transcription from the MCK-E box is inhibited in skeletal muscle cells treated with TPA, a phorbol ester, that mimics the effects of TNFα on cellular functions, including oxidative stress (Brenner et al., 1989; Henkel et al., 1993), with a superoxide-generating system (xanthine/xanthine oxidase), or with SIN-1, a donor of NO (Fig. 5). The inhibitory effect of TPA on MCK-E box expression was blocked by methylene blue, which inhibits adduct formation (Houglum et al., 1990), and by nitro-L-argine.

More definitive evidence for a critical role of Jun-D in skeletal muscle differentiated function was obtained by the use of antisense Jun-D oligonucleotides and overexpression of Jun-D (Fig. 5). An antisense, but not a sense, Jun-D oligonucleotide blocked not only Jun-D expression but also MCK transcription activities in skeletal muscle cells. In addition, transfection of skeletal cells with a vector expressing Jun-D normalized the inhibition of MCK transcription induced by TPA. Collectively, these results indicate that Jun-D is necessary for the expression of the differentiated function of the skeletal muscle. Addition of recombinant Jun-D or Ref-1 was sufficient to normalize the impaired MCK-binding activities in both skeletal muscle nuclear extracts from cachectic mice and skeletal muscle cells treated with components of the oxidative cascade. Experiments in transgenic animals, including targeted deletion of Jun-D with and without expression of various deleted/mutated Jun-D transgenes, should clarify the role of Jun-D in muscle-differentiated functions.

The weight loss and muscle wasting of cachexia are common findings in patients with AIDS (when associated with secondary infections or cancer), cancer, and diseases characterized by chronic inflammation (Beutler and Cerami, 1986; Grunfeld and Feingold, 1992; Tracey and Cerami, 1993). At present, there is no satisfactory treatment for these complications, which contribute significantly to the morbidity and mortality of patients with cachexia (Grunfeld and Feingold, 1992; Tracey and Cerami, 1993). Our results suggest that oxidative stress and NO mediate the skeletal muscle abnormalities, including the decreased affinity of Jun-D to the MCK-E box, and the muscle wasting characteristic of cachexia (Fig. 6).

These results may provide insights into this biological enigma, as well as potential therapeutic approaches using antioxidants (Buck and Chojkier, 1996), NOS inhibitors (Kilbourn et al., 1990; Kobzik et al., 1994; Buck and Chojkier, 1996), and/or ubiquitin–proteasome pathway inhibitors (Palombella et al., 1994; Traenckner et al., 1994), for patients with cancer, AIDS, and chronic inflammatory diseases.

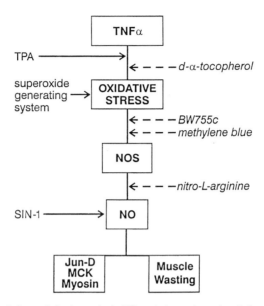

Figure 6. Oxidative stress induces skeletal muscle dedifferentiation and wasting. Induction of an oxidative stress cascade leads to decreased Jun-D activity and decreased expression of MCK and myogenin. The muscle wasting and/or dedifferentiation of skeletal muscle can be induced by TNFα, TPA, a superoxide -generating system (xanthine/xanthine oxidase), and SIN-1, and prevented by d-α-tocopherol, BW755c, methylene blue, and nitro-L-arginine. (Reproduced by permission of Oxford University Press from Buck and Chojkier, 1996.)

References

Akira S, Hirano T, Taga T and Kishimoto T (1990) Biology of multifunctional cytokines: IL 6 andrelated molecules (IL 1 and TNF). *FASEB J* 4: 2860–2867.

Bengal E, Ransone L, Scharfmann R, Dwarki VJ, Tapscott SJ, Weintraub H and Verma IM (1992) Functional antagonism between c-Jun and MyoD proteins: a direct physical association. *Cell* 68: 507–19.

Beutler B (ed.) (1992) *Tumor Necrosis Factors: The molecules and their emerging role in medicine.* Raven Press, New York.

Beutler B and Cerami A (1986) Cachectin and tumour necrosis factor as two sides of the same biological coin. *Nature* 320: 584–588.

Brennan T and Olson EN (1990) Myogenin resides in the nucleus and acquires high affinity for aconserved enhancer element on heterodimerization. *Genes Dev.* 4: 582–595.

Brenner DA, O'Hara M, Angel P, Chojkier M and Karin M (1989) Prolonged activation of jun and collagenase genes by tumour necrosis factor α. *Nature* 337: 661–663.

Brenner DA, Buck M, Feitelberg SP and Chojkier M (1990) Tumor necrosis factor α inhibits albumin gene expression in a murine model of cachexia. *J Clin Invest* 85: 248–255.

Buck M and Chojkier M (1996) Muscle wasting and dedifferentiation induced by oxidative stress in amurine model of cachexia, is prevented by inhibitors of nitric oxide synthesis and antioxidants. *EMBO J* 15: 1753–1765.

Buck M, Houglum K and Chojkier M (1996) Tumor necrosis factor α inhibits collagen $α_1$(I) gene expression and wound healing in a murine model of cachexia. *Am J Pathol* 149: 195–204.

Chaudhary AK, Munetaka N, Reddy GR, Yeola SN, Morrow JD, Blair IA and Marnett LJ (1994) Detection of endogenous malondialdehyde-deoxyguanosine adducts in human liver. *Science* 265: 1580–1582.

Cheng J, Turksen K, Yu QC, Schreiber H, Teng M and Fuchs E (1992) Cachexia and graft-*vs.*-host-disease-type skin changes in keratin promoter-driven TNF-α transgenic mice. *Genes Dev* 6: 1444–1456.

Costelli P, Carbó N, Tessitore L, Bagby GJ, Lopez-Soriano FJ, Argilés JM and Baccine FM (1993) Tumor necrosis factor-a mediates changes in tissue protein turnover in a rat cancer cachexia model. *J Clin Invest* 92: 2783–2789.

Dinarello CA, Cannon JG, Wolff SM, Bernheim HA, Beutler B, Cerami A, Figari IS, Palladino MA Jr and O'Connor JV (1986) Tumor necrosis factor (cachectin) is an endogenous pyrogen and induces production of IL-1. *J Exp Med* 163: 1433–1450.

Flores EA, Bistran BR, Pompselli JJ, Dinarello CA, Blackburn GL and Istfan NW (1989) Infusion of tumor necrosis/cachectin promotes muscle catabolism in the rat: A synergistic effect with interleukin 1. *J Clin Invest* 83: 1614–1622.

Fong Y, Moldawer LL, Marano MA, Wei H, Barber A, Manogue K, Tracey KJ, Kuo G, Fischman DA, Cerami A and Lowry SF (1989) Cachectin/TNF or IL-1α induces cachexia with redistribution of body proteins. *Am J Physiol* 256: R659–R665.

Goodman JC, Robertson CS, Grossman RG and Narayan RK (1990) Elevation of tumor necrosis factor in head injury. *J Neuroimmunol* 30: 213–217.

Grau GE, Taylor TE, Molyneaux ME, Wirima MB, Vassalli P, Hommel M and Lambert PH (1989) Tumor necrosis factor and disease severity in children with falciparum malaria. *N Engl J Med* 320: 1586–1591.

Grunfeld C and Feingold KR (1992) Metabolic disturbances and wasting in the acquired immunodeficiency syndrome. *N Engl J Med* 327: 329–337.

Henkel T, Machleidt T, Alkalay I, Krönke M, Ben-Neriah Y and Baeuerle PA (1993) Rapid proteolysis of IkB-α is necessary for activation of transcription factor NF-κB. *Nature* 365: 182–185.

Holvoet P, Perez G, Zhao Z, Brouwers E, Bernar H and Collen D (1995) Malondialdehyde-modified low density lipoproteins in patients with atherosclerotic disease. *J Clin Invest* 95: 2611–2619.

Houglum K, Filip M, Witztum J and Chojkier M (1990) Malondialdehyde and 4-hydroxynonenal protein adducts in plasma and liver of rats with iron overload. *J Clin Invest* 86: 1991–1998.

Kaushal S, Schneider JW, Nadal-Ginard B and Mahdavi V (1994) Activation of the myogenic lineage by MEF2A, a factor that induces and cooperates with MyoD. *Science* 266: 1236–1240.

Kilbourn RG, Gross SS, Jurban A, Adams J, Griffith OW, Levi R and Lodato RF (1990) N-methy-L-arginine inhibits tumor necrosis factor-induced hypotension: Implications for the involvement of nitric oxide. *Proc Natl Acad Sci USA* 87: 3629–3632.

Kirchberger M (1991) Excitation and contraction of skeletal muscle. *In*: JB West, (ed.): *Physiological Basis of Medical Practice*. Williams and Wilkins, Baltimore, MA, pp 62–102.

Kobzik L, Reid MB, Bredt DS and Stamler JS (1994) Nitric oxide in skeletal muscle. *Nature* 372: 546–548.

Li L, Zhou J, James G, Heller-Harrison R, Czech MP and Olson EN (1992a) FGF inactivates myogenic helix-loop-helix proteins through phosphorylation of a conserved protein kinase C site in their DNA-binding domains. *Cell* 71: 1181–1194.

Li L, Chambard J-C, Karin M and Olson EN (1992b) Fos and jun repress transcriptional activation by myogenin and MyoD: The amino terminus of Jun can mediate repression. *Genes Dev* 6: 676–689.

Murre C, McCaw PS and Baltimore D (1989) A new DNA binding and motif in immunoglobulin enhancer binding, daughterless, myoD, and myc proteins. *Cell* 56: 777–783.

Oliff A, Defeo-Jones D, Boyer M, Martinez D, Kiefer D, Vuocolo G, Wolfe A and Socher SH (1987) Tumors secreting human TNFα/cachectin induce cachexia in mice. *Cell* 50: 555–563.

Palombella VJ, Rando OJ, Goldberg AL and Maniatis T (1994) The ubiquitin-proteasome pathway is required for processing the NF-κB1 precursor protein and the activation of NF-κB. *Cell* 78: 773–785.

Pfarr C, Mechta F, Spyrou G, Lallemand D, Carillo S and Yaniv M (1994) Mouse Jun-D negatively regulates fibroblast growth and antagonizes transformation by ras. *Cell* 76: 747–760.

Roubenoff R, Roubenoff RA, Cannon JG, Kehayias JJ, Zhuang H, Dawson-Hughes B, Dinarello CA and Rosenberg IH (1994) Rheumatoid cachexia: Cytokine-driven hypermetabolism accompanying reduced body cell mass in chronic inflammation. *J Clin Invest* 93: 2379–2386.

Scuderi P, Lam KS, Ryan KJ, Petersen E, Sterling KE, Finley PR, Ray CG and Sylmen DJ (1986) Raised serum levels of tumour necrosis factor in parasitic infections. *Lancet* ii: 1364–1365.

Sherry BA, Gelin J, Fong Y, Marano M, Wei H, Cerami A, Lowry SF, Lundholm KG and Moldawer LL (1989) Anticachectin/tumor necrosis factor-alpha antibodies attenuate development of cachexia in tumor models. *FASEB J* 3: 1956–1962.

Skapek SX, Rhee J, Spicer DB and Lassar AB (1995) Inhibition of myogenic differentiation in proliferating myoblasts by cyclin D1-dependent kinase. *Science* 267: 1022–1024.

Spiegelman BM and Hotamisligil GS (1993) Through thick and thin: Wasting, obesity, and TNFα. *Cell* 73: 625–627.

Starnes HF Jr Warren RS, Jeevanandam M, Gabrilove JL, Larchian W, Oettgen HF and Brennan MF (1988) Tumor necrosis factor and the acute metabolic response to tissue injury in man. *J Clin Invest* 82: 1321–1325.

Strassmann G, Fong M, Kenney JS and Jacob CO (1992) Evidence for the involvement of interleukin 6 in experimental cancer cachexia. *J Clin Invest* 89: 1681–1684.

Tapscott SJ, Thayer MJ and Weintraub H (1993) Deficiency in rhabdomyosarcomas of a factor required for myoD activity and myogenesis. *Science* 259: 1450–1453.

Todorov P, Cariuk P, McDevitt T, Coles B, Fearon K and Tisdale M (1996) Characterization of a cancer cachectic factor. *Nature* 379: 739–742.

Tracey KJ (1992) The acute and chronic pathophysiologic effects of TNF: Mediation of septic shock and wasting (cachexia). *In*: B Beutler (ed.): *Tumor Necrosis Factors: The molecules and their emerging role in medicine.* Raven Press, New York, pp 255–273.

Tracey KJ and Cerami A (1993) Tumor necrosis factor, other cytokines and disease. *Annu Rev Cell Biol* 9: 317–343.

Tracey KJ, Wei H, Manogue KR, Fong Y, Hesse DG, Nguyen HT, Kuo GC, Beutler B, Cotran RS, Cerami A and Lowry SF (1988) Cachectin/tumor necrosis factor induces cachexia, anemia, and inflammation. *J Exp Med* 167: 1211–1227.

Tracey KJ, Morgello S, Koplin B, Fahey TJIII, Fox J, Aledo A, Manogue KR and Cerami A (1990) Metabolic effects of cachectin/tumor necrosis factor are modified by site of production. *J Clin Invest* 86: 2014–24.

Traenckner EB-M, Wilk S and Baeuerle PA (1994) A proteasome inhibitor prevents activation of NF-NF-κB and stabilizes a newly phosphorylated form of IkB-a that is still bound to NF-NF-κB. *EMBO J* 13: 5433–5441.

Treier M, Staszewski LM and Bohmann D (1994) Ubiquitin-dependent c-Jun degradation *in vivo* is mediated by the delta domain. *Cell* 78: 787–798.

Voth R, Rossol S, Klein K, Hess G, Schutt KH, Schroder HC, Büschenfelde K-HMZ and Müller WEG (1990) Differential gene expression of IFN-a and tumor necrosis factor-a in peripheral blood mononuclear cells from patients with AIDS related complex and AIDS. *J Immunol* 144: 970–975.

Waage A, Halstensen A and Espevik T (1987) Association between tumour necrosis factor in serum and fatal outcome in patients with meningococcal disease. *Lancet* i: 355–357.

Wing SS and Goldberg AL (1993) Glucocorticoids activate the ATP-ubiquitin-dependent proteolytic system in skeletal muscle during fasting. *Am J Physiol* 264: E668–E676.

Wong GHW, Elwell JH, Oberley LW and Goeddel DV (1989) Manganous superoxide dismutase is essential for cellular resistance to cytotoxicity of tumor necrosis factor. *Cell* 58: 923–931.

Xanthoudakis S and Curran T (1992) Identification and characterization of Ref-1, a nuclear protein that facilitates AP-1 DNA-binding activity. *EMBO J* 11: 653–665.

Xanthoudakis S, G Miao F Wang YCE Pan and Curran T (1992) Redox activation of Fos-Jun DNA binding activity is mediated by a DNA repair enzyme. *EMBO J* 11: 3323–3335.

Yoneda T, Alsina MA, Chavez JB, Bonewald L, Nishimura R and Mundy GR (1991) Evidence that tumor necrosis factor plays a pathogenetic role in the paraneoplastic syndromes of cachexia, hypercalcemia, and leukocytosis in a human tumor in nude mice. *J Clin Invest* 87: 977–985.

Free radicals and antioxidants in the pathogenesis of alcoholic myopathy

V.R Preedy[1], M.E. Reilly[1], D. Mantle[2] and T.J. Peters[1]

[1]Department of Clinical Biochemistry, King's College School of Medicine and Dentistry, Bessemer Road, London SE5 9PJ, UK
[2]Neurochemistry Department, Regional Neurosciences Centre, Newcastle General Hospital, Newcastle Upon Tyne, NE4 6BE, UK

Summary. Between one-third and two-thirds of all alcohol abusers have impairments in either muscle function, histology and/or muscle biochemistry, i.e. alcohol-induced muscle disease (AIMD). The chronic form of these lesions (chronic alcoholic myopathy, CAM) are generally characterised by selective atrophy of type II fibres. Affected subjects with CAM loose up to one third of the entire musculature. Clinical studies reveal a lack of correlation with either liver dysfunction, presence of neuropathy or general malnutrition. The activities of plasma carnosinase, an enzyme that cleaves the putative imidazole dipeptide anti-oxidant carnosine, is reduced in myopathic alcoholics, although the pathogenic basis of this is unknown. Plasma selenium and α–tocopherol are also reduced in patients with CAM. Overall, these data implicate free radical-mediated reactions in the pathogenesis of the myopathy, though detailed biochemical analysis in the clinical setting is lacking.
Alcohol-feeding studies in laboratory animals can reproduce the muscle lesions in CAM and have provided a greater insight into the mechanisms responsible for AIMD. In rats, certain anatomically-distinct skeletal muscles are strikingly rich in both type I and type II fibres, e.g. the soleus and plantaris, respectively. There is greater antioxidant capacity in type I fibre predominant muscles which may explain their resilience to ethanol toxicity. However, there are exceptions to the above, since the imidazole dipeptides are higher in type II fibre-rich muscles. In chronic ethanol fed rats, muscle levels of α-tocopherol are unaltered and dosing for 5 days with α-tocopherol is ineffective in ameliorating the protein synthetic lesions induced by acute ethanol administration. Furthermore, the imidazole dipeptide contents are not altered in plantaris muscle of ethanol-fed rats. In rats dosed acutely with ethanol the protein carbonyl concentrations (an index of oxidative damage to muscle) are reduced, with concomitant falls in protein synthesis. Overall, the data suggests that either (i) reactive oxygen species (ROS) are not directly responsible for the myopathy due to ethanol or (ii) AIMD is multifaceted, wherein the ROS are only a component of the lesion.

Introduction

Skeletal muscle comprises 40% of whole-body weight and is therefore a major contributor to whole-body protein metabolism. Thus, alcohol's effects on this tissue are not inconsequential and this is reflected in the fact that alcoholics frequently have disturbances in whole-body nitrogen economy; about half of chronic alcoholics display difficulties in gait and reduced muscle strength. Loss of lean tissue diminishes the patients capacity to respond to infections and trauma (Rennie, 1993).

Although estimates of ethanol misuse vary, there appears to be a significant proportion of patients with alcoholic muscle disease in all clinical populations, although assessed by diverse diagnostic criteria. Thus, between one-third and two-thirds of chronic alcoholics have skeletal muscle myopathy (reviewed in Preedy et al., 1994a, b). The diameter of fast-twitch or type II skeletal muscle fibres (with anaerobic, glycolytic metabolism) are particularly reduced. Type IIb fibres, which have few or no mitochondria, are more affected than the intermediate type IIa fibres, whereas slow-twitch or type I fibres (aerobic, oxidative) are relatively resilient (Slavin et al., 1983; Martin et al., 1985). Inflammation and fibrosis are not commonly found in biopsies (though there may be enhancement in lipid deposition; Sunnasy et al., 1993). There is no overt membrane damage, as reflected by plasma creatine kinase activities, except when an acute myopathy is super-imposed on the chronic myopathy (Del Villar Negro et al., 1984; Martin et al., 1985). The mid-arm circumference is reduced in affected patients and whole-body muscle mass, as assessed by the urinary creatinine/height index, is also reduced (Duane and Peters, 1988a). Thus alcoholic myopathy affects all muscle groups although the disease has been classified as a proximal myopathy. Acute clinical presentations (i.e. rhabdomyolysis with increased risk of renal failure) are infrequent (affecting less than 5% of chronic ethanol misusers) (Martin et al., 1985).

The myopathy is not the result of neurological involvement, raised cortisol levels, or liver disease (reviewed in Preedy et al., 1994a, b). Plasma carnosinase activities are, however, reduced in myopathic alcoholics (Duane and Peters, 1988b). The normal biochemical and physiological functions of carnosinase (which hydrolyses the dipeptide carnosine to β-alanine and histidine) are unknown, but it is claimed that carnosine is an important intracellular buffering agent as well as an antioxidant, particularly in type II fibres (Boldyrev et al., 1988; Quinn et al., 1992). The antioxidant properties of carnosine and other imidazole dipeptides have been challenged, albeit for *in vitro* studies (Aruoma et al., 1989).

The nutritional status (assessed as folate, pyridoxine, riboflavin, thiamine, vitamin B_{12} and vitamin D, and general nutritional intakes) are similar in alcoholics with myopathy compared with those with no skeletal muscle involvement (reviewed in Preedy et al., 1994a, b). Serum α-toco-pherol (Tab. 1) and selenium (Tab. 2) concentrations are, however, low in myopathic alcoholics compared with either non-misusing controls or non-myopathic alcoholics (Tabs 1 and 2) (Ward and Peters, 1992). Similar conclusions are obtained for α-tocopherol when data are expressed as a ratio to plasma lipid (α-tocopherol is transported by both low-density (LDL) and very-low-density lipoproteins (VLDL), although the latter are of less significance). There was no signi-

Table 1. Plasma α-tocopherol, β-carotene and retinol in alcohol misuse

	Concentration (μM)	Statistics (p)
α-Tocopherol		
Controls (non-misusers)	18.30 ± 1.27	
Non-myopathic alcohol misusers	18.90 ± 3.16	NS
Myopathic alcohol misusers	8.30 ± 0.98	$p<0.01$
β-Carotene		
Controls (non-misusers)	0.45 ± 0.06	
Non-myopathic alcohol misusers	0.26 ± 0.03	NS
Myopathic alcohol misusers	0.22 ± 0.05	NS
Retinol		
Controls (non-misusers)	3.80 ± 0.60	
Non-myopathic alcohol misusers	4.30 ± 0.63	NS
Myopathic alcohol misusers	2.30 ± 0.31	NS

All data are mean $\pm$ SEM of 10 observations in each group. Statistics (p values) pertain to differences from controls (non-misusers). NS, not significant; $p>0.05$. From Ward and Peters (1992).

ficant correlation of plasma selenium levels with severity of liver disease (categorized as either fatty, fibrotic or cirrhotic) (Ward and Peters, 1992). In contrast, plasma α-tocopherol was reduced with increasing severity of hepatic dysfunction, especially with cirrhosis (Ward and Peters, 1992). This effect of α-tocopherol and selenium appears to be specific, as neither β-carotene, retinol, copper and zinc are significantly altered in myopathic compared with non-myopathic alcoholics (Tabs 1 and 2) (Ward and Peters, 1992). The supposition that reduced selenium and α-tocopherol are not a consequence of a generalized malnutrition is supported by measurement of red cell transketolase (thiamine, vitamin B_1), glutathione reductase (riboflavin, vitamin B_2), aspartate aminotransferase (pyridoxine, vitamin B_6), folic acid and vitamin B_{12}; none of these indices is significantly different in myopathic compared to non-myopathic alcoholics (Ward and Peters, 1992).

Both selenium and α-tocopherol are dietary antioxidants. There is a correlation between reduced plasma selenium and depressed glutathione peroxidase activities (Lang et al., 1987). It is not known whether the reduced plasma α-tocopherol and selenium of myopathic alcoholics are a "cause or effect" although studies in laboratory rats have provided greater insight into this disease (see below).

Effects of selenium and α-tocopherol deficiencies on skeletal muscle

The mechanisms whereby α-tocopherol and selenium deficiencies affect skeletal muscle are unknown, although, in general, reduced levels of antioxidants may occur. Both α-tocopherol and selenium deficiencies induce skeletal muscle myopathies, although there are few studies on

Table 2. Plasma selenium, copper and zinc in alcohol misusers

	Concentration (μM)	Statistics (p)
Selenium		
Controls (non-misusers)	1.39±0.14	
Non-myopathic alcohol misusers	1.12±0.16	NS
Myopathic alcohol misusers	0.66±0.09	$p < 0.01$
Copper		
Controls (non-misusers)	19.30±1.64	
Non-myopathic alcohol misusers	14.90±0.95	NS
Myopathic alcohol misusers	17.50±1.14	NS
Zinc		
Controls (non-misusers)	12.00±0.41	
Non-myopathic alcohol misusers	9.20±0.85	NS
Myopathic alcohol misusers	8.30±0.70	NS

All data are mean±SEM of 10 observations in each group. Statistics (p values) pertain to differences from controls (non-misusers). NS, not significant; $p > 0.05$. From Ward and Peters (1992).

selenium deficiency as such. In many instances, the effects of selenium deficiency have been combined with α-tocopherol deficiency. For example, Van Vleet (1975) fed dogs diets deficient in both selenium and α-tocopherol and showed that, after about 2 months, marked skeletal muscle abnormalities were evident, including white streaks (indicative of myodegeneration) and necrosis with mixtures of intact, degenerated and regenerating fibres. In addition there were marked cardiac lesions localized to the subendocardial regions. Muscle weakness was also a distinguishing feature of chronic selenium and α-tocopherol deficiency, as well as anorexia (Van Vleet, 1975). The white streaks described by Van Vleet (1975) are a component of the condition *white muscle disease*, a degenerative disease seen in farm animals, such as calves, lambs and foals. This arises as a result of both α-tocopherol and selenium deficiency, possibly as result of poor soil selenium and/or enhanced oxidation of feedstuffs or poor diet (Higuchi et al., 1989; Hoshino et al., 1989; Osame et al., 1990).

The hypothesis to explain the myopathy

The assumption that reactive oxygen species (ROS) induces the myopathy by causing damage to the muscle is too simplistic. If free radicals are involved, then they must encompass a relationship with protein turnover to explain the reduced muscle protein content. The hypothesis that ROS are involved in alcohol-induced muscle disease (AIMD) is not new because over a decade ago, Garcia-Bunuel (1984) hypothesized that the ethanol-induced damage to skeletal (and heart) muscle results from ROS. To explain the mechanisms of the myopathy, two criteria must be met (and to a certain extent must be classified as scientifically non-negotiable):
1. The myopathy must be explained by defects in protein turnover, i.e. either protein synthesis and/or protein degradation.
2. Any mechanism must include the element of differential sensitivity, i.e. the preferential effects on type II fibres.

Animal studies in alcohol misuse

The myopathic lesions seen in alcohol abusers can be reproduced in laboratory rats fed ethanol as 35% of total energy. The weight of plantaris muscle, which contains a predominance of type II fibres, is more reduced than soleus muscle which contains a predominance of type I fibres. Even detailed analysis of the few type II fibres within the soleus indicates that the myopathy preferentially affects the fast-twitch, anaerobic fibres (Preedy et al., 1994a, b). This lends support to the concept of analysing anatomically distinct muscles containing a predominance of specific fibre types in the elucidation of the mechanisms of alcoholic myopathy. Certainly, Radák et al. (1994) have argued that, in terms of antioxidant systems, rats and humans both respond to exercise in a similar manner.

Effects on protein synthesis

The effects of ethanol on skeletal muscle protein synthesis have been detailed elsewhere (Preedy et al., 1994a, b). Very simply, ethanol reduces skeletal muscle protein synthesis in acutely and chronically ethanol-dosed rats: Protein synthesis is usually measured as a fractional rate, i.e. k_s, defined as the percentage of tissue proteins renewed each day or %/day. The fall in protein synthesis is greater in plantaris, compared with soleus muscle, although the differences are small. This small difference will nevertheless have cumulative implications. The use of metabolic inhibitors has facilitated studies into the role of acetaldehyde in mediating ethanol's deleterious effects. Thus, 4-methylpyrazole inhibits alcohol dehydrogenase reducing acetaldehyde formation; cyanamide inhibits acetaldehyde dehydrogenase raising acetaldehyde levels after a dose of ethanol. Use of these metabolic inhibitors in acute dosing studies has shown that both alcohol and acetaldehyde are pertubants of protein synthesis, contradicting the notion that alcohol itself is the causative agent (Preedy et al., 1992).

Antioxidant status in skeletal muscles

Skeletal muscle contains a variety of defence systems to counteract the damaging effects of ROS. Type I skeletal muscle fibres (i.e. the protected fibres in AIMD) have a higher antioxidant capacity and contain larger amounts of cytosolic and mitochondrial superoxide dismutase, catalase, glutathione peroxidase and α-tocopherol than muscles with predominantly type II-fibres (Asayama et al., 1986, 1987; Laughlin et al., 1990). This may reflect the fact that free radical formation occurs via the respiratory chain.

Table 3. Changes in protein synthesis in different tissues after an acute bolus of ethanol

	Mean α-tocopherol concentration (μg/g)	Synthesis rates in control rats[a] (k_s, %/day)	Mean change in k_s (%/day) after ethanol	Statistics (p)
Tissue				
Plantaris	5.8	16.9±0.5	−30	<0.001
Gastrocnemius	6.7	14.8±0.4	−29	<0.001
Soleus	9.3	22.5±0.7	−22	<0.001
Heart	17.9	18.4±0.3	−21	<0.001
Liver	15.0	86.2±2.2	−10	NS

Data (mean ± SEM, $n = 6$–9) are compiled from various studies using rats weighing about 150 g. Rats were injected intraperitoneally with ethanol at a dose of 75 mmol/kg body weight, and protein synthesis was measured after 2.5 h. k_s, fractional rate of protein synthesis (percentage of the tissue protein pool renewed each day).
[a] Control rats are those rats injected with saline in the same experimental study investigating the effects of an acute bolus of ethanol.
Statistics (p values) refer to differences from saline-injected control rats. NS, not significant; $p > 0.05$.

Table 4. The effect of chronic alcohol feeding for 6 weeks on levels of α-tocopherol in gastrocnemius muscle

	Control	Alcoholic	Statistics (p)
Muscle weight (mg wet weight)	1980±40	1620±80	<0.01
α-Tocopherol concentrations (μg/g wet weight)	4.8±0.8	6.8±1.0	NS
Total α-tocopherol contents (μg/muscle)	9.4±1.6	11.3±1.5	NS

All data are means±SEM of four to six pairs of rats. Rats were fed ethanol-containing diets as 36% of total energy for 6 weeks, and controls were fed identical amounts of the same diet in which ethanol was replaced by isocaloric glucose. Statistics (p values) refer to differences from glucose-fed controls and alcohol-fed rats. Muscle was represented by the gastrocnemius. NS, not significant; $p > 0.05$. From M.E. Reilly, H. Ansell, T.J. Peters and V.R. Preedy (unpublished data).

In this review the data for α-tocopherol have been presented in various muscles of the rat (Tab. 3, M.E. Reilly, H. Ansell, T.J. Peters and V.R. Preedy, unpublished data). The validity of these measurements is confirmed by the fact that gastrocnemius α-tocopherol levels in these studies (6.7 μg/g wet weight) are similar to the mean levels of 6.8 μg/g gastrocnemius muscle, in the rat studies of Reznick et al. (1992). The present authors' studies show that α-tocopherol levels in plasma (4.3 μg/ml) are lower than in muscle. In individual muscles, the following values were obtained (data as μg/g wet weight): plantaris (type II), 5.8; gastrocnemius (mixture of types I and II), 6.7; soleus, (type I) 9.3; heart (oxidative or super type I), 17.9. The liver had levels of 15 μg/g, which is similar to those in heart (M.E. Reilly, H. Ansell, T.J. Peters and V.R. Preedy, unpublished data). There is a relationship between rates of protein synthesis, the susceptibility to ethanol-induced reductions in protein synthesis and tissue α-tocopherol levels (Tab. 3). However, in rats fed ethanol for 6 weeks, there is no overt reduction in skeltal muscle α-tocopherol concentrations, despite the reductions in muscle weights (M.E. Reilly, H. Ansell, T.J. Peters and V.R. Preedy, unpublished data) (Tab. 4).

Protective effects of α-tocopherol and selenium

Various studies have indicated that supplementation of α-tocopherol prevents damage to tissue by ROS, for example, the effects of thyroxine-induced lipid peroxidation in oxidative muscle (Asayama et al., 1989) and the effect of exercise in oxidative damage to proteins (Reznick et al., 1992). These studies indicate that α-tocopherol supplementation may be beneficial to muscle, although this is not consistent. Thus, dietary supplementation of rats with α-tocopherol for 5 weeks has no effect on exercise-induced injury to skeletal muscle as defined by maximal tetanic force, the number of intact fibres, elevations in muscle glucose-6-phosphate dehydrogenase activities (an index of inflammation resulting from invasion of phagocytic cells) and plasma creatine kinase activities (Warren et al., 1992).

The studies above draw attention to the fact that a number of indices are employed to assess the effects of damage by ROS. Although thiobarbituric acid-reactive substances (TBARS – an index of lipid peroxidation) are frequently measured, they really pertain to lipid changes. Carbonyl assays are useful in determining damage to proteins themselves. Reznick et al. (1992) determined the carbonyl protein composition in gastrocnemius muscles of rats subjected to exhaustive exercise (1–1.5 h). They showed that protein carbonyl content in rats fed a normal diet was 2.14 ± 0.13 nmol/mg protein (all data as mean $\pm$ SEM, $n = 3–6$). In rats fed a vitamin E supplementation for 4 weeks, the carbonyl content of muscle was reduced to 1.45 ± 0.03 nmol/mg protein ($p < 0.001$ compared with control rats fed the normal diet). In rats fed the normal diet but exercised, the carbonyl concentrations was increased to 2.51 ± 0.06 nmol/mg protein but, in α-tocopherol-supplemented rats, exercise altered the carbonyl concentration to 1.57 ± 0.05 nmol/mg protein ($p < 0.001$ compared with 2.51 ± 0.06) (Reznick et al., 1992). From these studies it can be inferred that carbonyl content is a good index of protein oxidation.

In rats treated with a single bolus of ethanol, there is a reduction in muscle protein synthesis for up to 24 h, compared with saline-injected controls (Tab. 5). At 24 h muscle carbonyl concentrations are reduced, indicating that enhanced ROS generation is not a major effect of experimental muscle injury caused by ethanol, albeit in the post-ethanol phase (all the ethanol disappears 24 h after an acute bolus) (Tab. 6) (M.E. Reilly, D. Mantle, T.J. Peters and V.R. Preedy, unpublished data).

If the reductions in skeletal muscle protein synthesis were the result of a decrease in the availability of α-tocopherol, then supplementation with α-tocopherol should prevent this phenomenon. This has recently been tested by the authors' group. Rats were supplemented with α-tocopherol at a dose of 30 mg/kg per day for 5 days (M.E. Reilly, D. Mantle, T.J. Peters and V.R. Preedy, unpublished data). On day 6, rats were injected with alcohol (75 mmol/kg body weight, i.p.; protein synthesis measured after 2.5 h). Despite this treatment regimen, the ethanol still reduced skeletal muscle protein synthesis in α-tocopherol-supplemented rats (M.E. Reilly, D. Mantle, T.J. Peters and V.R. Preedy, unpublished data). It could be argued that the failure to observe the preventive effects of α-tocopherol on protein synthesis was the result of: (1) the studies being carried out acutely; and (2) tissue α-tocopherol levels not being adequately increased by the treatments. However, with respect to (1) Phoenix et al. (1989) showed that damage (defined in terms of creatine kinase efflux in isolated muscles) could be prevented by α-toco-

Table 5. Time course of ethanol on gastrocnemius muscle protein synthesis after an acute bolus

+		
Time of ethanol exposure		
0 (0.15 M NaCl)	13.3 ± 0.5	–
20 min	13.9 ± 1.9	NS
1 h	9.9 ± 0.2	<0.01
2.5 h	9.6 ± 0.9	<0.01
6 h	7.0 ± 0.8	<0.01
24 h	4.2 ± 0.5	<0.001

Data are mean $\pm$ SEM (4–8) from M.E. Reilly, D. Mantle, T.J. Peters and V.R. Preedy (unpublished data). Statistics (p values) pertain to differences from saline injected controls. k_s, fractional rate of protein synthesis (percentage of the tissue protein pool renewed each day). Rats were injected with ethanol at a dose of 75 mmol/kg body weight, i.p. NS, not significant; $p > 0.05$.

Table 6. Effect of ethanol on gastrocnemius muscle protein synthesis and carbonyl protein at 24 h

	Carbonyl protein (nmol/mg protein)	k_s (%/day)
Control	1.51±0.13	9.2±0.4
Ethanol	0.94±0.2	4.2±0.5
Statistics (P)	<0.05	<0.001

Data are mean±SEM (4–8) from M.E. Reilly, D. Mantle, T.J. Peters and V.R. Preedy (unpublished data). Statistics (p values) pertain to differences from saline-injected controls and ethanol-treated rats (75 mmol/kg body weight, i.p.). k_s, is the fractional rate of protein synthesis (percentage of the tissue protein pool renewed each day). Both these groups of rats were subjected to pair feeding.

pherol within 1 h of administration. In other words, α-tocopherol has the potential to act acutely. With respect to (2) the treatment protocol described above increased plasma and muscle α-tocopherol by about 40% ($p>0.05$; not significant) and in the liver by about 110% ($p<0.01$). There is very little literature on the levels of α-tocopherol necessary to prevent muscle damage. However, the argument that endogenous levels of muscle α-tocopherol in *ad libitum* fed rats are optimal and supplementation will have little effect, is not supported by Reznick et al. (1992), who showed that α-tocopherol supplementation (10 000 IU/kg diet; normal feed has a content of 30 IU/kg diet) of normal rats prevented oxidative damage to proteins, albeit after 4 weeks of treatment. In our studies, α-tocopherol treatment alone increased the synthesis rates of proteins relative to RNA (k_{RNA}, an index of translational efficiency) suggesting that supplementation may have some beneficial effects (M.E. Reilly, H. Ansell, T.J. Peters and V.R. Preedy, unpublished data). Studies are under way to ascertain the effects of more long-term (i.e. 4–6 weeks α-tocopherol supplementation) effects on ethanol-induced defects in muscle protein synthesis.

Imidazole dipeptides

The imidazole dipeptides occur in high concentrations in muscles, particularly those containing a predominance of type II fibres. They act as intracellular buffering agents and carnosine in particular is claimed to be a potent antioxidant. The fact that carnosinase activity is reduced in myopathic alcoholics suggests that the myopathy may be related to the tissue levels of carnosine. Although there may be a neurological basis for the changes in plasma carnosinase activities (Butterworth et al., 1995), reduced activities are also found in other myopathies such as hypothyroidism and muscular dystrophy (Bando et al., 1986; Wassif et al., 1993, 1994). It is thus not unreasonable to suggest that reduced tissue carnosine levels are implicated in the myopathy. This was investigated in laboratory rats fed ethanol-containing diets for 6 weeks (Tab. 7) (Ward and Preedy, 1992). Although there was a marked reduction in the weight of type II fibre-rich plantaris, carnosine and anserine concentrations (μmol/g wet weight) were increased (not shown) although the total contents of these dipeptides (μmol per muscle) were unaltered (Tab. 7) (Ward and Preedy, 1992). This does not support the original hypothesis.

Table 7. The effect of chronic alcohol feeding on levels of imidazole dipeptides

	Dipeptide content (μmol/muscle)		
	Control	Ethanol	Statistics (p)
Soleus			
Carnosine	0.05±0.01	0.05±0.01	NS
Anserine	0.04±0.01	0.03±0.01	<0.05
Carnosine plus anserine	0.09±0.01	0.07±0.01	<0.05
Plantaris			
Carnosine	0.15±0.01	0.16±0.01	NS
Anserine	0.15±0.01	0.13±0.01	NS
Carnosine plus anserine	0.29±0.02	0.29±0.02	NS

All data are means±SEM of four to six pairs of rats. Rats were fed ethanol-containing diets as 36% of total energy for 6 weeks, and controls were fed identical amounts of the same diet in which ethanol was replaced by isocaloric glucose. Statistics (p values) refer to differences from glucose-fed controls and alcohol-fed rats. NS, not significant; $p > 0.05$. From Ward and Preedy (1992).

The concept that the slow-twitch muscle has an inherent defence mechanism against ROS as a result of the high concentration of antioxidants is supported by the observation that most myopathic lesions are type II fibres predominant. However, there are instances when type I fibres are preferentially affected. In chronic hypoxic stress (6 months of intermittent simulation of altitudes of 4000 metres), a significant increase in TBARS occurs in type I fibre-predominant soleus, coupled with decreased amounts and activities of the mitochondrial isoform of superoxide dismutase (MnSOD: Radák et al., 1994). This may reflect reduced availability of oxygen to the mitochondria which leads to down-regulation of MnSOD activity. The specificity of the hypoxic effect was supported by the observation that increased levels of TBARS and reduced levels of MnSOD were not observed in tibialis anterior, classified by the authors as a type II muscle, of treated rats. Tibialis anterior has also been used to represent type II fibres or fast-twitch muscle in numerous other studies, for example, Wehrle et al. (1994). The activities of the non-mitochondrial isoform of superoxide dismutase (CuZnSOD), glutathione peroxidase and catalase in both soleus and tibialis anterior were not significantly altered (Radák et al., 1994).

Conclusion

Alcohol-induced lesions in skeletal muscle arise as a consequence of reduced protein synthesis although the mechanisms are poorly understood. Tentatively, a mechanism for the involvement of ROS has been forwarded and supported by clinical studies showing reduced serum levels of α-tocopherol and selenium. These observation are not consistent with the observation that (1) α-tocopherol loading studies are ineffective in ameliorating the effects of ethanol on protein synthesis and (2) protein carbonyl content is reduced when there is a marked fall in muscle protein synthesis in ethanol-treated rats. There is a clear need to extend these studies to investigate the assessment of oxidative damage to muscle itself, and to examine the relationship between long-term deficiencies of α-tocopherol and susceptibility to alcohol-induced muscle lesions.

Acknowledgments
Part of this work was carried out in the Rayne Institute, 123 Coldharbour Lane, London SE5 9NU. M.E. Reilly is supported by the Joint Research Committee (JRC) of Kings College School of Medicine and Dentistry.

References

Aruoma OI, Laughton MJ and Halliwell B (1989) Carnosine, homocarnosine and anserine: could they act as antioxidants *in vivo*? *Biochem J* 264: 863–869.

Asayama K, Dettbarn WD and Burr IM (1986) Differential effects of denervation on free-radical scavenging enzymes in slow and fast muscle of the rat. *J Neurochem* 46: 604–609.

Asayama K, Dobashi K, Hayashibe H, Megata Y and Kato K (1987) Lipid peroxidation and free radical scavengers in thyroid dysfunction in the rat: a possible mechanism of injury to heart and skeletal muscle in hyperthyroidism. *Endocrinology* 121: 2112–2118.

Asayama K, Dobashi K, Hayashibe H and Kato K (1989) Vitamin E protects against thyroxine-induced acceleration of lipid peroxidation in cardiac and skeletal muscles in rats. *J Nutr Sci Vitaminol* 35: 407–418.

Bando K, Ichihara K, Shimotsuji T, Toyoshima H, Koda K, Hayashi C and Miyai K (1986) Reduced serum carnosinase activity in hypothyroidism. *Annal Clin Biochem* 23: 190–194.

Boldyrev AA, Dupin AM, Siambela M and Stvolinsky SL (1988) The level of natural antioxidant glutathione and histidine-containing dipeptides in skeletal muscles of developing chick embryos. *Comp Biochem Physiol* 89B: 197–200.

Butterworth RJ, Wassif WS, Gerges A, Sherwood RA, Peters TJ and Bath PMW (1995) Serum carnosinase activity is reduced in acute stroke (abstract). *Cerebrovasc Dis* 5: 246.

Del Villar Negro A, Angulo J M and Rivera-Pomar JM (1984) Skeletal muscle changes in chronic alcoholic patients. A conventional, histochemical ultrastructural and morphometric study. *Acta Neurol Scand* 70: 185–196.

Duane P and Peters TJ (1988a) Nutritional status in alcoholics with or without skeletal muscle myopathy. *Alcohol Alcoholism* 23: 271–277.

Duane P and Peters TJ (1988b) Serum carnosinase activities in patients with alcoholic chronic skeletal muscle myopathy. *Clin Sci* 75: 185–190.

Garcia-Bunuel L (1984) Lipid peroxidation in alcoholic myopathy and cardiomyopathy. *Med Hypoth* 13: 217–231.

Higuchi T, Ichijo S, Osame S and Ohishi H (1989) Studies on serum selenium and tocopherol in white muscle disease of foal. *Jpn J Vet Sci* 51: 52–59.

Hoshino Y, Ichijo S, Osame S and Takahashi E (1989) Studies on serum tocopherol, selenium levels and blood glutathione peroxidase activities in calves with white muscle disease. *Jpn J Vet Sci* 51: 741–748.

Lang JK, Gohil K, Packer L and Burk RF (1987) Selenium deficiency, endurance exercise capacity, and antioxidant status in rats. *J Appl Physiol* 63: 2532–2535.

Laughlin MH, Simpson T, Sexton WL, Brown OR, Smith JK and Korthius RJ (1990) Skeletal muscle oxidative capacity, antioxidant enzymes, and exercise training. *J Appl Physiol* 68: 2337–2343.

Martin F, Ward K, Slavin G, Levi J and Peters TJ (1985) Alcoholic skeletal myopathy, a clinical and pathological study. *Quart J Med* 55: 233–251.

Osame S, Ohtani T and Ichijo S (1990) Studies on serum tocopherol and selenium levels and blood glutathione peroxidase activities in lambs with white muscle disease. *Jpn J Vet Sci* 52: 705–710.

Phoenix J, Edwards RH and Jackson MJ (1989) Inhibition of Ca^{2+}-induced cytosolic enzyme efflux from skeletal muscle by vitamin E and related compounds. *Biochem J* 257: 207–213.

Preedy VR, Keating JW and Peters TJ (1992) The acute effects of ethanol and acetaldehyde on rates of protein synthesis in Type I and Type II fibre-rich skeletal muscles of the rat. *Alcohol Alcoholism* 27: 241–251.

Preedy VR, Peters TJ, Patel VB and Miell JP (1994b) Chronic alcoholic myopathy: transcription and translational alterations. *FASEB J* 8: 1146–1151.

Preedy VR, Salisbury JR and Peters TJ (1994a) Alcoholic muscle disease: features and mechanisms. *J Pathol* 173: 309–315.

Quinn PJ, Boldyrev AA and Formazuyk VE (1992) Carnosine: its properties, functions and potential therapeutic applications. *Molec Aspects Med* 13: 379–444.

Radák Z, Lee K, Choi W, Sunoo S, Kizaki T Oh-ishi S, Suzuki K, Taniguchi N, Ohno H and Asano K (1994) Oxidative stress induced by intermittent exposure at a simulated altitude of 4000 m decreases mitochondrial superoxide dismutase content in soleus muscle of rats. *Eur J Appl Physiol* 69: 392–395.

Rennie MJ (1993) Lean tissue wasting in critically ill patients – is it preventable. *Brit J Intensive Care* 3: 139–147.

Reznick AZ, Witt E, Matsumoto M, Packer L (1992) Vitamin E inhibits protein oxidation of skeletal muscle of resting and exercised rats. *Biochem Biophys Res Commun* 189: 801–806.

Slavin G, Martin F, Ward P, Levi J and Peters T (1983) Chronic alcohol excess is associated with selective but reversible injury to type 2B muscle fibres. *J Clin Pathol* 36: 772–777.

Sunnasy D, Cairns SR, Martin F, Slavin G and Peters T J (1983) Chronic alcoholic skeletal myopathy: a clinical, histological and biochemical assessment of muscle lipid. *J Clin Pathol* 36: 778–784.

Van Vleet JF (1975) Experimentally induced vitamin E-selenium deficiency in the growing dog. *J Am Vet Med Assn* 166: 769–774.

Ward RJ and Peters TJ (1992) The antioxidant status of patients with either alcohol-induced liver damage or myopathy. *Alcohol Alcoholism* 27: 359–365.

Ward RJ and Preedy VR (1992) Imidazole dipeptides in experimental alcohol-induced myopathy. *Alcohol Alcoholism* 27: 633–639.

Warren JA, Jenkins RR, Packer L, Witt EH and Armstrong RB (1992) Elevated muscle vitamin E does not attenuate eccentric exercise-induced muscle injury. *J Appl Physiol* 72: 2168–2175.

Wassif WS, Preedy VR, Summers B, Duane P, Leigh N and Peters TJ (1993) The relationship between muscle fibre atrophy factor, plasma carnosinase activities and muscle RNA and protein composition in chronic alcoholic myopathy. *Alcohol Alcoholism* 28: 325–331.

Wassif WS, Sherwood RA, Amir A, Idowa B, Summers B, Leigh N and Peters TJ (1994) Serum carnosinase activities in central nervous system disorders. *Clin Chim Acta* 22: 57–64.

Wehrle U, Dusterhoft S and Pette D (1994) Effects of chronic electrical stimulation on myosin heavy chain expression in satellite cell cultures derived from rat muscles of different fiber-type composition. *Differentiation* 58: 37–46.

Drug-induced muscle damage

G.A. Brazeau

Department of Pharmaceutics, College of Pharmacy, University of Florida Box 100494 J.H.M.H.C., Gainesville, Florida 32610, USA

Summary. In recent years, numerous reports in the clinical and basic science literature have suggested that systemic drug administration can cause damage to skeletal muscle. The extent of skeletal muscle damage can range from that which is minor and self-limiting following drug discontinuation, to muscle weakness which may or may not be ameliorated following drug discontinuation, to rhabdomyolysis which, if not treated, can lead to acute renal failure and even death. The mechanisms of drug-induced toxicity are numerous and can range from disruption of intracellular calcium homeostasis, to depletion of cellular energy sources, to free radical mediated mechanisms. This chapter will examine specifically those drugs or substances of abuse which have been shown to cause toxicity to skeletal muscle.

Skeletal muscle damage, myotoxicity, myopathy, myalgia and rhabdomyolysis

Lane and Mastaglia (1978) provided an excellent description of the many features of drug-induced muscular syndromes. These authors discussed the toxicity of drugs on skeletal muscle by dividing the effects into the following types: (1) focal myopathy; (2) muscle fibrosis with contracture; (3) acute/subacute painful proximal myopathy; (4) acute rhabdomyolysis, subacute/chronic painless proximal myopathy; (5) myasthenic syndromes; (6) polymyositis/dermatomyositis syndromes; (7) myotonic syndrome; and (8) malignant hyperpyrexia. Le-Quintrec and Le-Quintrec (1991) have proposed an alternative classification scheme of drug-induced muscle damage based upon histological classification into: (1) vacuolar muscle damage; (2) mitochondrial myopathy; or (3) a necrotizing myopathy. The reader is referred to these two references for more details about these two different classification systems and the drugs in the various groups.

A review of this and more recent literature reveals, however, that the terminology used to describe the toxic effects of drugs on skeletal muscle is confusing, therefore clouding the interpretation and comparison between reported studies. In this work, "drug-induced muscle damage" or "toxicity" or "myotoxicity" is broadly defined to mean a reversible or irreversible change in the anatomy, biochemistry, and/or physiology of the muscle specifically associated with drug administration. The key issue in determining the presence of drug-induced muscle damage is whether this condition is secondary to another primarily pathological condition suffered by the patient, (e.g. AIDS). Consequences of skeletal muscle damage are myalgia (namely pain in skeletal muscle tissue) or muscle weakness and/or atrophy. Drug-induced muscle damage can occur either after systemic drug administration (e.g. oral, intravenous injection, intravenous infusion) or directly after an intramuscular injection. In contrast, myopathy is defined as disease involving skeletal muscle (e.g. mitochondrial myopathy, HIV-induced myopathy, muscular dystrophy), and consequently is independent of an externally administered substance of abuse or therapeutic agent. The confusion arises in using the term "myopathy" associated with drug-induced muscle damage (e.g. alcoholic myopathy).

The general mechanisms underlying the development of drug-induced skeletal muscle damage are similar to those responsible for muscle disease; however, in many cases drug-induced skeletal muscle damage is self-limiting once the noxious or toxic agent has been removed from the systemic circulation. In contrast, it is difficult to arrest or retard the progression of disease-induced alterations in skeletal muscle structure or function. Rhabdomyolysis, the most severe muscle toxicity which can be fatal if not treated aggressively, is defined as a process of total muscle tissue dissolution and destruction which causes the release of myoglobin into the systemic circulation. If not treated, the increased serum myoglobin can block the kidney glomeruli, resulting in acute renal failure. This condition is also characterized by very high levels of muscle-derived enzymes in serum.

Presentation and diagnosis of possible drug-induced muscle damage

The clinical presentation of drug-induced muscle damage often includes an elevation in levels of muscle enzymes in the serum (namely creatine kinase, lactate dehydrogenase, aldolase, or myoglobin), generalized muscle weakness as measured by decreased function or mobility, and muscle pain (which may be difficult to discriminate from other pain sources). A key indication of possible drug-induced muscle damage is that these serum enzyme levels decrease after discontinuation of the suspected agent. Although the elevation in the serum levels of muscle-specific enzymes through electrophoresis is often used to determine muscle damage resulting from either drug administration or disease, it must be remembered that muscle damage may or may not be associated with an elevation of these enzymes. In alcoholic men and women about 50% of these patients who underwent muscle biopsy fulfilled the histological criteria for mild-to-moderate muscle damage or myopathy; however, serum creatine kinase levels were elevated in only a quarter to one-third of the patients (Urbano-Marquez et al., 1989; Urbano-Marquez et al., 1995). Furthermore, aldolase levels were increased in 20% of the men and about 50% of the women, whereas about 10% of all patients showed increased myoglobin levels (Urbano-Marquez et al., 1995).

Possible reasons that can account for the non-significant increases in the levels of muscle-derived enzymes in serum include; (1) the half-life of the investigated enzyme in serum for a given individual relative to the time and extent of the toxicity or the blood draw, and/or (2) the presence of other endogenous substances in the serum, which inhibit the activity of these enzymes (often measured using spectrophotometric, kinetic, or colorimetric assays), thus causing the reported levels to be lower than the true levels. For example, assuming that a drug had an immediate toxic effect on rodent skeletal muscle after systemic administration, leading to a rapid rise in creatine kinase, and that collection of a blood sample was made about 15 h after that toxic exposure, the level of this serum enzyme would be about 5% of the original amount released into serum given that the half-life of rat serum creatine kinase activity is reported to be about 2.5 h (Page, 1992). Second, it is possible that endogenous inhibitors or the presence of drug in the serum could interfere with the enzyme activity (Clarkson and Ebbeling, 1988; Brazeau and Fung, 1989).

In clinical reports of drug-induced muscle damage, patients often report cramps and pain, proximal or generalized muscle weakness which interferes with normal daily functioning, contraction and induration of muscles, inability or difficulty with walking and/or the inability to rise from a seated position (Lane and Mastaglia, 1978; DeAngelis et al., 1991; Urbano-Marquez et al., 1995; Suzuki et al., 1997). It may be useful to measure specific muscle function and strength (e.g. deltoid) or conduct electromyographic (EMG) studies, particularly in those patients who will be taking those drugs long-term that may cause damage to skeletal muscle. Muscle damage is best confirmed histologically using paraffin or cryostat sections. Investigators often report their findings as either necrotic and/or degenerative/regenerative. Histological studies present little difficulty in animal studies where the investigator can specifically conduct these analyses on slow-twitch oxidative muscles (e.g. soleus), fast-twitch glycolytic muscles (e.g. extensor digitorum longus muscles) or mixed muscles (tibialis anterior or gastrocnemius). In human studies, muscle biopsies, although not entirely commonplace in clinical studies, are becoming increasingly important as this methodology is improved to obtain better samples with less patient distress and/or discomfort. Tests that have been used to determine muscle damage include hematoxylin and eosin, modified Gomori trichrome, and the periodic acid–Shiff reaction for paraffin sections and myofibrillar ATPase at pH 9.4, succinate dehydrogenase, NADH diaphorase, non-specific esterase activity, and oil-red O (Peña et al., 1990; Urbano-Marquez et al., 1995) for histochemistry using crysotat sections. The presence of phagocytes in the muscle tissue can be identified using acid phosphatase (Peña et al., 1990). Electron microscopy is also an important tool to understand possible mechanisms. Ultrathin sections can be stained with uranyl acetate and lead citrate for visualization to determine ultrastructural changes (Peña et al., 1990; Rago et al., 1994). The diagnosis of muscle damage or myopathy can be made using the criteria proposed by Mastaglia and Walton (1982).

In animal studies, muscle function tests *in situ* and *in vitro* are useful methods to access the degree to which a particular drug has caused damage to skeletal muscle function. Contractile properties in the skeletal muscle can easily be conducted *in situ* or *in vitro* using extensor digitorum longus, tibialis anterior, soleus, and diaphragm muscles. Parameters that can be investigated in these types of studies include maximal twitch tension, optimal length, peak tetanic tension, isometric and isotonic force measurements, and fatigue index. A limitation with the *in vitro* studies is the lack of adequate oxygenation to the entire muscle, which becomes more severe with higher incubation temperatures. The soleus, extensor digitorum longus, and diaphragm muscles are ideal for these *in vitro* experiments because their geometry permits adequate gaseous diffusion within an organ bath as illustrated by their widespread use in *in vitro* experiments (Fitts et al., 1984; Brown et al., 1992; Powers et al., 1996).

Drugs reported to cause skeletal muscle damage with systemic administration

Table 1 is a list of the most commonly reported drugs and substances of abuse that cause muscle damage after oral or intravenous administration, as determined by increased serum enzymes or histological evaluation, combined with or without an assessment of muscle function and strength. This particular review will only focus on those drugs or substances for which there is a substantial

Table 1. Drugs or substances of abuse commonly reported to cause muscle damage with systemic administration

Drug	References
Alcohol	Urbano-Márquez et al. (1989)
	Pendergast et al. (1990)
	Urbano-Márquez et al. (1995)
	Preedy et al. (1994a)
Fibrates	
Clofibrate	Godoy et al. (1992)
Bezafibrate	Vita et al. (1993)
Chloroquine	Parodi et al. (1985)
	Sugita et al. (1987)
	Avina-Zubieta et al. (1995)
	Velasco et al. (1995)
Cocaine	Pagala et al. (1991)
	Pagala et al. (1993)
	Brazeau et al. (1995)
Colchicine	Goodman and Murray (1953)
	Angeven (1957)
	Naldus et al. (1977)
	Murray et al. (1983)
	Collot et al. (1984)
	Kuncl et al. (1987)
	Levy et al. (1991)
Emetine (ipecac)	Sugie et al. (1984)
	Palmer and Guay (1985)
	Halbig et al. (1988)
	Thyagarajan et al. (1993)
Heroin	Richter et al. (1971)
	Schwartzfarb et al. (1977)
	Gibb and Shaw (1985)
	Peña et al. (1990)
HMG-CoA reductase inhibitors	Jingami (1994)
Lovastatin/gemfibrozil	Chucrallah et al. (1992)
Pravastatin/gemfibrozil	Winklund et al. (1993)
Lovastatin/itraconazole	Neuvoen and Jalava (1996)
Lovastatin	Waclawik et al. (1993)
Simvastatin	Veerkamp et al. (1996)
Simvastatin/pravastatin	Pierno et al. (1995)
Simvastatin/pravastatin/lovasatatin	Smith et al. (1991)
Simvastatin/pravastatin	Fukami et al. (1993)
HMG-CoA reductase inhibitors/cyclosporin	Smith et al. (1991)
Leuprolide acetate	Crayton et al. (1991)
Suramin	Rago et al. (1994)
Steroids	Bowyer et al. (1985)
(prednisone, methylprednisolone,	Dekhuijzen and Decramer (1992)
dexamethasone)	Lacomis et al. (1993)

Table 1. (continued)

Drug	References
	vanBalkom et al. (1994)
	Koehler (1995)
	Decramer et al. (1996)
	Nava et al. (1996)
Vincristine/dexamethasone	DeAngelis et al. (1991)
Valproate	Papadimitriou and Servidei (1991)
Zidovudine	Gertner et al. (1989)
	Dalakas et al. (1990)
	Groopman (1990)
	Arnaudo et al. (1991)
	Chalmers et al. (1991)
	Mhiri et al. (1991)
	Weissman et al. (1992)
	Lewis et al. (1992)
	Schroder et al. (1992)
	Manji et al. (1993)
	Modica-Nalolitano (1993)
	Peters et al. (1993)
	Rachlis and Fanning (1993)
	Spadaro et al. (1993)

body of literature. Based on a review of the literature, the lipid-lowering agents, the steroids, zidovudine (azidothymidine or AZT), and alcohol have received the most attention with respect to their potential to cause muscle damage. Table 2 lists other compounds that have been associated

Table 2. Drugs or substances of abuse less commonly reported to cause muscle damage with systemic administration

Drug	References
Cyclosporin A/colchichine	Yaminishi et al. (1993)
Diuretic therapy	Shintani et al. (1991)
Griseofulvin	Deo et al. (1994)
Morphine/dihydrocodeine	Blain et al. (1985)
Methimazole	Suzaki et al. (1997)
Neuromuscular blocking agents	Gooch (1995)
(pancuronium, vecuronium)	Elliot and Bion (1995)
	Zprielipp et al. (1995)
D-Penicillamine	Lund and Nielsen (1983)
Phenytoin	Harney and Glasberg (1983)
Pivampicillin	Rose et al. (1992)

with the development of muscle toxicity. For additional information in this area, readers should consult Lane and Mataglia (1978) or Le-Quintrec and Le-Quintrec (1991). The drugs discussed in this review and/or listed in Tables 1 and 2 should not be confused with other therapeutic agents (e.g. local anesthetics, cephalosporins) which cause pain and/or localized muscle damage after direct intramuscular injection (Yagiela et al., 1982; Gaertner et al., 1987; Arnold et al., 1988; Brazeau et al., 1992).

Mechanism(s) of drug-induced muscle damage

There are several general mechanism(s) that may underlie the development of drug-induced muscle damage. The following are the possible mechanisms:
- Disruption/alteration in intracellular calcium homeostasis
- Disruption/alteration in intracellular energy sources
- Ischemia–reperfusion
- Oxidative stress and free radical generation
- Direct sarcolemma membrane disruption and/or solubilization
- Drug-induced hypokalemia and hypochloremia
- Drug-induced polymyositis or dermatomyositis
- Secondary to drug-induced neuropathy

Additional information on the toxicology of muscle has been provided by Harris and Blain, 1990. Oxidative stress, reactive oxygen species (ROS), and free radicals may be directly responsible for muscle damage via their interactions with lipids causing lipid peroxidation, with proteins causing the formation of protein carbonyls, and with DNA causing the formation of DNA adducts. Alternatively, the formation of free radicals may be secondary to other pathological changes (e.g. disruption of intracellular energy sources). The mechanisms responsible for oxidative stress and the generation of free radicals in drug-induced muscle damage will be discussed in subsequent sections.

It is possible that the toxic effects of some myotoxic drugs or compounds may be a direct function of their ability to disrupt the sarcolemma (outer membrane) causing an alteration in intracellular ion regulation, increased levels of cytosolic calcium, and activation of enzymes responsible for degradative processes (e.g. proteases, phospholipases – see below) (Brazeau and Fung, 1990). Alternatively, other agents may cause myotoxic effects through a disruption of the calcium ion channels and/or pumps in the sarcolemma or sarcoplasmic reticulum, leading to increases in cytosolic calcium from extracellular sources or a redistribution from intracellular sites of calcium sequestration such as the sarcoplasmic reticulum or the mitochondria (Wrogemann and Peña, 1976; Duncan, 1978; Publicover et al., 1978; Benoit et al., 1980). Increased cytosolic calcium has been reported to enhance protein degradation via calcium-dependent proteases, to alter muscle histology, and to activate phospholipases (e.g. phospholipase A_2) (Jackson et al., 1984; Duarte et al., 1992; Belcastro et al., 1996). Toxicity to skeletal muscle is potentiated because this activated phospholipase leads to the release of membrane-bound fatty acids and the formation of lysophospholipids. Furthermore, the activation of phospholipase A_2 leads to the release of membrane-bound arachidonic acid, which is subsequently converted to bioactive molecules such as prostaglandins and leukotrienes by lipoxygenase and cyclo-oxygenase enzymes, respectively. Lysophospholipids, prostaglandins and leukotrienes are known to disrupt skeletal muscle membranes which will, in turn, further amplify intracellular cytosolic calcium levels (Duncan and Jackson, 1987; Jackson et al., 1987).

Drug-induced muscle damage may occur if the toxic agent affects the cellular energetics of the muscle fibre. Byrne et al. (1985) reported an irreversible failure of the isometric twitch tension and the induction of a severe progressive contracture in rodent muscle after the infusion of the uncoupling agent, 2,4-dinitrophenol, intra-arterially into the rat hindlimb. This process was accompanied by low levels of ATP and phosphocreatine, together with lactate accumulation in the muscle. As suggested by this study, if a drug or compound depletes intracellular ATP levels through either a defect in the respiratory chain or an uncoupling of the respiratory chain to the phosphorylation system, this could result in disruption in the functioning of the ion pumps responsible for maintaining intracellular cytosolic calcium homoeostasis and lead to decreased muscle function and/or damage (as described above). In addition, a depletion in the levels of ATP has also been shown to result in increased proteolysis in muscle fibres (Fagan et al., 1992).

Other possible mechanisms of drug-induced muscle damage may be secondary to a toxic effect on adjacent vasculature and nerve tissue. Drug-induced muscle damage may result from an ischemia–reperfusion-like injury if the drug causes vasoconstriction in the muscle-associated vasculature (e.g. cocaine, zidovudine, alcohol). The anoxia in this situation, followed by the subsequent reperfusion, could expose muscle fibres to free radicals generated directly from the tissue or from neutrophils during the reperfusion process. Finally, the close correlation between muscle and nerve tissue requires the investigator to rule out the possibility that the toxic effect on muscle tissue is secondary to the effects on nerves or blood vessels. It has been suggested that certain local anesthetics, when administered locally, are primarily toxic to the muscles, whereas other compounds are toxic to other muscle components, particularly blood vessels and nerves (Foster and Carlson, 1980).

Drug-induced myotoxicity can be the result of alterations in serum electrolytes or depletion of carnitine, for example, drug-induced muscle damage may be secondary to hypokalemia and/or hypochloremia as a result of diuretic therapy (Shiintani et al., 1991). Muscle damage may be secondary to inflammatory processes. Polymyositis and dermatomyositis are muscle disorders that are characterized by lymphocytic infiltration which leads to regeneration and vacuolar changes (Bohan et al., 1977; Hochberg et al., 1986; Crayton et al., 1991).

Drug-induced muscle and role of free radicals

Skeletal muscle, like other tissue, may be suspectible to the toxic effects of free radicals which can cause the formation of lipid peroxides, protein carbonyls, or DNA adducts. The role of free radicals as a possible mechanism underlying the development of muscular dystrophy, malignant hyperthermia, alcoholic myopathy, inflammatory myopathies, and ischemia-reperfusion injury has been discussed by Jackson and O'Farrell (1993). For example, recent studies have suggested the presence of elevated oxyradical production in muscular dystrophies as a function of the lack of dystrophin in the *mdx* mouse (Hauser et al., 1995). Furthermore, it has been suggested that oxidative stress and the formation of mitochondrial DNA deletions may be responsible for the sarcopenia or muscle wasting associated with the ageing process (Weindruch, 1995; Wei et al., 1996). It would not be unexpected for drug-induced skeletal muscle damage to be mediated via oxidative stress or ROS (which in turn can lead to the formation of other free radicals) because: (1) skeletal muscle at rest accounts for a large share of the body's total O_2 consumption; (2) muscles, in particular working or exercising muscles, are exposed to higher levels of O_2 consumption; (3) muscles may be more susceptible to ischemia–reperfusion injury because of the nature of sympathetic intervention of muscle vasculature (susceptibility to constriction), combined with the fact that skeletal muscle is particularly susceptible to phospholipid oxidation during reperfusion injury (Lindsay et al., 1988); and (4) the decreased repaired capacity of skeletal muscle (compared with other more mitotically active tissues) (Weindruch, 1995). In a recent study, a free radical and/or lipid peroxidation-based mechanism has been proposed for the toxic effects of the diisopropylphosphofluoridate-induced muscle necrosis (Yang and Dettbarn, 1996). Likewise, a free radical-based mechanism has been proposed for the toxicity of alcohol, cocaine, and zidovudine on skeletal muscle (see subsequent discussion of these individual compounds).

Zidovudine (AZT)

Over the past 10 years, zidovudine has clearly been the most extensively studied drug with respect to its ability to cause toxicity to skeletal muscle. One difficulty in interpreting these studies has focused on whether the drug versus the disease process in HIV patients is responsible for the development of myopathy in the muscle (Manji et al., 1993). Gherardi (1994) proposed a classification system of the muscle involvement in HIV patients which included: (1) an HIV myopathy that meets the critieria of a polymyositis in most patients; (2) a zidovudine toxicity to muscle which appears to be a reversible effect on the mitochondria; (3) an HIV-wasting syndrome and AIDS-associated cachexias; (4) opportunistic infections and tumoral infiltration; and (5) vasculitic processes and iron pigment deposits. There are numerous reports of zidovudine-induced muscle damage that is dose and time dependent and has been characterized by proximal weakness and muscle tenderness elevated creatine kinase and aldolase levels, abnormal EMG activity, widespread focal necrosis, and ragged-red fibres indicative of abnormal mitochondria (Helbert et al., 1988; Gertner et al., 1989; Dalakas et al., 1990; Groopman, 1990; Panegyres et al., 1990; Chalmers et al., 1991; Chen et al., 1992; Schroder et al., 1992; Peters et al., 1993; Rachlis and Fanning, 1993; Spadaro et al., 1993).

Studies on zidovudine-induced muscle damage suggest that the underlying mechanism is mitochondrial myopathy. This is characterized by depletion of mitochondrial DNA (mtDNA), mtRNA enzymatic defects in the respiratory chain system (e.g. NADH cytochrome *c* reductase, cytochrome *c* oxidase, citrate synthase), delayed resynthesis of phosphocreatine and impaired synthesis of ATP and accumulation of lipid droplets (Arnuado et al., 1991; Mhiri et al., 1991; Pezeshkpour et al., 1991; Lewis et al., 1992; Weissman et al., 1992; Modica-Napolintano, 1993; Semino-Moro et al., 1994; Sinnwell et al., 1995). The toxicity to muscle cells has been shown to be reduced with treatment by L-carnitine (Semino-Mora et al., 1994). Alternative hypotheses to explain zidovudine-induced muscle damage include a secondary additional effect associated with vascular change leading to ischemia (Chairot et al., 1995) and a decrease in the proliferation of muscle cells (Herzberg et al., 1992).

Lipid lowering agents: HMG-CoA reductases, fibrate derivatives and valproate

In the last few years, no other drug class, apart from zidovudine, has been studied as extensively as the lipid-lowering agents regarding their potential to cause damage to skeletal muscle. In particular, these studies have focused on the hydroxymethylglutaryl coenzyme A (HMG-CoA) reductases (namely lovastatin, simvastatin, fluvastatin, and pravastatin), the fibrate derivatives (namely clofibrate and bezafibrate, and gemfibrozil), and valproate. It is unclear whether muscle toxicity is markedly increased when a HMG-CoA reductase is taken concurrently with a fibrate derivative or with cyclosporin A (Smith et al., 1991; Churcrallah et al., 1992; Wiklund et al., 1993; Hutchesson et al., 1994; Jingami, 1994; Smit et al., 1995). It has been suggested that combined bezafibrate/simvastatin (Hutchesson et al., 1994) and fluvastatin/gemfibrozil (Smit et al., 1995) for the treatment of hyperlipidemia has no adverse effects on skeletal muscle. Furthermore, it has been demonstrated by Spence et al. (1995) that there is no pharmacokinetic interaction between patients who are taking fluvastatin together with gemfibrozil. Alternatively, it has been recommended that combined pravastatin/gemfibrozil (Wiklund et al., 1993), as well as lovastatin/gemfibrozil (Churcrallah et al., 1992), should not be used in this therapy as a result of the potential for muscle damage. Furthermore, it has been suggested that the incidence of HMG-CoA reductase-induced muscle toxicity is enhanced to about one in three in patients who are receiving immunosuppressant therapy (e.g. cyclosporin A) (Smith et al., 1991). This increased toxicity in the presence of cyclosporin A was subsequently confirmed in rodent studies in which there was a marked potentiation of the toxicity (75–100% incidence of muscle toxicity) as evidenced by myofiber necrosis, interstitial edema, and inflammatory infiltration in primarily type IIb (fast-twitch glycolytic fibres) (Smith et al., 1991). The pharmacokinetic evaluation in these studies indicated that cyclosporin A increased the HMG-CoA reductase AUC (area under the plasma concentration–time curve) and muscle drug levels. It was hypothesized that this effect may be mediated through a cholestasis associated with this combined drug therapy. Likewise, as a result of its ability to inhibit the metabolism of lovastatin, the antifungal agent itraconazole should not be used in combination therapy or the dose of lovastatin should be reduced to avoid the potential development of drug-induced muscle damage (Newonen and Jalava, 1996).

In case reports of HMG-CoA reductase toxicity to muscle tissue, the clinical picture ranges from asymptomatic elevated serum creatine kinase levels to acute rhabdomyolysis (Bradford et al., 1993; Veerkamp et al., 1996). The mechanisms that have been postulated to cause this muscle damage include: (1) inhibitory effects on growth and differentiation (Veerkamp et al., 1996); (2) inadequate synthesis of coenzyme Q and heme A in the inner mitochondrial membrane, leading to degeneration of membranous organelles and impairment of energy production (Waclawik et al., 1993); and (3) a reduction in skeletal muscle ubiquinone levels (Laaksonen et al., 1995). The toxicity is enhanced with increased doses (Smith et al., 1991; Pierno et al., 1995). Furthermore, studies indicate that the risk for developing muscle toxicity is higher for the more lipophilic

HMG-CoA reductases (e.g. simvistatin) compared with those that are more hydrophilic (e.g. pravastatin) because of their greater potential for crossing the cellular membranes (Fukami et al., 1993; Pierno et al., 1995; Veerkamp et al., 1996).

There are isolated case reports of clofibrate-, bezafibrate-, and valproate-induced muscle toxicity (Papadimitriou and Servidei, 1991; Godoy et al., 1992; Vita et al., 1993). Vita et al. (1991) concluded that the toxic effects of bezafibrate were a function of the drug acting directly on the sarcolemma, leading to discontinuities and initiating cell necrosis *versus* an effect on glycolytic and mitochondrial enzymes. The report by Papimitriou and Servidei (1991) highlighted a case where valproate administration precipitated the development of a pre-exisiting multiple acyl-CoA dehydrogenase deficiency myopathy, as determined by ragged-red fibres, neutral lipid storage, decreased total and free carnitine, and a 40% decrease in the levels of short-medium and long-chain acyl-CoA dehydrogenases.

Steroids

Steroids also make up one of the most studied group of drugs inducing muscle damage. Steroid-induced damage to muscle tissue (which may be associated with neuropathy) has been reported to occur when these agents are used to treat patients for status asthmaticus (Bowyer et al., 1985; Lacomis et al., 1993), chronic obstructive pulmonary disease (COPD) (Decramer et al., 1996), neuro-oncology (Koehler, 1995), myasthenia gravis (Panegyres et al., 1993), and rheumatoid arthritis (Caldwell and Furst, 1991). The toxic effects of steroids, in particular prednisone, methylprednisolone, dexamethasone, and triamcinolone, have been shown to affect both peripheral and ventilatory muscles. Furthermore, it has been suggested that this condition has substantial morbidity in patients with status asthmaticus and seems to be associated with reduced survival of patients when compared with patients with a similar degree of airflow obstruction in COPD (Lacomis et al., 1993; Decramer et al., 1996). Steroid-induced muscle damage can be generally characterized into two types: (1) those associated with high-dose steroids versus (2) those associated with chronic moderate-to-high-dose steroids (Dekhuijen and Decramer, 1992; Van-Balkom et al., 1994; Decramer et al., 1996; Nava et al., 1996).

The toxic effects on muscle associated with acute high-dose steroid therapy include generalized muscle atrophy, diffuse muscle weakness, and rhabdomyolysis, whereas for chronic conditions intermediate-to-high doses result in proximal muscle weakness (Dekuijzen and Decramer, 1992; van-Balkom et al., 1994; Decramer et al., 1996; Nava et al., 1996). Muscle biopsy in patients with COPD revealed variation in peripheral muscle fiber diameter, angular trophic muscle fibers, and diffuse necrotic and basophilic fibers that predominantly affect type IIb fibers (fast-twitch glycolytic) (Decramer et al., 1996). Similar findings were reported in patients with status asthmaticus (Lacomis et al., 1993). Rodent studies have shown atrophy of type IIb (fast-twitch glycolytic) in peripheral muscle tissue and type IIa (fast-twitch glycolytic oxidative fibers) in diaphragm muscle. The levels of serum creatine kinase or lactate dehydrogenase were not elevated or only moderately elevated (Lacomis et al., 1993; Decramer et al., 1996). In contrast, steroid-induced muscle toxicity appears to be associated with increased creatine excretion in the urine (Decramer et al., 1996). The mechanism of steroid-induced muscle damage has not been determined to date; there is uncertainty whether this is a primary effect or secondary to decreased nutritional status, resulting in steroid-induced muscle wasting. Fernandez-Sola et al. (1993) reported that 11 asthmatic patients with low-dose glucocorticoid treatment did not exhibit changes in muscle weakness or evidence of histological changes in muscle; however, the regulatory enzymes of glycogen metabolism were modified compared with those of controls. At this stage, nutritional deprivation does not account entirely for the development of the toxic effects of steroids on peripheral and ventilatory muscle (Nava et al., 1996).

Chemotherapy

Chemotherapeutic agents may be associated with the development of both myopathy (muscle damage) and neuropathy. The determination of the agent responsible for the toxic effect on skeletal muscle and/or nerve is difficult to interpret because chemotherapy often involves combination drug therapy. DeAngelis et al. (1991) reported the development of moderate-to-severe signs and symptoms of neuropathy and/or myopathy after combination and high-dose dexamethasone for the treatment of intermediate- and high-grade non-Hodgkin's lymphomas. Vincristine has previously been reported to cause necrotizing myopathy with neuropathy (Lane and Mastaglia, 1978; Le-Quintrec and Le-Quintrec, 1991). Likewise, steroids have also been shown to cause myopathy (see above). Muscle weakness, most apparent in the hands and feet, developed in all patients. Furthermore, when some patients received half the dose of dexamethasone, there was no weakness in the deltoid muscle compared with those who received the high dose. A similar effect was noted in quadriceps, iliopsoas, and hamstring muscles. Through an examination of various motor functions and skills, DeAngelis and co-workers (1991) concluded that two different patterns can be used to discriminate clinically between myopathy caused by dexamethasone versus vincristine toxicity. Vincristine toxicity (neuropathy) seems to impair fine-motor tasks, whereas dexamethasone myopathy seems to affect the patient's ability to rise from a seated position or climb stairs via a proximal muscle weakness.

Other chemotherapeutic agents that have been known to cause muscle toxicity include leuprolide acetate for prostate cancer and the antiparasitic drug suramin which has been used to treat metastatic cancer (Crayton et al., 1991; Rago et al., 1994). Crayton et al. (1991) reported a patient who developed diffuse muscle weakness and myalgias associated with fever, sweats, shortness of breathe, anorexia, and malaise after the administration of leuprolide acetate. Creatine kinase levels were markedly elevated, but these were predominant the isoenzyme fraction associated with cardiac muscle. A muscle biopsy revealed necrosis, degeneration, and regeneration of muscle fibres with inflammatory cells. The symptoms resolved with short-term steroid therapy. The administration of suramin for the treatment of metastatic cancer was associated with hyperphosphatemia and mitochondria myopathy, determined by histochemical, biochemical, and ultrastructural findings. Muscle weakness continued despite phosphate repletion therapy and resolved with drug discontinuation. High-dose paclitaxel (Taxol) therapy has also been noted to cause damage to muscle, either alone or in combination with cisplatin (Rowinsky et al., 1993a, b).

Emetine (ipecac)

Thyagarajan et al. (1993) reported that the antiemetic emetine, found in ipecac, caused a necrotizing muscle damage process as determined by EMG examination in a patient who ingested up to 200 ml/week for 3 months. The patient reported severe neck and limb weakness and diffuse body ache. Serum creatine kinase levels were slightly elevated; however, muscle biopsy reported vacuolar degeneration with myofibrolysis and cytoplasmic body formation. The systems resolved gradually after discontinuation of the ipecac. This study is consistent with previous studies by Palmer and Guay (1985) who reported the presence of muscle damage predominantly in type I fibers, and Halbig et al. (1988) and Sugie et al. (1984) who reported "Z band streaming" the formation of cytoplasmic bodies and sarcotubular abnormalities.

Colchicine

Colchicine is an example in which the damage to the muscle is secondary to neuropathy (Le-Quintrec and Le-Quintrec, 1991; Levy et al., 1991). The development of this muscle damage associated with neuropathy has been reported to occur after colchicine use for a time period ranging from 6 months to several years (Kuncl et al., 1987). Alternatively, other patients exhibited muscle toxicity associated with acute intoxication (Naldus et al., 1977; Murray et al., 1983). Patients with colchicine toxicity to skeletal muscle usually have mild chronic renal insufficiency. Serum creatine kinase levels may be elevated 10–20 times above normal and some patients were unable to rise from a chair or to lift objects above their shoulders. Muscle biopsy from these patients revealed lysosome vacuolar muscle damage with accumulation of autophagic vacuoles and lysosomes in the absence of necrosis. Muscle strength improved and creatine kinase levels decreased over a 4- to 6-week period after discontinuation of the drug. The development of this muscle damage may be secondary to the nerve damage, however, other studies have suggested that colchicine may have direct effects on skeletal muscle (Goodman and Murray, 1953; Angeven, 1957; Collot et al., 1984).

Chloroquine

The antimalarial agent chloroquine is similar to colchicine in that the toxic effects on skeletal muscle may be secondary to neurologic involvement (Le-Quintrec and Le-Quintrec, 1991; Avina-Zubieta et al., 1995). Avina-Zubieta et al. (1995) reported an incidence of 1 in 100 patient-years in the development of muscle damage after the administration of the antimalarial agents, chloroquine and hydroxychloroquine. The damage in skeletal muscle was seen in those patients who had received chloroquine for 12 and 18 months at a maximum daily dose of 250 mg. These patients exhibited difficulty in rising from a seated position, walking, and raising their leg against gravity. There was no increase in the serum levels of muscle-derived enzymes. One of the three patients exhibited a type II fibre atrophy with rare rimmed vacuoles; however, these biopsies were con-ducted about 26 days after discontinuation of the drug. The symptoms resolved upon discontinuation of the chloroquine. This report by Avina-Zubieta et al. (1995) presents an excellent review of the isolated English language reports of chloroquine-induced muscle damage. A neuro-myopathy has also been reported for a patient taking chloroquine for the treatment of systemic lupus erythematosus (Parodi et al., 1985). Chloroquine-induced muscle damage and neuropathy have been confirmed in rodent studies by Velasco et al. (1995) and Sugita et al. (1987). Velasco et al. (1995) reported that soleus muscles (type I fibres) exhibited a vacuolar myopathy, whereas extensor digitorum longus muscles (type II) did not exhibit any indication of vacuolation. However, these latter muscle (type II fibers) seemed to exhibit neuropathic changes. Sugita et al. (1987) have reported that chloroquine myopathy, characterized by autophagic vacuole formation and increases in lysosomal enzymes (e.g. cathepsins B and L), could be inhibited by the concurrent administration of a cysteine protease inhibitor EST.

Alcohol

A common feature of chronic ethanol abuse in humans is the presence of alcoholic myopathy. The reader is referred to the chapter by V.R. Preedy et al. in this volume for further details on this particular drug-induced muscle damage. Two distinct type of myopathies have been reported in alcoholics: (1) a fulminate acute rhabdomyolysis reported in 1955 by Hed et al., and (2) a chronic proximal myopathy first described in the mid-1960s (Ekbom et al., 1964). Alcoholic myopathy has been reported to occur to varying degrees in 20–60% of patients in alcohol treatment facilities (Farris et al., 1967; Martin and Peters, 1985; Urbano-Márquez et al., 1989). Recent studies have suggested that women may be more susceptible to the toxic effects of ethanol on cardiac muscle than men (Urbano-Márquez et al., 1995). This chronic proximal myopathy can be characterized by marked muscle weakness and diminished muscle mass leading to decreased muscle function, even in detoxified alcoholics (Urbano-Márquez et al., 1989; Pendergast et al., 1990; Urbano-Márquez et al., 1995). In its most severe form, the myopathy may involve 25% of skeletal muscle mass (Cook et al., 1992). The diagnosis of alcoholic myopathy utilizes clinical symptoms (e.g. pain, the presence of weakness and decreased performance), electrophysiological measurements, muscle histological and histochemical evaluation, elevated serum levels of cytosolic enzymes, myoglobinemia, and myoglobinuria.

This myopathy, in human and animal models, appears primarily to involve anaerobic (fast-twitch, type II) fibres, whereas aerobic (slow-twitch, type I) fibres remain relatively unaffected (Hanid et al., 1981; Slavin et al., 1983; Preedy and Peters, 1988; Trounce et al., 1990; Preedy et al., 1994a, b). Numerous mechanisms to explain the development of chronic ethanol-induced myopathy have been postulated including: (1) a myopathy secondary to ethanol-induced general nutritional deficiency, as well as to specific deficiencies including hypophosphatemia and/or hypokalemia; (2) prolonged ischemia as a result of ethanol-induced vasoconstriction; (3) a direct effect of ethanol on muscle extracellular and intracellular membranes; (4) a direct toxic effect of ethanol on transport processes in both extracellular and intracellular membranes; (5) ethanol-induced free radical formation and lipid peroxidation; (6) ethanol-mediated changes in carbo-hydrate or lipid metabolism; (7) the influence of ethanol on protein turnover; and (8) an effect on translation and/or transcription (Knochel et al., 1975; Garcia-Bunuel, 1984; Ohnishi, 1985;

Ohnishi et al., 1985; Preedy and Peters, 1988; Trounce et al., 1990; Held, 1991; Cook et al., 1992; Ward and Peters, 1992; Preedy et al., 1994a, b; Amaladevi et al., 1995; Pagala et al., 1995).

Cocaine

The increasing number of clinical reports describing cocaine-induced rhabdomyolysis leading to acute renal failure, and the publicized deaths of athletes who used cocaine as an ergogenic agent, have highlighted our lack of knowledge about the mechanisms responsible for cocaine's toxicity on cardiac and skeletal muscle (Cregler and Mark, 1986; Herzlich et al., 1988; Pogue and Nurse, 1989; Turbat-Herrar, 1994). Pagala et al. (1993) demonstrated an increased release of creatine kinase from rat soleus *in vitro*, but not from extensor digitorum longus muscle when incubated in the presence of cocaine over a 4-hour period. Furthermore, Pagala et al. (1991) reported the cocaine reduces skeletal muscle function by reducing muscle and nerve membrane excitability, without effects on neuromuscular transmission, excitation–contraction coupling, or contractility.

At this time, the mechanism of cocaine-induced rhabdomyolysis and/or muscle damage has not been elucidated. In subsequent work, Brazeau et al. (1995), in attempts to determine whether cocaine has a direct effect on muscle, reported that the release of creatine kinase from the isolated extensor digitorum longus muscle was statistically significant only when muscles were exposed to 1 mM cocaine for a period of 30 min. These findings suggest that the muscle damage in cocaine-induced rhabdomyolysis may be mediated by an indirect action rather than a direct action on the muscle fibres. Rather it seems that this myotoxicity might be mediated via a secondary effect. One possibility is that chronic cocaine ingestion could lead to repeated ischemic events mediated by vasoconstrictor properties of cocaine and its metabolites. Cocaine-induced vasoconstriction has been implicated as a possible mechanism in myocardial infarction, cerebrovascular accidents, obstetric complications, and rhabdomyolysis (Inaba et al., 1978; Cregler and Mark, 1986; Turbat-Herrara, 1994). Recent studies have suggested that cocaine's toxic effects on the heart and brain seem to be mediated via vasoconstriction and decreased blood flow (Madden and Powers, 1990; Kurth et al., 1993; Nunez et al., 1994). Damage to the muscle tissue would result from the subsequent reperfusion and generation of free radicals. If these free radicals are not inactivated by radical-scavenging systems in the muscle and associated vasculature, this could result in damage to the muscle fibres via the formation of lipid peroxides, protein carbonyls, or DNA adducts. This hypothesis was further suggested when sedentary BalbC mice injected with a single intravenous dose of cocaine compared with physiological saline in controls, demonstrated creatine kinase levels that were significantly elevated fivefold, whereas TBARSs (an indication of lipid peroxidation) were elevated by 100% in gastrocnemius muscle of cocaine-treated animals that were compared with controls injected with physiological saline. The present findings suggest that lipid peroxidation may occur in mouse skeletal muscle after a single intravenous cocaine dose.

Miscellaneous drugs

Antithyroid drugs may cause damage to skeletal muscle. Suzuki et al. (1997) reported increased creatine kinase levels and muscle pain and cramps in patients who were receiving methimazole for chronic hyperthyroidism caused by Graves' disease. These investigators hypothesized that the major reason for these symptoms was the rapid decrease in thyroid hormones (namely tyroxine or T_4) in muscle tissues, which was caused by a hypothyroid state leading to increased release of creatine kinase. Furthermore, they recommended the addition of levothyroxine in conjunction with methimazole, rather than the discontinuation of methimazole in this condition.

Amiodarone has been reported to cause a vacuolar-type myopathy, secondary to neurological complications (Le-Quintrec and Le-Quintrec, 1991). In other case reports, phenytoin and pivampicillin have been suggested to cause skeletal muscle damage via elevations in creatine kinase, muscle weakness, and muscle biopsy (Harney and Glasberg, 1983; Rose et al., 1992). Patients, in whom phenytoin-induced muscle damage is reported, should be switched to alternative anticonvulsant therapy (Harney and Glasberg, 1983). In the case of pivamipicillin-induced muscle damage, carnitine replacement therapy was found to be successful as a treatment modality (Rose et al., 1992). Penicillamine, used in the treatment of rheumatoid arthritis, has also been reported to cause muscle damage, which may characterized as a polymyositis/dermatomyositis and presents as proximal muscle weakness, with elevated serum enzymes and histological evidence of necrosis/regeneration and inflammation (Lane and Mastaglia, 1978; Lund and Nieslen, 1983). A proximal myopathy has been reported to occur with griseofulvin therapy (Deo et al., 1994). As for the myotoxicity of combination therapy discussed previously for the HMG-CoA reductases, cyclosporin A, in combination with colchicine, has been reported to cause myalgia, muscle weakness, and elevated creatine kinase levels. The authors attributed this effect to cyclosporin A because symptoms eased off with discontinuation of the cyclosporin, but at the same time there was a reduction in the dose of colchicine (Yamanishi et al., 1993).

Neuromuscular blocking agents (e.g. pancuronium, vecuronium), which are often used in routine care of critically ill patients, have also been associated with the development of muscle damage or dysfunction after prolonged use (Elliot and Bion, 1995; Gooch, 1995; Prielipp et al., 1995). Muscle weakness may continue long after the discontinuation of these therapeutic agents (Prielipp et al., 1995). Patient presentation may include decreased muscle tone, reduced strength, and elevated creatine kinase levels (Gooch, 1995). The mechanism underlying the development of this myopathy is not understood at this time; however, toxicity may be exacerbated by the concurrent administration of steroids or aminoglycoside antibiotics (Elliot and Bion, 1995; Gooch, 1995).

Besides cocaine, other drugs of abuse may be associated with muscle toxicity after systemic administration. Heroin has been shown to cause muscle damage in animals and humans (Richter et al., 1971; Schwartzfarb et al., 1977; Gibb and Shaw, 1985; Peña et al., 1990). The difficulty in interpreting case reports of cocaine-induced muscle damage has been in determining whether the myopathy was a function of the heroin itself or the agents that are frequently used as adulterants. Schwartzfarb et al. (1977), who conducted biopsies in the muscles of a heroin addict, reported the presence of intense necrosis in soleus muscle, whereas rectus femoris or gastrocnemius muscles showed few changes. Likewise, Peña et al. (1990) reported similar findings in soleus versus tibialis anterior muscle of rodents who were administered intraperitoneal heroin over a 3-month period. These soleus muscles were observed to be in the process of degeneration/regeneration with considerable involvement of macrophages and phagocytosis. A possible reason for these disparate results in different muscles was a greater degree of vascularization in soleus muscle (predominately slow-twitch oxidative) which may mean that it is exposed to higher drug levels (Peña et al., 1990). A possible mechanism underlying the development of this muscle damage is the metabolism of heroin to morphine and other opiates which provoke altered cell metabolism, membrane transport, and cellular energy (Simon, 1971). Likewise, Blain et al. (1985) reported three cases of opiate self-poisoning with morphine or dihydrocodeine, who developed acute muscle damage as measured by elevated serum creatine kinase and myoglobin levels, and acute renal failure in one patient.

References

Amaladevi B, Pagala S, Pagala M, Namba T and Grob D (1995) Effect of alcohol and electrical stimulation on leakage of creatine kinse from isolated fast and slow muscles of the rat. *Alcohol Clin Exp Res* 19: 147–152.

Angeven JR (1957) Nerve destruction by colchicine in mice and golden hamsters. *J Exp Zool* 136: 363.

Arnaudo E, Dalakas M, Shanke S, Moraes CT, DiMauro S and Schon EA (1991) Depletion of muscle mitochondrial DNA in AIDS patients with zidovudine-induced myopathy. *Lancet* 337: 508–510.

Arnold J, Berger A and March L (1988) A comparative evaluatin of pain following intramuscular administration of three parenteral antimicrobials to healthy volunteers. *Curr Therap Res* 43: 1082–1088.

Avina-Zubieta JA, Johnson ES, Suarez-Almazor ME and Russell AS (1995) Incidence of myopathy in patients treated with antimalarials. A report of three cases and a review of the literature. *Brit J Rheumatol* 34: 166–170.

Belcastro AN, Alibisser TA, Littlejohn B (1996) Role of calcium-activated neutral protease (calpain) with diet and exercise. *Can. J Appl Physiol* 21: 328–346.

Benoit PW, Yagiela JA and Fort NF (1980) Pharmacological correlation between local anesthetic-induced myotoxicity and disturbances of intracellular calcium disturbances. *Toxicol Appl Pharmacol* 52: 187–198.

Blain PG, Lane RJ, Bateman DN and Rawlins MD (1985) Opiate-induced rhabdomyolysis. *Hum Toxicol* 4: 71–74.

Bohan A, Jones P, Bowman R and Pearson C (1977) A computer-assisted analysis of 153 patients with polymyositis and dermatomyositis. *Medicine* 56: 255–286.

Bowyer SL, LaMothe MP and Hollister JR (1985) Steroid myopathy: Incidence and detection in a patient population with asthma. *J Allergy Clin Immunol* 76: 234–342.

Bradford RH, Downton M, Chremos AN, Langendorfer A, Stinnett S, Nash DT, Mantell G and Shear CL (1993) Efficacy and tolerability of lovastatin in 3390 women with moderate hypercholesterolemia. *Ann Intern Med* 118: 850–855.

Brazeau GA and Fung H-L (1989) *In vitro* assay interferences of creatine kinase activity. *Biochem J* 257: 619–621.

Brazeau GA and H-L Fung (1990) Mechanisms of creatine kinase release from isolated rat skeletal muscles damaged by propylene glycol and ethanol. *J Pharm Sci* 79: 393–397.

Brazeau GA, Arthurs S and Mott J (1992) No correlation between pain and *in vitro* myotoxicity after intramuscular injection of three cephalosporins. *Curr Therap Res* 51: 839–843.

Brazeau GA, McArdle A and Jackson MJ (1995) Effects of cocaine on leakage of creatine kinase from skeletal muscle: *in vitro* and *in vivo* effects. *Life Sci* 57: 1569–1578.

Brown M, Ross TP and Holloszy JO (1992) Effects of aging and exercise on soleus and extensor digitorum longus muscles of female rats. *Mech Age Dev* 63: 69–77.

Byrne E, Hayes DJ, Shoubridge EA, Morgan-Hughes JA and Clark JB (1985) Experimentally induced defects in mitochondrial metabolism in rat skeletal muscle. *Biochem J* 229: 101–108.

Caldwell JR and Furst DE (1991) The safety and efficacy of low-dose corticosteroids for rhematoid arthritis. *Semin Arthritis Rheum* 21: 1–11.

Chairot P, Le Maguet F, Autheir FJ, Labes D, Poron F and Gherardi R (1995) Cytochrome c oxidase deficiency in zidovudine myopathy affects perifascucular muscle fibers and arterial smooth muscle cells. *Neuropathol Appl Neurobiol* 21: 540–547.

Chalmers AC, Greco CM and Miller RG (1991) Prognosis in AZT myopathy. *Neurology* 41: 1181–1184.

Chen SC, Barker SM, Mitchell DH, Stevens SM, O'Neill P and Cunningham AL (1992) Concurrent zidovudine-induced myopathy and hepatoxicity in patients treated for human immunodeficiency (HIV) virus. *Pathology* 24: 109–11.

Churcrallah A, De-Girolami U, Freeman R and Federman M (1992) Lovastatin/gemfibrozil myopathy: a clinical, histochemical and ultrastructural study. *Eur Neurol* 32: 293–296.

Clarkson PM and Ebbeling C (1988) Investigation of serum creatine kinase variability after muscle-damaging exercise. *Clin Sci* 75: 257–261.

Collot M, Louvard D and Singer SJ (1984) Lysosomes are associated with mnicrotubules and not with intermediated filaments in cultured fibroblasts. *Proc Natl Acad Sci USA* 81: 788–792.

Cook EB, Adebiyi LAY, Preedy VR, Peters TJ and Palmer TN (1992) Chronic effects of ethanol on muscle metabolism in the rat. *Biochim Biophys Acta* 1180: 207–214.

Crayton H, Bohlmann T, Sufit R and Graziano FM (1991) Drug-induced polymyositis secondary to lueprolide acetate (Lupron) therapy for prostate cancer. *Clin Exp Rheumatol* 9: 525–528.

Cregler LL and Mark H (1986) Medical complications of cocaine abuse *N Engl J Med* 315: 1495–1500.

Dalakas MC, Illa I, Pezeshkpour GH, Laukaitis JP, Cohen B and Griffin JL (1990) Mitochondrial myopathy caused by long-term zidovudine therapy. *N Engl J Med* 322: 1098–1105.

DeAngelis LM, Gnecco C, Taylor L and Warrell RP (1991) Evolution of neuropathy and myopathy during intensive vincristine/corticosteroid chemotherapy for non-Hodgkin's lymphoma. *Cancer* 67: 2241–2246.

Decramer M, deBock V and Dom R (1996) Functional and histological picture of steroid-induced myopathy in chronic obstructive pulmonary disease. *Am J Respir Crit Care Med* 153: 1958–1964.

Dekhuijzen PN and Decramer M (1992) Steroid-induced myopathy and its significance to respiratory disease: a know disease rediscovered. *Eur Respir J* 5: 997–1003.

Deo A, Mehta HG, Biniyala R, Pathare S, Mehta PJ and Mehtalia SD (1994) Proximal myopathy associated with griseofulvin therapy. *J Assoc Physicians India* 42: 85.

DeSmet Y (1993) Status asthmaticus. Acute myopathy induced by cortisone and neuropathy during resuscitation. *Rev Neurol Paris* 149: 573–576.

Duarte JA, Soares JM, Appell HJ (1992) Nifedipine diminishes exercise-induced muscle damage in mouse. *Int J Sport Med* 13: 274–277.

Duncan CJ (1978) Role of intracellular calcium in promoting muscle damage: a strategy for controlling the dystrophic condition. *Experientia* 34: 1531–1535.

Duncan CJ and Jackson MJ (1987) Different mechanisms mediate structural changes and intracellular enzyme efflux following damage to skeletal muscle. *J Cell Sci* 87: 183–188.

Ekbom K, Hed R, Kirstein L and Astrom KE (1964) Muscular affections in chronic alcoholism. *Arch Neurol* 10: 449–458.

Elliot JM and Bion JF (1995) The use of neurmuscular blocking agents in intensive care practice. *Acta Anaesthesiol Scand Suppl* 106: 70–82.

Fagan JM, Wajnberg EF, Culbert L and Waxman L (1992) ATP depletion stimulates calcium-dependent protein breakdown in chick-skeletal muscle. *Am J Physiol* 262: E637–643.

Farris AA, Reyes MG and Abrams BB (1967) Subclinical alcoholic myopathy. Electromyographic and biopsy study. *Am Neurol* 92: 102–196.

Fernandez-Sola J, Cusso R, Picado C Vernet M, Grau JM and Urbano-Marquez A (1993) Patients with chronic glucocorticoid treatment develop changes in muscle glycogen metabolism. *J Neurol Sci* 117: 103–106.

Fitts RH, Troup JP, Witzmann FA and Holloszy JO (1984) The effect of aging and exercise on skeletal muscle function. *Mech Age Dev* 27: 161–172.

Foster AH and Carlson BM (1980) Myotoxicity of local anesthetics and regeneration of damaged muscle fibers. *Anesth Analg* 58: 727–736.

Fukami M, Maeda N, Fukushige J, Kogure Y, Shimada Y, Ogaea T and Tsujita Y (1993) Effects of HMG-CoA reductase inhibitors on skeletal muscle of rabbits. *Res Exp Med Berl* 193–263–273.

Gaertner DJ, Boschert KR and Schoeb TR (1987) Muscle necrosis in syrian hamsters resulting from intramuscular injections of ketamine and xylazine. *Lab Animal Sci* 37: 80–83.

Garcia-Bunuel L (1984) Lipid peroxidation in alcoholic myopathy. *Med Hypotheses* 13: 217–231.

Gertner E, Thurn JR, Williams DN, Simpson M, Balfour HH, Rhame F and Henry K (1989) Zidovudine-associated myopathy. *Am J Med* 86: 814–818.

Gherardi RK (1994) Skeletal muscle involvement in HIV-infected patients. *Neuropathol Appl Neurobiol* 20: 232–237.

Gibb WRG and Shaw IC (1985) Myoglobinuria due to heroin abuse. *J R Soc Med* 78: 862–863.

Godoy JM, Nicaretta DH, Balassiano SL and Skacel M (1992) *Arq Neuropsiquiatr* 50: 123–125.

Gooch JL (1995) AAEM case report #29: prolonged paralysis after neuromuscular blockade. *Muscle Nerve* 18: 937–942.

Goodman GC and Murray MR (1953) Influence of colchicine on the form of skeletal muscle in tissue culture. *Proc Soc Exp Biol Med* 84: 668–672.

Groopman JE (1990) Zidovudine intolerance. *Rev Infect Dis* 12: S500–S506.

Halbig L, Gutmann L, Goebel HH, Brick JF and Schochet S (1988) Ultrastructural pathobiology in emetine-induced myopathy. *Acta Neuropathol (Berlin)* 75: 577–582.

Hanid A, Slavin G, Mair W, Sowter C, Ward P, Webb J and Levi J (1981) Fibre type changes in striated muscle of alcoholics. *J Clin Pathol* 34: 991–995.

Harney J and Glasberg MR (1983) Myopathy and hypersensitivity to phenytoin. *Neurology* 33: 790–791.

Harris JB and Blain PG (1990) Introduction to the toxicology of muscle. *Bailliére's Clin Endocrinol Metabol* 4: 665–686.

Hauser E, Höger H, Widhlam K and Lubec G (1995) Oxyradical damage and mitochondrial enzyme activities in the mdx mouse. *Neuropediatrics* 26: 260–262.

Hed R, Larsson H and Wahlgren F (1955) Acute myoglobinuria in alcoholism. *Acta Med Scand* 152: 459–463.

Helbert M, Fletcher T, Peddle B, Harris JRW and Pinching AJ (1988) Zidovudine-associated myopathy. *Lancet* 2: 689–690.

Held IR (1991) Ribosomal RNA activity and protein in skeletal muscles of chronic ethanol-feed rats. *Alcohol* 9: 79–82.

Herzberg NH, Zorn I, Zwart R, Portegies P and Bolhuis PA (1992) Major growth reduction and minor decrease in mitochondrial enzyme activity in cultured human muscle cells after exposure to zidovudine. *Muscle Nerve* 15: 706–710.

Herzlich BD, Arsura EL, Palag M and Grob D (1988) Rhabdomyolysis related to cocaine abuse. *Ann Intern Med* 109–335–336.

Hochberg M, Feldman D and Stevens MD (1986) Adult onset olymyositis/dermatomyositis: An analysis of clinical and laboratory features and survival in 76 patients with a review of the literature. *Semin Arth Rheum* 15: 168–178.

Hutchesson AC, Moran A and Jones AF (1994) Dual bezafibrate-simvastatin therapy for combined hyperlipideamia. *J Clin Pharm Ther* 19: 387–389.

Inaba TT, Stewart and Kalow W (1978) Metabolism of cocaine in man. *J Clin Pharm Ther* 23: 547–552.

Jackson MJ and O'Farrell S (1993) Free radical and muscle damage. *Brit Med J* 49: 630–641.

Jackson MJ, Jones DA and Edwards RHT (1984) Experimental skeletal muscle damage: the nature of calcium activated degenerative processes. *Eur J Clin Invest* 14: 369–374.

Jackson MJ, Wagenmakers AJM and Edwards RHJ (1987) Effect of inhibitors of arachidonic acid metabolism on efflux on intracellular enzymes from skeletal muscle following experimental damage. *Biochem J* 241: 403–407.

Jingami H (1994) HmG-CoA redcuatse inhibitor for therapy of patients with hyperlipoprotenemia. *Nippon Rinsho* 52: 3271–3281.

Knochel JP, Bilbrey GL, Fuller TJ and Carter NW (1975) The muscle cell in chronic alcoholism: The possible role of phosphate depletion in alcoholic myopathy. *Ann N Y Acad Sci* 252: 274–286.

Koehler PJ (1995) Use of corticosteroids in neuro-oncology. *Anti-Cancer Drugs* 6: 19–33.

Kuncl RW, Duncan G, Watson D, Alderson K, Rogawaski MA and Peper M (1987) Colchichine myopathy and neuropathy. *N Engl J Med* 316: 1562–1568.

Kurth CD, Monitto C, Albuquerque ML, Feuer P, Anday E and Shaw L (1993) Cocaine and its metabolites constrict cerebral arterioled in newborn pigs. *J Pharmacol Exp Ther* 265: 587–591.

Laaksonen R, Jokelainen K, Sahi T, Tikkanen MJ and Himberg JJ (1995) Decreases in serum ubiquinone do not result in reduced muscle levels in muscle tissue during short-term simvastatin treatment in humans. *Clin Pharmacol Therap* 57: 62–66.

Lacomis D, Smith TW and Chad DA (1993) Acute myopathy and neuropathy in status asthmaticus: Case report and literature review. *Muscle Nerve* 16: 84–90.

Lane RJM and Mastaglia FL (1978) Drug-induced myopathies in man. *Lancet* 27: 562–566.

Lewis W, Gonzalez B, Chomyn A and Papoian T (1992) Zidovudine induces molecular, biochemical and ultrastructural changes in skeletal muscle mitochondria. *J Clin Invest* 89: 1354–1360.

Levy M, Spino M and Read SE (1991) Colchicine: a state-of-the-art review. *Pharmacotherapy* 11: 196–211.

Le-Quintrec JS and Le-Quintrec JL (1991) Drug-induced myopathies. *Bailliére's Clin Rheumatol* 5: 21–38.

Lindsay TP, Walker PM, Mickle DAG and Romaschin AD (1988) Measurement of hdyroxy-conjugated dienes after ischemia-reperfusion in canine skeletal muscle. *Am J Physiol* 254: H578–H583.

Lund HI and Nielsen M (1983) Penicillamine-induced dermatomyositis. A case history. *Scand J Rheumatol* 12: 350–352.

Madden JA and Powers RH (1990) Effect of cocaine and cocaine metabolites on cerebral arteries *in vitro*. *Life Sci* 47: 1109–1114.

Manji H, Harrison MJG, Roudn JM, Jones DA, Connolly S, Fowler CJ, Williams I and Weller IVD (1993) Muscle disease, HIV and zidovudine: the spectrum of muscle disease in HIV-infected individuals treated with zidovudine. *J Neurol* 240: 479–488.

Martin F and Peters TJ (1985) Alcoholic muscle disease. *Alcohol* 20: 125–136.

Mastaglia FL and Walton J (1982) *Skeletal Muscle Pathology*. Churchill Livingstone, New York.

Mhiri C, Baudrimont M, Bonne G, Geny C, Degoul F, Marsac C, Roullet E and Gherardi R (1991) Zidovudine myopathy: a distinctive disorder associated with mitochondrial dysfunction. *Ann Neurol* 29: 606–614.

Modica-Napolitano JS (1993) AZT causes tissue specific inhibition of mitochondrial bioenergetic function. *Biochem Biophys Res Commun* 194: 170–177.

Murray SS, Kramlinger KG, McMichan JC and Mohr DN (1983) Acute toxicity after excessive ingestion of colchicine. *Mayo Clin Proc* 58: 528–532.

Naldus RM, Rodvien R and Mielke CH (1977) Colchicine toxicity. A multisystem disease. *Arch Intern Med* 137: 394–396.

Nava S, Gayan-Ramierz CG, Rollier H, Bisschop A, Dom R, deBock V and Decramer M (1996) Effects of acute steroid administration on ventilatory and peripheral muscles in the rat. *Am J Respir Crit Care Med* 153: 1888–1896.

Neuvonen PJ and Jalava KM (1996) Itraconazole drastically increases plasma concentrations of lovastatin and lovastatin acid. *Clin Pharmacol Ther* 60: 54–61.

Nunez BD, Mialo L, Wang Y, Nunez MM, Sellke FW Ross JN, Susulic V, Carrozza GY, Paik GY Carrozza JP and Morgan JP (1994) Cocaine-induced microvascular spasm in Yucatan miniture swine. *Circ Res* 74: 281–290.

Ohnishi ST (1995) Chronic alcohol ingestion alters the calcium permeability of sarcoplasmic reticulum of rat skeletal muscle. *Memb Biochem* 6: 33–47.

Ohnishi ST, Waring AJ, Fang S-RG, Horiuchi K and Ohnishi T (1985) Sarcoplasmic reticulum membrane of rat skeletal muscle is disordered with chronic alcoholic ingestion. *Memb Biochem* 6: 49–63.

Pagala MKD, Venkatachari SAT, Herzlich B, Ravindran K, Namba T and Grob D (1991) Effect of cocaine on responses of mousephrenic nerve-diaphragm preparation. *Life Sci* 48: 795–802.

Pagala M, Amaladevi B, Azad D, Pagala S, Herzlich B, Namba T and Grob D (1993) Effect of cocaine on leakage of creatine kinase from isolated fast and slow muscles of the rat. *Life Sci* 52: 751–756.

Pagala M, Ravindran K, Amaladevi B, Namba T and Grob B (1995) Effect of ethanol on function of the rat heart and skeletal muscle. *Alcohol Clin Exp Res* 19: 676–684.

Page SF (1992) *Determination of post-translationally modified creatine kinase – MM as a means of assessing skeletal muscle damage.* Thesis, University of Liverpool.

Palmer EP and Guay AT (1985) Reversible myopathy secondary to abuse with ipecac in patients with major eating disorders. *N Engl J Med* 313: 1457–1459.

Panegyres PK, Squier M, Mills KR and Newsom-Davis J (1993) Acute myopathy associated with large parenteral dose of corticosteroid in myasthenia gravis. *J Neurol Neurosurg Psychiat* 56: 702–704.

Panegyres PK, Papadimitriou JM, Hollingsworth PN, Armstrong JA and Kakulas BA (1990) Vesicular changes in the myopathies of AIDS. Ultrastructural observations and their relationships to zidovudine treatment. *J Neurol Neurosurg Psychiat* 53: 649–655.

Papadimitriou A and Servidei S (1991) Late onset-lipid storage myopathy due to multiple acyl coA dehydrogenase deficiency triggered by valproate. *Neuromusc Disord* 1: 247–252.

Parodi A, Regesta G and Rebora A (1985) Chloroquine-induced myopathy. Report of a case. *Dermatologica* 171: 203–205.

Peña J, Aranda C, Luque E and Vaamonde R (1990) Heroin-induced myopathy in rat skeletal muscle. *Acta Neuropathol* 80: 72–76.

Pendergast DR, York JL and Fisher NM (1990) A survey of muscle function in detoxified alcoholics. *Alcohol* 7: 361–366.

Peters BS, Winer J, Landon DN, Stotter A and Pinching AJ (1993) Mitochondrial myopathy associated with chronic zidovudine therapy in AIDS. *Quart J Med* 86: 5–15.

Pezeshkpour G, Illa I and Dalakas MC (1991) Ultrastructural characteristics and DNA Immunocytochemistry in Human Immunodeficiency. *Hum Pathol* 22: 1281–1288.

Pierno S, De Luca A, Tricafico D, Roselli A, Natuzzi F, Ferrannini E, Laico M and Camerino DC (1995) Potential risk of myopathy by HMG-CoA reductase inhibitors: a comparison of pravastatin and simvastatin effects on membrane electrical properties of rat skeletal muscle fibers. *J Pharmacol Exp Ther* 275: 1490–1496.

Powers SK, Criswell D, Herb RA, Demirel H and Dodd S (1996) Age-related changes in diaphragmatic maximal shortening velocity. *J Appl Physiol* 80: 445–451.

Preedy VR and Peters TJ (1988) Acute effects of ethanol on protein synthesis in different muscles and muscle protein fractions of the rat. *Clin Sci* 74: 461–466.

Preedy VR, Peters TJ, Patel VB and Miell JP (1994a) Chronic alcoholic myopathy: transcriptional and translational alterations. *FASEB J* 8: 1146–1151.

Preedy VR, Salisbury JR and Peters TJ (1994b) Alcoholic muscle disease: features and mechanisms. *J Pathol* 173: 309–315.

Prielipp RC, Coursin DB, Wood KE and Murray MJ (1995) Complications associated with sedative and neuromuscular blocking agents in critically ill patients. *Crit Care Clin* 11: 983–1003.

Publicover SJ, Duncan CJ and Smith JL (1978) The use of A23187 to demonstrate the role of intracellular calcium in causing ultrastructural damage in mammalian muscle. *J Neuropathol Exp Neurol* 37: 544–557.

Rachlis A and Fanning MM (1993) Zidovudine toxicity: clinical features and management. *Drug Safety* 8: 312–320.

Rago RP, Miles JM, Sufit RL, Springgs DR and Wilding G (1994) Suramin-induced muscle weakness from hypophosphatemia and Mitochondrial Myopathy. *Cancer* 73: 1954–1959.

Richter RW, Challenor YB, Pearson J, Kagen LJ, Hamilton LL and Ramsey WH (1971) Acute myoglobinuria associated with heroin addiction. *JAMA* 216: 1172–1176.

Rose SJ, Stokes TC, Patel S, Cooper MB, Betteridge DJ and Payne JE (1992) Carnitine deficiency associated with long-term pivampicillin treatment: the effect of a replacement therapy regime. *Postgrad Med J* 68: 932–934.

Rowinsky EK, Chaudhry V, Forastiere AA, Sartorius SE, Ettinger DS, Grochow LB, Lubejko BG, Cornblath DR and Donehower RC (1993a) Phase I and pharmacologic study of paclitaxel and cisplatin with granulocyte colony-stimulating factor: neuromuscular toxicity is dose-limiting. *J Clin Oncol* 11: 2010–2020.

Rowinsky EK, Chaudhry V, Cornblath DR and Donehower RC (1993b) Neurotoxicity of Taxol. *Monogr Natl Cancer Inst* 15: 107–115.

Schroder JM, Bertram M, Schnabel R and Pfaff U (1992) Nuclear and mitochondrial changes of muscle fibers in AIDS after treatment with high doses of zidovudine. *Acta Neuropathol Berlin* 85: 39–47.

Schwartzfarb L, Singh G and Marcus D (1977) Heroin-associated rhabdomyolysis with cardiac involvement. *Arch Int Med* 137: 1255–1257.

Semino-Mora MC, Leon-Monzon ME and Dalakas MC (1994) Effect of L-carnitine on the zidovudine-induced destruction of human myotubules. Part I: L-carnitine prevents the myotoxicity of AZT *in vitro. Lab Invest* 71: 102–112.

Shiintani S, Shiligai T and Tsukagoshi H (1991) Marked hypokalemic rhabdomyolysis with myoglobinuria due to diuretic therapy. *EurJ Neurol* 31: 396–398.

Simon EJ (1971) The effects of narcotics on cells in tissue culture. *In*: D Clouet (ed.): *Narcotic Drug: Biochemical Pharmacology.* Plenum Press, New York, pp 248–259.

Sinnwell TM, Sivakumar K, Soueidan S, Jay C, Frank JA and McLaughlin AC (1995) Metabolic abnormalities in skeletal muscle of patients receiving zidovudine therapy observed by ^{31}P *in vivo* magnetic resonance spectroscopy. *J Clin Invest* 96: 126–131.

Slavin G, Martin F, Ward P Levi J and Peters J (1983) Chronic alcohol excess is associated with selective but reversible injury to type 2B muscle fibers. *J Clin Pathol* 36: 772–777.

Smit JWA, Jansen GH, deBruiin TWA and Erkelens DW (1995) Treatment of combined hyperlipidemia with fluvastatin and gemfibrozil, alone or in combination, does not induce muscle damage. *Am J Cardiol* 76: 126A–128A.

Smith PF Eydelloth RS, Grossman SJ, Stubbs RJ, Schwartz MS, Germershausen JI, Vyase KP, Kari PH and MacDonald JS (1991) HMG-CoA reducatse inhibitor-induced myopathy in the rat: cyclosporin A interaction and a mechanism studies. *J Pharmacol Exp Ther* 257: 1225–1235.

Spadaro M, Tilia G, Massara Mc Damiani A, Parisi L Tomelleri G, D'Offizi G and Morocutti C (1993) Myopathy in long-term AZT therapy: clinical, electrophysiological and biopsy study in 67 HIV$^+$ subjects. *Ital J Neurol Sci* 14: 369–374.

Spence JD, Munoz CE, Hendricks L, Latchinian L and Khouri HE (1995) Pharmacokinetics of the combination of fluvastatin and gemfibrozil. *Am J Cardiol* 76: 80A–83A.

Sugie H, Russin R and Verity MA (1984) Emetine myopathy: two case reports with pathobiochemical analysis. *Muscle Nerve* 7: 54–59.

Sugita H, Higuchi I, Sano M and Ishiura S (1987) Trial of a cysteine proteinase inhibitor, EST, in experimental chloroquine myopathy. *Muscle Nerve* 10: 5165–523.

Suzuki S, Ichikawa K, Nagai M, Mikoshiba M, Mori J, Kaneko A, Sekine R, Asanuma N, Hara M, Nishii Y et al.(1997) Elevation of serum creatine kinase during treatment with antithryoid drugs in patients with hyperthyroidism due to Graves Disease. *Arch Intern Med* 157: 693–696.

Thyagarajan D, Day BJ, Wodak J, Gilligan B and Dennett X (1993) Emetine myopathy in a patient with an eating disorder. *Med J Aust* 159: 757–760.

Trounce I, Byrne E and Dennett X (1990) Biochemical and morphological studies of skeletal muscle in experimental chronic alcoholic myopathy. *Acta Neurol Scand* 82: 386–391.

Turbat-Herrara EA (1994) Myglobinuric acute renal failure associated with cocaine use. *Ultrastruct Patholol* 18: 127–131.

Urbano-Márquez A Ramon E, Navarro-Lopez F, Grau JM, Mont L and Rubin E (1989) The effects of alcoholism on skeletal and cardiac muscle. *N Engl J Med* 320: 409–415.

Urbano-Márquez A, Estruch R, Fernández-Solá, J, Nicolás JM, Paré JC and Rubin E (1995) The greater risk of alcoholic cardiomyopathy and myopathy in women compared to men. *JAMA* 274: 149–154.

Van-Balkom RH, vander-Heijden HF, van-Herwaarden CL and Dekhuijzen PN (1994) Corticosteroid-induced myopathy of the respiratory muscles. *Netherlands J Med* 45: 114–122.

Veerkamp JH, Smit JWA Benders AAGM and Oosterhof A (1996) Effects of HMG-CoA reductase inhibitors on growth and differentiation of cultured rate skeletal muscle cells. *Biochim Biophys Acta* 1315: 217–222.

Velasco E Finol HJ and Marquez A (1995) Toxic and neurogenic factors in chloroquine myopathy fibre selectivity. *J Submicrosc Cytol Pathol* 27: 451–457.

Vita G, Toscano A, Mileto G, Pitrone F, Ferro MT, Gagliardi E, Bresolin N, Fortunato F and Messina C (1993) Bezafibrate-induced myopathy: no evidence for defects in muscle metabolism. *Eur Neurol* 33: 168–172.

Waclawik AJ, Lindal S and Engel AG (1993) Experimental lovastatin myopathy. *J Neuropathol Exper Neurol* 52: 542–549.

Ward RJ and Peters TJ (1992) The antioxidant status of patients with either alcohol-induced liver damage or myopathy. *Alcohol Alcoholism* 27: 359–365.

Wei Y-H, Kao S-H and Lee H-C (1996) Simultaneous increase of mitochondrial DNA deletions and lipid peroxidation in aging. *Ann N Y Acad Sci* 786: 24–43.

Weindruch R (1995) Interventions based on the possibility that oxidative stress contributes to sarcopenia. *J Gerontol* 50A: 157–161.

Weissman JD, Constantinitis I, Hudgins P and Wallance DC (1992) ^{31}P magnetic resonance spectroscopy suggests impaired mitochondrial function in AZT-treated HIV-infected patients. *Neurology* 42: 619–623.

Wiklund O, Angelin B, Bergman M, Berglund L, Bondjers G and Carlsson A (1993) Pravastatin and gemfibrozil alone and in combination for the treatment of hypercholesterolemia. *Am J Med* 94: 13–20.

Wrogemann K and Peña SDJ (1976) *Lancet* ii: 672–274.

Yagiela JA, Benoit PW and Fort NF (1982) Mechanism of epinephrine enhancement of lidocaine-induced skeletal muscle necrosis. *J Dental Res* 61: 686–690.

Yamanishi Y, Ishibe Y, Taooka Y, Mukuzono H, Aoi K and Yamana S (1993) A case of cyclosporin A-induced myopathy. *Ryumachi* 33: 63–67.

Yang ZP and Dettbarn W-D (1996) Diisopropylphosphorofluoridate-induced cholinergic hyperactivity and lipid peroxidation. *Toxicol Pharmacol* 138: 48–53.

Free radicals and diseases of animal muscle

J.R. Arthur

Division of Micronutrient and Lipid Metabolism, Rowett Research Institute, Bucksburn, Aberdeen, AB21 9SB, Scotland, UK

Introduction

There are many diseases of both skeletal and cardiac muscle which occur in animals. These include congenital dystrophies, viral-induced myopathies and nutritional disorders, the last often being associated with micronutrients. The diagnosis and pathology of animal muscle diseases are well documented in many comprehensive reviews and are not discussed in detail in this chapter (Bradley and Fell, 1981; Ohlendieck, 1996). Many muscle disorders involve breakdown or necrosis of muscle and/or nerve tissues, a process that is likely to be associated with production of free radicals. However, it is often difficult to determine whether these free radicals are the cause, rather than the consequence, of the disease (Duthie and Arthur, 1993). Thus, supplementation with antioxidant nutrients, which may decrease signs of free radical activity in diseased animals, may be suppressing the symptoms of the disorder rather than causing it. However, environmental stresses that favour *in vivo* free radical formation often exacerbate nutritional myopathies in animals and thus support a role for free radicals in the underlying disease process (Kennedy et al., 1987; Arthur, 1988; Smith and Allen, 1997; Walsh et al., 1993b). This chapter concentrates on two main types of muscle disorders, namely those caused by combined selenium and vitamin E deficiencies and those associated with malignant hyperthermia. In both these types of disorder, free radical activity is likely to be an underlying factor in the disease process, particularly when it is associated with deficiencies in cell antioxidant systems.

Myopathies associated with selenium and vitamin E deficiencies

Several adverse effects have been associated with selenium and vitamin E deficiency in cattle, sheep and pigs; these include poor reproductive performance, retained placenta, low growth rates, liver necrosis, and most frequently, skeletal myopathies and cardio-myopathies (Tab. 1). The involvement of selenium and vitamin E deficiencies in these latter diseases was first recognized in the late 1950s. Economic losses caused by myopathies, impaired growth and premature deaths initiated the institution of many selenium and vitamin E supplementation programmes although, at this time, very little was known about the metabolic functions of selenium and vitamin E (Blaxter, 1962). However, since recognition of the essentiality of selenium in glutathione peroxidases and the involvement of vitamin E in the protection of membrane fatty acids, free radicals have been hypothezised as a significant cause of the myopathies (Arthur, 1988; Walsh et al., 1993a). Thus

Table 1. Selenium- and/or vitamin E-responsive diseases in farm animals

Disease	tissue
Myopathies heart/ skeletal (white muscle disease)	Muscle
Exudative diathesis	Capillaries
Pancreatic necrosis	Pancreas
Hepatosis dietetica	Liver
Growth (ill thrift)	Muscle
Poor reproduction	Ovary/testis
Retained placenta	Uterus/muscle

For further details of the above conditions see Blaxter (1962); Arthur (1982); Combs and Combs (1986); Putnam and Comber (1987); Vanvleet and Ferrans (1992); Hansen and Deguchi (1996).

selenium and vitamin E deficiencies are thought to compromise the antioxidant systems which maintain the structure and function of cell membranes, particularly those stabilizing lipids and sulphhydryl-containing proteins.

Two major forms of clinical myopathy occur in selenium- and vitamin E-deficient ruminants; the first, the "congenital" form, occurs in newborn lambs and calves, being associated with stillbirths or deaths within the first few days of life, particularly after "stresses" such as feeding or exercise. The second "delayed" form of the disease affects older animals and may occur in animals up to 2 years of age (McDowell et al., 1996) The heart is less often involved in the delayed disease, which therefore causes fewer fatalities than the congenital form. In cattle and sheep, additional stresses such as changes in diet, environment and transport may predispose towards disease (Arthur, 1988). Such factors may increase endogenous production of free radicals and thus strengthen a role for these molecules in the pathogenesis of the disorder. Before further consideration of the role of selenium deficiency and the free radicals in myopathies, it is necessary to summarize the many biological functions of selenium. The recent discovery of many new roles for selenium has allowed a more accurate interpretation of the relationships between the micronutrient and vitamin E in antioxidant systems.

Biological functions of selenium

The biological functions of selenium are primarily based on the activities and properties of 30 or more selenoproteins thought to occur in mammals. Twelve selenoproteins, each with distinct properties as well as unique intracellular and tissue distributions, have been characterized by purification and/or cloning. These fulfil a range of metabolic functions (Tab. 2). The four glutathione peroxidases function to remove potentially injurious peroxides, which can initiate free radical chain reactions and thus cause damage to cell membranes. The glutathione peroxidases are found in different organs and cell fractions and probably have complementary functions (Hoekstra, 1975; Thomas et al., 1990; Maiorino et al., 1995; Imai et al., 1996). Extracellular glutathione peroxidase (eGSHPx) has the potential to metabolize peroxides in body fluids and extracellular

Table 2. Mammalian selenoproteins

Enzyme/protein	Function
Glutathione peroxidases	
Cytosolic	Antioxidant
Phospholipid hydroperoxide	
Extracellular	
Gastrointestinal	
Iodothyronine deiodinases	
Type I	Thyroid hormone
Type II	Metabolism
Type III	
Thioredoxin reductase	Redox/antioxidant
Selenoprotein P	Antioxidant?
Selenoprotein W	Redox/antioxidant?
Selenophosphate synthetase metabolism	
Sperm capsule selenoprotein	Structural?
Selenium-binding proteins 14, 56 and 58 kDa	Unknown

The above selenium-containing proteins are discussed in greater detail in the following reviews and original publications: Burk and Hill (1993; Combs and Combs (1986); Arthur and Beckett (1994); Croteau et al. (1995); Croteau et al. (1996); Arthur et al. (1996); Kim et al. (1997).

spaces, whereas the other three glutathione peroxidases have intracellular functions. Cytosolic glutathione peroxidase (cGSHPx) is primarily located in the soluble fraction of cells and was the first selenium-containing enzyme to be discovered (Rotruck et al., 1973). Thus, it was hypothesised that the basis of the interaction between selenium and vitamin E deficiencies was the impairment of peroxide metabolism in the cytosol and free radical scavenging in the cell membrane (Hoekstra, 1975). However, in many instances, an almost complete absence of cGSHPx activity in liver or muscle could not be associated with any increases in free radical activity (Reiter and Wendel, 1983; Arthur et al., 1987). Thus, it is now uncertain whether loss of this enzyme activity plays a major part in the interactions between selenium and vitamin E deficiencies.

The discovery of phospholipid hydroperoxide glutathione peroxidase (phGSHPx), a protein associated with cell membranes, has provided a more plausible basis for the interactions between selenium and vitamin E-dependent antioxidant systems (Thomas et al., 1990). Crucially, phGSHPx can metabolise phospholipid hydroperoxides that cGSHPx cannot. Cytosolic glutathione peroxidase can only metabolise long-chain fatty acid hydroperoxides after they have been released from phospholipids by phospholipase A_2 (Grossmann and Wendel, 1983). Although this latter mechanism may provide an antioxidant role for cGSHPx, it is less likely to be important than that of the membrane-associated phGSHPx. The compartmentalisation of antioxidant activities of the three most widely distributed glutathione peroxidases is summarized in Figure 1. Thus, in cooperation with vitamin E located in the lipid of the cell membranes, the glutathione peroxidases are able to provide antioxidant protection to the cell. In addition to cGSHPx, eGSHPx and phGSHPx which are widely distributed throughout the body, gastrointestinal glutathione peroxidase (giGSHPx) has been cloned and expressed, and has very similar properties to cGSHPx. The mRNA for giGSHPx is found predominantly in the intestinal tract and the specific functions of the enzyme are unclear (Chu et al., 1993; Chu and Esworthy, 1995). The function is, however, unlikely to be related to myopathies unless it is involved in prevention of absorption of dietary peroxides, which could act as an oxidative stress in selenium- and vitamin E-deficient tissues.

Although changes in glutathione peroxidase activities can be used to explain the relationships between selenium and vitamin E, other selenoproteins, which may influence this process, have been detected in skeletal muscle. In particular, selenoprotein W has been purified and cloned and contains one selenocysteine coded for by an "in frame" TGA codon (Vendeland et al., 1995; Yeh et al., 1997). The functions of muscle selenoprotein W have not been defined, although it is postulated to have an antioxidant role. This would be consistent with the redox properties of selenocysteine at physiological pH and the binding of glutathione molecules to selenoprotein W

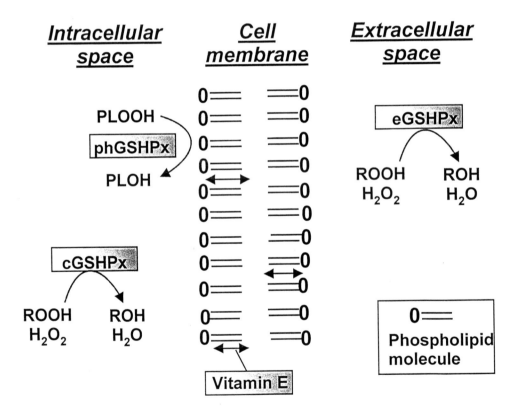

Figure 1. Sites of action and substrates of glutathione peroxidases and relationship to vitamin E. PLOOH, phospholipid hydroperoxide; PLOH, phospholipid alcohol; ROOH, lipid hydroperoxide; ROH, lipid alcohol; H_2O_2, hydrogen peroxide.

during its purification from muscle (Beilstein et al., 1996). Selenoprotein W was originally identified by *in vivo* labelling of muscle from lambs with ^{75}Se, animals with selenium- and vitamin E-responsive nutritional myopathy (white muscle disease) having decreased levels of muscle selenoprotein W. Thus loss of the protein was thought to be involved in the pathological changes associated with the disease (Combs and Combs, 1986).

Selenium is also essential for three deiodinase enzymes which regulate the synthesis and further metabolism of 3,3',5-triiodothyronine (T_3) – the metabolically active form of the thyroid hormone (Arthur and Beckett, 1994; Arthur et al., 1996). The mRNA for type II deiodinase has been detected in human muscle and even a low enzyme activity per gram of tissue would make a significant contribution to whole-body T_3 production (Salvatore et al., 1996). If deiodinase activity occurs in muscle of cattle and sheep, it would influence T_3 levels and thus the tissue's basal metabolism. However, there is no evidence to connect thyroid metabolism with the incidence of myopathy in sheep and cattle.

Thus, selenium is likely to be involved in the antioxidant systems in muscle and, in particular, in those that lower free radical formation by decreasing the likelihood of non-enzymatic degradation of peroxides and metal ion-catalyzed formation of hydroxyl or similar reactive chemical species. Although all cell constituents may be vulnerable to attack by free radicals, those components associated with membranes, particularly the unsaturated lipids, are at particular risk. Oxidation of the fatty acids produces toxic by-products including nonenal and other aldehydes as well as causing major disruption to the membrane structure (Walsh et al., 1993b). In the context of the initiation of disease, selenium deficiency will allow accumulation of potentially injurious peroxides, which may propagate free radical activity and damage the essential components of cells. In addition, vitamin E deficiency will also allow "survival of radicals" and thus potentially damaging chain reactions in the lipid membranes of the cell. As well as damage to the structural components of membranes, vitamin E deficiency will also allow oxidation of essential thiol groups in membrane enzymes, which maintain the correct inflow and outflow of metal ions from the cell. Imbalances in distribution and intracellular accumulation of Ca^{2+} are a characteristic of many selenium/vitamin E-based myopathies and are consistent with oxidative damage to the cell membrane (Arthur, 1988).

If free radical activity, following combined selenium and vitamin E deficiencies, underlies muscle problems seen in animals, it follows that the disease should be associated with oxidation of cell components and will be exacerbated by "oxidative stress". The term oxidative stress is used loosely in this context to include: direct and indirect stresses, which can range from toxicity of free transition metal ions to increased metabolism involving oxygen such as during and after excessive exercise. In addition to the direct generation of free radicals, the structure of membranes can provide a very variable substrate for free radical attack. Thus, a high percentage of polyunsaturated fat in membranes can be a form of "oxidative stress" by providing the ideal substrate for lipid peroxidation.

Nutritional myopathy and free radicals

Nutritional myopathy is almost always a consequence of combined selenium and vitamin E deficiencies and can be treated with a supplement of either micronutrient (Blaxter, 1962). The association of the condition with low antioxidant status is consistent with a crucial role for free radicals in the pathogenesis of the disease. However, a low selenium and vitamin E status will not always cause myopathy and further stresses are often needed to precipitate the disease (Arthur, 1988). As discussed in the previous section, loss of selenium-containing glutathione peroxidases would favour conditions in which free radicals are formed. Thus, in muscle from selenium- and vitamin E-deficient cattle, there are elevated indices of lipid peroxidation – a process mediated through free radical reactions (Walsh et al., 1993b). These indices, thiobarbituric acid-reactive substances (TBARS), ascorbate-induced TBARS, ascorbate-induced hexanal and ion-induced 4-hydroxy-nonenal, were not increased in selenium-deficient calves but only in those that were both selenium and vitamin E deficient (Walsh et al., 1993b). Although TBARS, hexanal and nonenal are indirect measures of lipid peroxidation, they are indicative of increased free radical activity and, as this occurred before or at the time of onset of myopathy in cattle it is likely that they are associated with the origin of the disease. In addition, in selenium- and vitamin E-deficient rats there is increased free radical production in muscle tissue as determined by electron spin resonance (ESR) spectroscopy. This only occurs with combined selenium and vitamin E deficiency and emphasizes that free radical formation may be stopped either at the stage of peroxide formation or directly by radical scavenging in the membrane (Arthur et al., 1988).

The role of free radical activity in causing nutritional myopathy is further supported by the range of factors that are needed to precipitate the disease in selenium- and vitamin E-deficient animals. Cattle with a very low selenium status, even with combined vitamin E deficiency, do not invariably develop muscle disease. The acute focal myopathy and accumulation of calcium in muscle that can occur with these deficiencies is frequently linked to stresses, such as turnout of deficient animals from indoor housing to fresh open pasture (Arthur, 1988). This raises the possibility of either dietary or environmental stresses being essential for the onset of disease, for example, an intake of fresh grass would result in an increase in the intake of polyunsaturated fatty acids (Kennedy et al., 1987). That this causes disease is consistent with early studies in which high levels of cod liver oil were used to induce myopathy in vitamin E-deficient cattle (Blaxter, 1962). Polyunsaturated fatty acids are very susceptible to free radical attack, and any increased oxidative activity associated with impaired antioxidant systems or exercise in muscle of selenium- and vitamin E-deficient cattle would provide ideal circumstances for the initiation of myopathy (Rice et al., 1986; Arthur, 1988).

In addition to fat intake, the fatty acid composition of muscle can also modulate its susceptibility to selenium and vitamin E deficiency. Myopathy occurs spontaneously in selenium- and vitamin E-deficient cattle with high endogenous levels of polyunsaturated fatty acids in their tissues. Dependent on the type of diet consumed, linoleic acid, (18:2, ω6), can comprise between 8% and 30% of total fatty acids in muscle of cattle (Rice et al., 1986). When diets consumed by cattle are treated with sodium hydroxide to improve digestibility, this causes high levels of endogenous polyunsaturated fatty acids. Furthermore, alkali treatment destroys endogenous vitamin E in the diet, enhancing susceptibility to lipid peroxidation and thus the chance of spontaneous myopathy

(Walsh et al., 1993b). In contrast, animals with a very low selenium and vitamin E status may be resistant to treatment with polyunsaturated fatty acids if their endogenous fatty acids are predominantly saturated, as occurs with some commercial diets (Arthur and Duthie, 1994a). However, in combination with the stress of turnout from indoor housing to pasture, exercise and changes in temperature, consumption of grass causes nutritional myopathy (Arthur, 1988).

Exercise causes major increases in muscle oxidative metabolism which can result in free radical-mediated damage to lipids and proteins within the muscle (Reznick et al., 1992). Thus, stimulation of muscle metabolism in antioxidant-deficient animals will provide the conditions that are most likely to induce myopathy. Anecdotal evidence also associated magnesium deficiency with myopathy caused by selenium and vitamin E deficiency in farms. This is also consistent with a free radical-mediated mechanism causing myopathy because magnesium deficiency increases free radical formation in muscle of rats (Rock et al., 1995).

As metabolic rates may increase twentyfold during exercise, this can result in leakage of oxygen radicals from mitochondria into the cytosol. In antioxidant-deficient muscle, this rise in free radical concentration will exceed the capacity of the cell to detoxify the molecules and thus result in tissue damage (Reznick et al., 1992). Therefore, free radical production can be associated with pathogenesis of nutritional myopathy caused by selenium and vitamin E deficiency, particularly if exacerbated by exercise and high dietary and tissue levels of polyunsaturated fatty acids. These different processes are summarised in Figure 2 which emphasises that disease will only occur when the oxidative stress on the muscle of the animal exceeds the antioxidant capacity; this will vary with dietary selenium and vitamin E.

Malignant hyperthermia

Malignant hyperthermia (MH) is a disorder that used to cause significant economic losses for the swine production industry. Malignant hyperthermia is an inherited disorder which has been associated with a mutation in the gene for the ryanodine-sensitive calcium transporter (Duthie and Arthur, 1993). The muscle damage and release of enzymes such as pyruvate kinase and creatine kinase into plasma, which occur in MH, have many similarities to vitamin E deficiency. Thus, although the underlying defect of MH may be abnormal calcium transport, the pathogenesis of the disease may be dependent on free radical activity. Consistent with this in MH-susceptible pigs, there are increased indices of lipid peroxidation in plasma. In addition, tissues from MH-susceptible animals are more prone to be oxidized and produce free radicals which can be detected by spin-trapping and ESR spectroscopy (Duthie et al., 1990). Thus, although free radical activity is involved in the pathogenesis of MH in pigs, it is not the initial cause of the condition. The influence of antioxidants on biochemical indicators of free radical activity and muscle damage in MH-susceptible swine reveals an ability to modulate rather than prevent the syndrome (Duthie and Arthur, 1993).

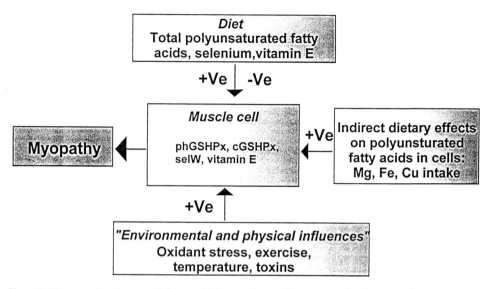

Figure 2. Dietary and environmental factors which can influence the pathogenesis of nutritional myopathy. –Ve acts to prevent myopathy, +Ve acts to induce myopathy.

Conclusions

There is good evidence that free radicals are involved in the initiation of muscle damage in selenium- and vitamin E-deficient animals. This is consistent with the physical and dietary stresses that exacerbate the myopathy also being potential stimulators of free radical activity. The ability of antioxidants to prevent myopathy is a direct effect on the disease process, whereas their effects on MH are a secondary amelioration of symptoms.

Acknowledgements
Work in the author's laboratory is supported by the Scottish Office Agriculture Environment and Fisheries Department (SOAEFD).

References

Arthur JR (1982) Nutritional inter-relationships between selenium and vitamin E. *Report of the Rowett Institute* 38: 124–135.
Arthur JR (1988) Effects of selenium and vitamin E status on plasma creatine kinase activity in calves. *J Nutr* 118: 747–755.
Arthur JR and Beckett GJ (1994) New metabolic roles for selenium. *Proc Nutr Soc* 53: 615–624.
Arthur JR and Duthie GG (1994a) Free radicals and trace elements in muscle disorders and sport. *In*: AE Favier, J Neve and P Faure (eds): *Trace Elements and Free Radicals in Oxidative Diseases*. AOCS Press, Champaign IL, pp 241–248.

Arthur JR, Morrice PC, Nicol F, Beddows SE, Boyd R, Hayes JD and Beckett GJ (1987) The effects of selenium and copper deficiencies on glutathione S-transferase and glutathione peroxidase in rat liver. *Biochem J* 248: 539–544.

Arthur JR, McPhail DB and Goodman BA (1988) Spin trapping of free radicals in homogenates of heart from selenium and vitamin E deficient rats. *Free Radical Res Commun* 4: 311–315.

Arthur JR, Bermano G, Mitchell JH and Hesketh JE (1996) Regulation of selenoprotein gene expression and thyroid hormone metabolism. *Biochem Soc Trans* 24: 384–388.

Beilstein MA, Vendeland SC, Barofsky E, Jensen ON and Whanger PD (1996) Selenoprotein W of rat muscle binds glutathione and an unknown small molecular weight moiety. *J Inorg Biochem* 61: 117–124.

Blaxter KL (1962) Vitamin E in health and disease in cattle and sheep. *Vitamins Hormones* 20: 633–643.

Bradley R and Fell BF (1981) Myopathies in animals. *In*: Sir J Walton (ed.): *Disorders of Voluntary Muscle*, 4th edn. Churchill Livingstone, London, pp 824–872.

Burk RF and Hill KE (1993) Regulation of Selenoproteins. *Annu Rev Nutr* 13: 65–81.

Chu FF and Esworthy RS (1995) The expression of an intestinal form of glutathione peroxidase (GSHPx-GI) in rat intestinal epithelium. *Arch Biochem Biophys* 323: 288–294.

Chu FF, Doroshow JH and Esworthy RS (1993) Expression, characterization, and tissue distribution of a new cellular selenium-dependent glutathione peroxidase, GSHPx-GI. *J Biol Chem* 268: 2571–2576.

Combs GF and Combs SB (1986) *The Role of Selenium in Nutrition*. Academic Press, New York.

Croteau W, Whittemore SL, Schneider MJ and St Germain DL (1995) Cloning and expression of a cDNA for a mammalian type III iodothyronine deiodinase. *J Biol Chem* 270: 16569–16575.

Croteau W, Davey JC, Galton VA and St Germain DL (1996) Cloning of the mammalian type II iodothyronine deiodinase – A selenoprotein differentially expressed and regulated in human and rat brain and other tissues. *J Clin Invest* 98: 405–417.

Duthie GG and Arthur JR (1993) Free radicals and calcium homeostasis: relevance to malignant hyperthermia. *Free Radical Biol Med* 14: 435–442.

Duthie GG, McPhail DB, Arthur JR, Goodman BA and Morrice PC (1990) Spin trapping of free radicals and lipid peroxidation in microsomal preparations from malignant hyperthermia susceptible swine. *Free Radical Res Commun* 8: 93–99.

Grossmann A and Wendel A (1983) Non-reactivity of the selenoenzyme glutathione peroxidase with enzymically hydroperoxidised phospholipids. *Eur J Biochem* 135: 549–552.

Hansen JC and Deguchi Y (1996) Selenium and fertility in animals and man – A review. *Acta Vet Scand* 37: 19–30.

Hoekstra WG (1975) Biochemical function of selenium and its relation to vitamin E. *Fed Proc* 34: 2083–2089.

Imai H, Sumi D, Sakamoto H, Hanamoto A, Arai M, Chiba N and Nakagawa Y (1996) Overexpression of phospholipid hydroperoxide glutathione peroxidase suppressed cell death due to oxidative damage in rat basophile leukemia cells (RBL-2H3). *Biochem Biophys Res Commun* 222: 432–438.

Kennedy S, Rice DA and Davidson WB (1987) Experimental myopathy in vitamin E and selenium depleted calves with and without added dietary poly-unsaturated fatty acid as a model for nutritional degenerative myopathy in ruminant cattle. *Res Vet Sci* 43: 384–394.

Kim IY, Guimaraes MJ, Zlotnik A, Bazan JF and Stadtman TC (1997) Fetal mouse selenophosphate synthetase 2 (SPS2): Characterization of the cysteine mutant form overproduced in a baculovirus-insect cell system. *Proc Natl Acad Sci USA* 94: 418–421.

McDowell L R, Williams SN, Hidiroglou N, Njeru CA, Hill G M, Ochoa L and Wilkinson NS (1996) Vitamin E supplementation for the ruminant. *Anim Feed Sci Technol* 60: 273–296.

Maiorino M, Aumann KD, Brigeliusflohe R, Doria D, Vandenheuvel J, McCarthy J, Roveri A, Ursini F and Flohe L (1995) Probing the presumed catalytic triad of selenium-containing peroxidases by mutational analysis of phospholipid hydroperoxide glutathione peroxidase (PHGPx). *Biol Chem Hoppe Seyler* 376: 651–660.

Ohlendieck K (1996) Molecular pathogenesis of diseases of the muscle. *Naturwissenschaften* 83: 555–565.

Putnam ME and Comber N (1987) Vitamin E. *Vet Rec* 121: 541–545.

Reiter R and Wendel A (1983) Selenium and drug metabolism-I. Multiple modulations of mouse liver enzymes. *Biochem Pharmacol* 32: 3063–3067.

Reznick AZ, Witt E, Matsumoto M and Packer L (1992) Vitamin-E inhibits protein oxidation in skeletal muscle of resting and exercised rats. *Biochem Biophys Res Commun* 189: 801–806.

Rice DA, Kennedy S, McMurray CH and Blanchflower WJ (1986) Differences in tissue concentrations of n-3 and n-6 fatty acids in vitamin E and selenium deficient cattle. *In*: CH McMurray and DA Rice (eds): *Proceedings of the 6th International Conference on Production Diseases in Farm Animals*. University of Belfast, pp 229–232.

Rock E, Astier C, Lab C, Vignon X, Gueux E, Motta C and Rayssiguier Y (1995) Dietary magnesium deficiency in rats enhances free radical production in muscle. *J Nutr* 125: 1205–1210.

Rotruck JT, Pope AL, Ganther HE, Swanson AB, Hafeman DG and Hoekstra WG (1973) Selenium: biochemical role as a component of glutathione peroxidase. *Science* 179: 588–590.

Salvatore D, Bartha T, Harney JW and Larsen PR (1996) Molecular biological and biochemical characterization of the human type 2 selenodeiodinase. *Endocrinology* 137: 3308–3315.

Smith GM and Allen JG (1997) Effectiveness of alpha-tocopherol and selenium supplements in preventing lupinosis-associated myopathy in sheep. *Aust Vet J* 75: 341–348.

Thomas JP, Maiorino M, Ursini F and Girotti AW (1990) protective action of phospholipid hydroperoxide glutathione peroxidase against membrane-damaging lipid peroxidation – *in situ* reduction of phospholipid and cholesterol hydroperoxides. *J Biol Chem* 265: 454–461.

Vanvleet JF and Ferrans VJ (1992) Etiologic factors and pathologic alterations in selenium-vitamin-E deficiency and excess in animals and humans. *Biol Trace Elem Res* 33: 1–21.

Vendeland SC, Beilstein MA, Yeh JY, Ream W and Whanger PD (1995) Rat skeletal muscle selenoprotein W: cDNA clone and mRNA modulation by dietary selenium. *Proc Natl Acad Sci USA* 92: 8749–8753.

Walsh DM, Kennedy DG, Goodall EA and Kennedy S (1993a) antioxidant enzyme activity in the muscles of calves depleted of vitamin E or selenium or both. *Brit J Nutr* 70: 621–630.

Walsh DM, Kennedy S, Blanchflower WJ, Goodall EA and Kennedy DG (1993b) Vitamin E and selenium deficiencies increase indices of lipid peroxidation in muscle tissue of ruminant calves. *Int J Vitam Nutr Res* 63: 188–194.

Yeh JY, Gu QP, Beilstein MA, Forsberg NE and Whanger PD (1997) Selenium influences tissue levels of selenoprotein W in sheep. *J Nutr* 127: 394–402.

Therapeutic trials of antioxidants in muscle diseases

M.J. Jackson and R.H.T. Edwards

Department of Medicine, University of Liverpool, Liverpool L69 3GA, UK

Introduction

Therapeutic trials of antioxidants in muscle diseases have a long history. One of the first clinical trials in patients with Duchenne muscular dystrophy (DMD) was to examine the effect of vitamin E on progression of the disease (Bicknell, 1940). Although this author was initially very hopeful of the benefit of these compounds, it latterly became clear that no true positive effects were seen. This study exemplifies the history of clinical trials in DMD, and other severe genetically determined muscle diseases, where initial data from uncontrolled and hopeful studies of possible therapies have been inevitably followed by carefully controlled trials that show no benefit of the compound being tested.

A critical review of the early trials and suggestions for the design of future trials was presented by Dubowitz and Heckmatt in 1980 and this is the benchmark against which all trials in this area should be judged. Subsequent trials in DMD have been reviewed by Toescu et al. (1994). This short review will concentrate on a critical re-examination of trials of antioxidants in Duchenne, Becker and myotonic muscular dystrophies. These are the muscle diseases for which there are some firm data indicating an involvement of free radicals in the degenerative process.

Background to trials of antioxidant therapy in muscle diseases

It has been recognized for a number of years that the most severe forms of the degenerative muscular dystrophies share many of the characteristics of vitamin E- or selenium-deficiency myopathy of animals and indeed the animal disorders have been incorrectly termed "nutritional muscular dystrophies" (Bradley and Fell, 1980). The recognition that the major functions of vitamin E and selenium within the body are probably as antioxidants, acting to inhibit the toxic effects of free radicals has prompted investigations of the possible role of free radicals in some of the human disorders, particularly myotonic muscular dystrophy and Duchenne muscular dystrophy.

The evidence for an involvement of free radical-mediated processes in the muscular dystrophies is largely based on extrapolation from animal work, studies of the antioxidant content of tissues and body fluids, and measurement of free radical reaction products in blood or tissues of patients. In essence, it appears that patients with DMD have evidence of increased lipid peroxidation in blood and muscle tissue (Kar and Pearson, 1979; Jackson et al., 1984; Mechler et al., 1984), and have some evidence for an increase in the activity of antioxidant enzymes in muscle tissue (Kar and Pearson, 1979; Austin et al., 1992), which is assumed to be an adaptive response to chronic increased oxidative stress in muscle.

The possibility that such changes reflect increased free radical activity occurring as a result of the process of muscle degeneration (rather then as a cause of it) has received little attention, although there is some suggestion that this may occur in DMD. Foxley et al. (1991) reported that indicators of lipid peroxidation were normal in muscles from the *mdx* mouse, a precise animal model of DMD, which does not show the same extent of muscle degeneration.

Similar but less numerous reports have appeared indicating a possible increase in oxidative stress in patients with myotonic dystrophy. In these patients, blood selenium levels may be low (Orndahl et al., 1982; Ihara et al., 1995) and serum lipid peroxides appear to be increased (Ihara et al., 1995).

Duchenne/Becker muscular dystrophy

The genetic defect responsible for Duchenne (DMD) and Becker muscular dystrophies (BMD) was identified in 1985 (Monaco et al., 1985), and localized to band Xp21 on the human X chromosome. The cloned gene was used to identify the protein product, dystrophin, which was shown to be absent or greatly diminished in muscle from DMD patients (Hoffman et al., 1987).

Dystrophin is found in skeletal, cardiac and smooth muscle, as well as brain (Karpati and Carpenter, 1988), and proteins from the small C-terminal transcripts of the same gene are found in other tissues (Chamberlain et al., 1993). Immunohistological staining, using antibodies raised against dystrophin, located the protein on the cytoplasmic side of the plasma membrane (Watkins et al., 1988; Cullen et al., 1990). Karpati and Carpenter (1988) initially proposed that dystrophin was orientated with its N-terminus attached to actin in the cytoskeleton, and its C-terminus anchored through the plasma membrane. Research groups in the USA (Ervasti and Campbell, 1991) and Japan (Yoshida and Ozawa, 1990) have proposed that dystrophin is a cytoplasmic protein which exists in association with a large oligomeric glycoprotein complex that spans the plasma membrane; this is now thought to be attached to the merosin (laminin M) component of the muscle extracellular matrix (Ibraghiminov-Baskrovnaya et al., 1992), although the precise molecular organization of the glycoprotein–dystrophin complex remains unclear (Ervasti and Campbell, 1991; Suzuki et al., 1994).

Lack of dystrophin appears to lead to the loss of associated proteins in *mdx* mouse and DMD muscle (Ervasti et al., 1990; Ohlendieck and Campbell, 1991). It has been suggested that it may be the loss of one or more of these associated proteins, or the whole complex, that leads to the degeneration seen in dystrophin-deficient muscle, because the specific loss of one of the associated glycoproteins, in the presence of apparently normal expression and localization of dystrophin, results in a myopathy similar to DMD (Matsumura et al., 1992; Sewry et al., 1994). Furthermore, recent data also indicate that patients with congenital muscular dystrophy have a specific reduction in merosin in muscle extracellular matrix (Hayashi et al., 1993), further supporting the idea that defects in various parts of this protein–glycoprotein structure lead to myopathy.

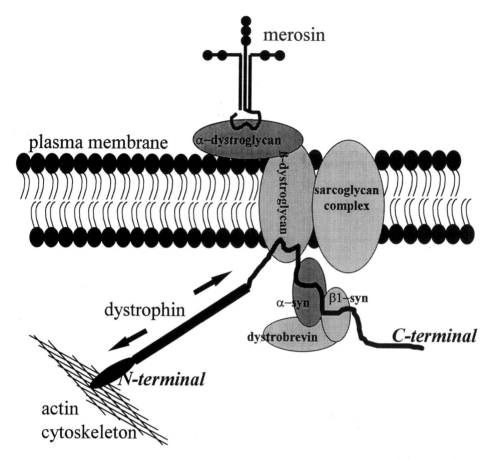

Figure 1. A schematic representation of the dystrophin–protein complex which appears to link the muscle cyto-skeleton to the extracellular matrix. The function of the complex is currently obscure. A lack of dystrophin gives rise to Duchenne muscular dystrophy and Becker muscular dystrophy is associated with the presence of abnormal dystrophin. Defects in other parts of the complex are associated with other less common muscular dystrophies. (Derived from McArdle and Jackson, 1997.)

A recently proposed structure for the dystrophin–protein complex (DPC) is shown in Figure 1. There is still considerable controversy about the function of this complex and the manner in which a lack or defect in dystrophin leads to muscle degeneration. A review of the theories has been presented by McArdle et al. (1995).

A potential link between these cell biological advances in the understanding of DMD and BMD and free radical metabolism has been identified by Brenman and co-workers (1995). They have shown that nitric oxide synthase (NOS) is absent from the membranes of fast-twitch fibres in DMD patients and *mdx* mice. They attribute this to an interaction between the DPC and an N-terminal domain of neuronal-type NOS which localizes the enzyme. Thus loss of whole or part of the DPC in DMD will lead to a lack of NOS in the muscle plasma membrane. Further studies have demonstrated a similar loss of NOS from BMD muscle (Chao et al., 1996). The functional consequences of this defect are currently unclear, but Brenman et al. (1995), have speculated that aberrant regulation of NOS may contribute to preferential degeneration of fast-twitch muscle fibres in DMD/BMD.

Trials of antioxidants in DMD/BMD

A large number of therapeutic trials of antioxidants have been undertaken in patients with DMD/BMD (Tab. 1). Although some authors have claimed benefit for their patients, all the well-controlled trials of antioxidants have shown no beneficial effects (Edwards et al., 1984).

Myotonic dystrophy

There have also been substantial advances in the understanding of the genetic and phenotypic basis for myotonic muscular dystrophy, although this field has not developed to the same extent as DMD/BMD.

The myotonic dystrophy gene codes for a protein homologous to serine/threonine protein kinases. This gene contains a repeated trinucleotide motif (CTG) in the 3'-untranslated region (Timchenko et al., 1995). In myotonic dystrophy patients, the CTG repeats are extremely unstable, varying in length from patient to patient and generally increasing in length in successive generations. There is a strong correlation between the size of the repeats and the age of onset of

Table 1. Antioxidant agents tested in patients with Duchenne/Becker muscular dystrophy

Agent	Stated result	Reference
Vitamin E	Benefit	Bicknell (1940)
	No benefit	Fitzgerald and McArdle (1941)
	No benefit	Bernske et al. (1960)
	No benefit	Edwards et al. (1984)
Vitamin E and selenium	No benefit	Gamstorp et al. (1986)
Selenium	No benefit	Jackson et al. (1989)
Superoxide dismutase ("Orgotein")	No benefit	Stern et al. (1982)
Allopurinol	Benefit	Thompson and Smith (1978)
	No benefit	Mendell and Weichers (1979)
	No benefit	Bertorini et al. (1985)
	no benefit	Griffiths et al. (1985)
Zinc	No benefit	Jackson and Edwards (1986)
Coenzyme Q_{10}	Benefit	Folkers and Simonsen (1995)

the disease. A considerable amount of research effort has been concentrated on trying to evaluate the pathophysiological effect of these genetic differences. Initial data indicated that the phenotypic defect might result from a steady-state reduction in the myotonic dystrophy protein kinase (DM protein kinase) and, although there is increasing evidence that myotonic dystrophy patients have reduced levels of the protein (Koga et al., 1994), there is now increasing doubt concerning the validity of this hypothesis (Tinchenko et al., 1995). Transgenic and knock-out mice studies indicate that simple loss or gain of expression of the protein kinase is not the only crucial requirement for development of the disease (Jansen et al., 1996).

The role of the DM protein kinase is also currently unknown, although some data indicate that it can modulate sodium channels in skeletal muscle, which is compatible with known changes in myotonic dystrophy (Mounsey et al., 1995). Currently, there appears to be no clear link between these accepted defects in myotonic dystrophy and free radical production or antioxidant metabolism.

Trials of antioxidants in myotonic dystrophy

A number of studies have examined the effect of selenium and vitamin E supplements in myotonic dystrophy patients. The first two studies by Orndahl and co-workers (1983, 1986) reported impressive beneficial results, but a third study by them (Orndahl et al., 1994) could find no evidence of beneficial effects. The sparse data from therapeutic trials are therefore contradictory. However, in the light of the new data indicating the underlying genetic and biochemical defects in myotonic dystrophy, it is difficult to find a logical rationale for further study of antioxidant therapy in this disorder.

Other disorders

Antioxidant therapy has been attempted, in an uncontrolled manner, in many other muscle disorders, but the lesson from studies in patients with DMD or BMD is that isolated subjective reports of beneficial effects should be interpreted cautiously. There is increasing evidence that antioxidant supplementation may be of benefit in some acquired muscle disorders such as alcoholic myopathy (V.R. Preedy et al., this volume), post-ischaemic muscle dysfunction (see Bushell et al., 1996), and the muscle degeneration associated with HIV infection. However, their use in patients with primary muscle disease does not seem to be warranted.

Acknowledgements
The authors would like to acknowledge the input of all their co-workers in this area and the Muscular Dystrophy Group of Great Britain and Northern Ireland for their financial support of much of the work described.

References

Austin L, de-Niese M, McGregor A, Arthur H, Gurusinge A and Gould MK (1992) Potential oxyradical damage and energy status in individual muscle fibres from degenerating muscle diseases. *Neuromusc Dis* 2: 27–33.

Bernske GM, Burton ARC, Gould EN and Levy D (1960) *Neurology* 35: 61–65.

Bertorini TE, Palmieri GMA, Griffin J, Chesney C, Pifer D et al. (1985) Chronic allopurinol and adenine therapy in Duchenne muscular dystrophy: effects on muscle function, nucleotide degradation and muscle ATP and ADP content. *Neurology* 35: 61–65.

Bicknell F (1940) Vitamin E in the treatment of muscular dystrophies and nervous diseases. *Lancet i*: 10–13.

Bradley R and Fell BF (1980) Myopathies in animals. *In*: JN Walton (ed.): *Disorders of Voluntary Muscle*, 4th edn. Churchill, London, pp 824–872.

Brenman JE, Chao DS, Xia H, Aldape K and Bredt DS (1995) Nitric oxide synthase complexed with dystrophin and absent from skeletal muscle sarcolemma in Duchenne muscular dystrophy. *Cell* 82: 743–752.

Bushell A, Klenerman L, Davis H, Grierson I and Jackson MJ (1996) Ischaemia-reperfusion-induced muscle damage: protective effect of corticosteroids and antioxidants in rabbits. *Acta Orthop Scand* 67: 393–398.

Chamberlain JS, Phelps SF, Cox GA, Maichele AJ and Greenwood AD (1993) PCR analysis of muscular dystrophy in mdx mice. *In*: T Partridge (ed.): *Molecular and Cell Biology of Muscular Dystrophy*. Chapman and Hall, London, pp 167–189.

Chao DS, Gorospe GR, Brenman JE et al. (1996) Selective loss of sarcolemmal nitric oxide synthase in Becker muscular dystrophy. *J Exp Med* 184: 609–618.

Cullen MJ, Walsh J, Nicholson LVB and Harris JB (1990) Ultrastructural localisation of dystrophin in human muscle using gold immunolabelling. *Proc R Soc Lond* 240: 197–210.

Dubowitz V and Heckmatt JZ (1980) Management of muscular dystrophy. *Brit Med Bull* 36: 139–144.

Edwards RHT, Jones DA and Jackson MJ (1984) An approach to treatment trials in muscular dystrophy with particular reference to agents influencing free radical damage. *Med Biol* 62: 143–147.

Ervasti JM and Campbell KP (1991) Membrane organisation of the dystrophin-glycoprotein complex. *Cell* 66: 1121–1131.

Ervasti JM, Ohlendieck K, Kahl SD, Gaver MG and Campbell KP (1990) Deficiency of a glycoprotein component of the dystrophin complex in dystrophic muscle. *Nature* 345: 315–319.

Fitzgerald G and McArdle B (1941) Vitamins E and B6 in the treatment of muscular dystrophy and motor neurone disease. *Brain* 64: 19–42.

Folkers K and Simonsen R (1995) Two successful double-blind trials with coenzyme Q10 (vitamin Q10) on muscular dystrophies and neurogenic atrophies. *Biochim Biophys Acta* 1271: 281–286.

Foxley A, Edwards RHT and Jackson MJ (1991) Enhanced lipid peroxidation in Duchenne muscular dystrophy may be secondary to muscle damage. *Biochem Soc Trans* 19: 180S.

Gamsdorp I, Gustavson KH, Helstrom O and Nordgren B (1986) A trial of selenium and vitamin E in boys with muscular dystrophy. *J Child Neurol* 1: 211–214.

Griffiths RD, Cady EB, Edwards RHT and Wilkie DR (1985) Muscle energy metabolism in Duchenne dystrophy studied by ^{31}P NMR: Controlled trials show no effect of allopurinol or ribose. *Muscle Nerve* 8: 760–767.

Hayashi YK, Engvall E, Arikaea-Hirasawa E, Goto K, Koga R, Nonaka I, Sugita H and Arahata K (1993) Abnormal localisation of laminin subunits in muscular dystrophies. *J Neurol Sci* 119: 53–64.

Hoffman EP, Brown RH and Kunkel LM (1987) Dystrophin, the protein product of the Duchenne muscular dystrophy locus. *Cell* 51: 919–928.

Ibraghiminov-Beskrovnaya O, Ervasti JM, Leville CJ, Slaughter CA, Sernett SW and Campbell KP (1992) Primary structure of dystrophin-associated glycoproteins linking dystrophin to the extracellular matrix. *Nature* 355: 696–702.

Ihara Y, Mori T, Namba A, Nobukini K, Sato K and Miyata S (1995) Free radicals, lipid peroxides and antioxidants in blood of patients with myotonic dystrophy. *J Neurol* 242: 119–122.

Jackson MJ and Edwards RHT (1986) Critical review of the role of zinc in muscular dystrophy. *Cardiomyology* 5: 31–38.

Jackson MJ, Jones DA and Edwards RHT (1984) Techniques for studying free radical damage in muscular dystrophy. *Med Biol* 62: 135–138.

Jackson MJ, Coakley J, Stokes M, Edwards RHT and Oster O (1989) Selenium metabolism and supplementation in patients with Duchenne muscular dystrophy. *Neurology* 39: 655–659.

Jansen G, Groenen PJ, Bachner D et al. (1996) Abnormal myotonic dystrophy protein kinase levels produce only mild myopathy in mice. *Nat Genet* 13: 316–324.

Kar NC and Pearson CM (1979) Catalase, superoxide dismutase, glutathione reductase and thiobarbituric acid-reactive products in normal and dystrophic human muscle. *Clin Chim Acta* 94: 277–280.

Karpati G and Carpenter S (1988) The deficiency of a sarcolemmal cytoskeletal protein (dystrophin) leads to the necrosis of skeletal muscle fibres in Duchenne-Becker dystrophy. *In*: LC Sellin, R Libelius and S Thesleff (eds): *Neuromuscular Junction*. Elsevier Science, Amsterdam, pp 429–436.

Koga R, Nakao Y, Kurano Y, Tsukahara T, Nakamura A, Ishiura S and Nonaka I (1994) Decreased myotonin-protein kinase in the skeletal and cardiac muscles in myotonic dystrophy. *Biochem Biophys Res Commun* 202: 577–585.

McArdle A and Jackson MJ (1997) The pathophysiology of Duchenne muscular dystrophy. *Med Biochem*; *in press*.

McArdle A, Edwards RHT and Jackson MJ (1995) How does dystrophin deficiency lead to muscle damage in muscular dystrophy: lessons from the *mdx* mouse. *Neuromusc Disord* 5: 445–456.

Matsumura K, Torne MS, Collin H, Azibi K, Chaonch M, Kaplan JC, Fardeau M and Kampbell KP (1992) Deficiency of the 50KD dystrophin-associated glycoprotein in severe childhood autosomal recessive muscular dystrophy. *Nature* 359: 320–322.

Mechler F, Imre S and Dioszeghy P (1984) Lipid peroxidation and superoxide dismutase in Duchennne muscular dystrophy. *J Neurol Sci* 63: 279–283.

Mendell JR and Wiechers DO (1979) Lack of benefit of allopurinol in Duchenne muscular dystrophy. *Muscle Nerve* 2: 53–56.

Monaco AP, Bertelson CJ, Middlesworth W, Colletti CA, Aldridge J, Fischbeck KH, Bartlett R, Pericak-Vance MA, Roses AD and Kunkel LM (1985) Detection of deletions spanning the Duchenne muscular dystrophy locus using a tightly linked DNA segment. *Nature* 316: 842–845.

Mounsey JP, Xu P, John JE, Horne LT, Gilbert J, Roses AD and Moorman JR (1995) Modulation of skeletal muscle sodium channels by human myotonin protein kinase. *J Clin Invest* 95: 2379–2384.

Ohlendieck K and Campbell KP (1991) Dystrophin-associated proteins are greatly reduced in skeletal muscle from mdx mice. *J Cell Biol* 115: 1685–1694.

Orndahl G, Rindby A and Selin E (1982) Myotonic dystrophy and selenium. *Acta Med Scand* 211: 493–499.

Orndahl G, Rindby A and Selin E (1983) Selenium therapy of myotonic dystrophy. *Acta Med Scand* 213: 237–239.

Orndahl G, Seliden U, Hallin S, Wetterqvist H, Rindby A and Selin E (1986) Myotonic dystrophy treated with selenium and vitamin E. *Acta Med Scand* 219: 407–414.

Orndahl G, Grimby G, Grimby A, Johansson G and Wilhelmsen L (1994) Functional deterioration and selenium-vitamin E treatment in myotonic dystrophy. *J Intern Med* 235: 205–210.

Sewry CA, Sansome A, Matsumura K, Campbell KP and Dubowitz V (1994) Deficiency of the 50 kDa dystrophin-associated glycoprotein and abnormal expression of utrophin in two south Asian cousins with variable expression of severe childhood autosomal recessive muscular dystrophy. *Neuromusc Disord* 4: 121–129.

Stern LZ, Ringel SP, Ziter FA et al. (1982) Drug trial of superoxide dismutase in Duchenne's muscular dystrophy. *Arch Neurol* 39: 342–346.

Suzuki A, Yoshida M, Hayashi K, Mizuno Y, Hagiwara Y and Ozawa E (1994) Molecular organisation at the glycoprotein-complex-binding site of dystrophin. Three dystrophin-associated proteins bind directly to the carboxyl-terminal portion of dystrophin. *Eur J Biochem* 220: 283–292.

Thompson WHS and Smith I (1978) X-linked recessive (Duchenne) muscular dystrophy and purine metabolism: effects of oral allopurinol and adenylate. *Metabolism* 27: 151–163.

Timchenko L, Monkton DG and Caskey CT (1995) Myotonic dystrophy: an unstable CTG repeat in a protein kinase gene. *Semin Cell Biol* 6: 13–19.

Toescu V, Edwards RHT and Jackson MJ (1994) Treatment of Duchenne muscular dystrophy: History and future directions. *Basic Appl Myol* 4: 217–226.

Watkins SC, Hoffman EP, Slater HS and Kunkel LM (1988) Immunoelectron microscopic localisation of dystrophin in myofibres. *Nature* 333: 863–866.

Yoshida M and Ozawa E (1990) Glycoprotein complex anchoring dystrophin to sarcolemma. *J Biochem* 108: 748–752.

Subject index

MCBU
Molecular and Cell Biology Updates

A. Mackiewicz, University School of Medical Sciences, Poznan, Poland
P.B. Sehgal, New York Medical College, Valhalla, NY (Eds)

Molecular Aspects of Cancer and its Therapy

1998. Approx. 300 pages. Hardcover.
ISBN 3-7643-5724-X
Due in July 1998

This book highlights recent progress in the molecular, cellular and immunological mechanisms that contribute to the pathophysiology of cancer and the design of therapeutic modalities based upon these molecular insights. Areas of particular emphasis include cancer immunology and the immunotherapy of cancer, the role of cytokines in modulating the social behaviour of cancer cells, the genetic alterations that characterize human cancer and metastasis, and a consideration of the more experimental approaches to cancer therapy, including gene therapy using expression vectors for cytokines and their receptors, antisense RNA therapy, and anti-idiotypic antibody immunization.

This volume serves to introduce the general reader as well as the cancer specialist to personalized perspectives of particular topics in cancer research by leading research groups in the field. The combination of a „reviews"-approach with a more research-oriented approach in discussions of specific research topics provides a stimulating and forward-looking volume which serves to update selected aspects of cancer research today. This combination will be useful to both the beginner as well as the more advanced biomedical scientist.

For orders originating from all over the world except USA and Canada
Birkhäuser Verlag AG
P.O. Box 133
CH-4010 Basel / Switzerland
Fax: +41/61/205 07 92
e-mail: orders@birkhauser.ch

For orders originating in the USA and Canada
Birkhäuser
333 Meadowland Parkway
USA-Secaurus, NJ 07094-2491
Fax: +1 201 348 4033
e-mail: orders@birkhauser.com

Birkhäuser

Free Radical Research • Cell Biology • Biochemistry

A.E. Favier, CHU Albert Michallon, Grenoble, France
J. Cadet, CEA, Grenoble, France
B. Kalyanaraman, Medical College of Wisconsin, Milwaukee, WI, USA
M. Fontecave, J.L. Pierre, LEDSS II, St. Martin d'Hères, France (Eds)

Analysis of Free Radicals in Biological Systems

1995. 312 pages. Hardcover.
ISBN 3-7643-5137-3

The main aim of the book is to provide a comprehensive survey on recent methodological aspects of the measurement of damage within cellular targets, information which may be used as an indicator of oxidative stress.

In the introductory chapters, emphasis is placed on the chemical properties of reactive oxygen species and their role in the induction of cellular modifications together with their links to various diseases. The central part of the book is devoted to the description of selected methods aimed at monitoring the production of free radicals in cellular systems. In addition, several assays are provided to assess the chemical damage induced by reactive oxygen species in critical cellular-targets in vitro and in humans in vivo.

Thus both practical aspects and general considerations, including discussions on the applications and limitations of the assays, are critically reviewed. One of the major features of the book is the description of new experimental methods. These include the measurement of oxydized DNA bases and nucleosides, new techniques for the determination of LDL oxidation using spin-trap agents, and the use of salicylate as an indicator of oxidative stress. In addition, more classical though significantly improved techniques devoted to the measurement of hydroperoxides and aldehydes are described.

The book will serve a large scientific community including biologists, chemists, and clinicians working on the chemical and biological effects of oxidative stress. It may also be of interest to investigators in the fields of drug, cosmetic and new food research.

For orders originating from all over the world except USA and Canada
Birkhäuser Verlag AG
P.O. Box 133
CH-4010 Basel / Switzerland
Fax: +41/61/205 07 92
e-mail: orders@birkhauser.ch

For orders originating in the USA and Canada
Birkhäuser
333 Meadowland Parkway
USA-Secaurus, NJ 07094-2491
Fax: +1 201 348 4033
e-mail: orders@birkhauser.com

M.-F. Schulz-Aellen, Institutions Universitaire de Gériatrie et de Psychiatrie de Genève, Switzerland

Aging and Human Longevity

1997. 284 pages. 20 illus. Hardcover.
ISBN 3-7643-3875-X
Softcover.
ISBN 3-7643-3964-0

This book combines a scientific and medical description of aging with a critical review of ways to prolong life. The first part gives an overview of the complex biological mechanisms of aging and of the consequences of tissue and system aging in humans. The role of genetic and environmental factors that influence the rate of aging in several species is discussed.

The second part of this book evaluates the various means, including life styles, behavioral variables, medical interventions and vitamin supplementation that may slow down the physiological and psychological effects of aging. Biotechnology and gene therapy are also becoming a part of medical interventions to prolong life. This book is aimed at readers with some knowledge of biology and medicine, as well as to a larger audience eager to know more about how to live a healthy, happy and productive life in their old age.

For orders originating from all over the world except USA and Canada
Birkhäuser Verlag AG
P.O. Box 133
CH-4010 Basel / Switzerland
Fax: +41/61/205 07 92
e-mail: orders@birkhauser.ch

For orders originating in the USA and Canada
Birkhäuser
333 Meadowland Parkway
USA-Secaurus, NJ 07094-2491
Fax: +1 201 348 4033
e-mail: orders@birkhauser.com

MCBU
Molecular and Cell Biology Updates

P. Bannasch, Deutsches Krebsforschungszentrum, Heidelberg, Germany
D. Kanduc, University of Bari
S. Papa, University of Bari
J.M. Tager, University of Amsterdam, The Netherlands (Eds)

Cell Growth and Oncogenesis

1998. 312 pages. Hardcover.
ISBN 3-7643-5727-4

Rapid progress has been made in our understanding of the molecular mechanisms of cell growth and oncogenesis during the past decade. This book comprises recent results on the regulation of cell growth in normal and neoplastic tissues by growth factors including hormones, and by the activation and inactivation of oncogenes and tumor suppressor genes, respectively. Special attention has been given to the presentation of the frequently neglected close correlation between changes in signal transduction and metabolism pathways during oncogenesis.

For orders originating from all over the world except USA and Canada
Birkhäuser Verlag AG
P.O. Box 133
CH-4010 Basel / Switzerland
Fax: +41/61/205 07 92
e-mail: orders@birkhauser.ch

For orders originating in the USA and Canada
Birkhäuser
333 Meadowland Parkway
USA-Secaurus, NJ 07094-2491
Fax: +1 201 348 4033
e-mail: orders@birkhauser.com